Progress in Systems and Control Theory

Volume 23

Series Editor
Christopher I. Byrnes, Washington University

I. Csiszár
Gy. Michaletzky

Stochastic Differential and Difference Equations

Birkhäuser
Boston • Basel • Berlin

Imre Csiszár
Mathematical Institute of the
Hungarian Academy of Sciences
Realtanoda u. 13-15
Budapest 1053
Hungary

György Michaletzky
Eötvös Loránd University
Department of Probability Theory
and Statistics
Múzeum krt. 6-8
Budapest 1088, Hungary

Library of Congress Cataloging-in-Publication Data

Stochastic differential and difference equations / Imre Csiszár and
  György Michaletzky, editors.
      p.   cm. -- (Progress in systems and control theory ; v. 23)
      Papers from the Conference on Stochastic Differential and
  Difference Equations held Aug. 21-24, 1996, in Györ, Hungary.
      Includes bibliographical references.
      ISBN 0-8176-3971-3 (hc : alk. paper). -- ISBN 3-7643-3971-3 (Basel
  : hc : alk. paper)
      1. Stochastic differential equations--Congresses.   2. Stochastic
  difference equations--Congresses. I. Csiszár, Imre, 1938-   .
  II. Michaletzky, György, 1950- . III. Conference on Stochastic
  Differential and Difference Equations (1996 : Györ, Hungary)
  IV. Series.
  QA274.23.S8   1997
  519.2--dc21
                                                            97-20913
                                                            CIP

Printed on acid-free paper

© 1997 Birkhäuser Boston

Birkhäuser ®

ISBN 0-8176-3971-3
ISBN 3-7643-3971-3
Typeset by the Editors in LaTeX.
Printed and bound by Edwards Brothers, Ann Arbor, MI.
Printed in the U.S.A.

9 8 7 6 5 4 3 2 1

Table of Contents

Preface

The *Conference on Stochastic Differential and Difference Equations* held at Győr, Hungary, August 21–24, 1996 was organized jointly by Eötvös Loránd University, Budapest and Kossuth Lajos University, Debrecen, with the sponsorship of the Hungarian Regional, the International Executive and the European Regional Committees of the Bernoulli Society as a satellite event to the 4th World Congress of the Bernoulli Society, August 26–31, 1996, Vienna, Austria.

It is noteworthy that the meeting had a strong international flavour with 76 participants from 21 countries, including 6 each from Japan and the USA. The core of the conference consisted of the 14 invited lectures, delivered by distinguished experts in their research fields. The majority of contemporary research areas have been covered in these lectures.

The list of the invited speakers included T. Duncan, M. Fukushima, T. Funaki, I. Gyöngy, R. Khasminskii, I. Kubo, H. Kunita, A. Lindquist, D. Nualart, R. Ober, M. Pavon, G. Picci, T. SubbaRao, M. Zakai.

Invited lectures were presented in plenary sessions, while the contributed papers were presented in two parallel sessions. The first session was devoted to various problems of stochastic partial differential equations (SPDE) and related random fields. The second session covered discrete and continuous time parameter ARMA processes and stochastic differential equations in general.

The Széchenyi István College in Győr provided the venue of the event, seemingly the satisfaction of the participants.

The Welcome party and the Gala Dinner, with its fabulous cuisine and Hungarian wines, provided the guests with gracious and pleasing entertainments. These occasions proved to be golden opportunities to meet old friends and make new ones from all over the world.

Budapest, Febr. 25, 1997

Csiszár Imre	Michaletzky György	Márkus László
chairman	chairman	secretary

Győr. The Conference Venue

Győr is called town of the waters, because it lies at the confluence of the Danube, Raba, Rabca and Marcall Rivers. It has made history over many centuries. In the 5th century B.C. Győr was inhabited by Celtic and Illyrian tribes. The Romans took over from them but kept the Illyrian name of Arrabona, which means, dwellings by the River Rab". Arrabona - thanks to its strategic location - soon became an important town, and fortress of Pannonia Province. After the fall of the Roman Empire the Avars ruled the town and they fortified the Roman "Castrum" by ring-shaped mounds. This is how the settling Hungarians found the castle, and from the ring (ring - gyuru in Hungarian) they gave it its name. The Hungarians also recognised the strategic importance of the place. Already during the reign of the first king Stephen 1st, it had become an episcopal seat, turning very soon into a cathedral town. In 1271 King Stephen 5th elevated Győr to the rank of free royal town and gave the right to tax traffic in transit. During the Turkish invasion strategic importance increased dramatically as the Turks threatened Vienna. By the help of treachery the Turks could capture Győr, but their rule lasted only 4 years. Legend has it that after capturing Győr the Turks put an iron rooster and half moon on the tower of the castle, saying that Hungarians will recapture the town only when the rooster crows and the half moon becomes full. The Hungarian general Pallfy Miklos used this legend and took back the castle by ruse. This is how the iron rooster became part of the city arms. In 1809 Emperor Napoleon, after defeating the Austro-Hungarian army near Győr blasted the bastions of the castle, and it has never been rebuilt. From the 18th century a robust economic prosperity characterised the town. By the 19th century only Budapest surpassed Győr as a Hungarian grain marketing centre. But industry soon eclipsed trade as the economic strong-point, and today the city is best known as a manufacturer of trains and trucks.

As befits a dynamic city of more than 125,000 people, Győr is pervaded by modern buildings, though the historical monuments are all around. Outstanding among the eye-catching new structures is the Kisfaludy Theatre, named after the local poet and playwright Karoly Kisfaludy. It is the home of the highly regarded Győr ballet company, which specialized in modern dance. The Cathedral was founded in the 11th century, but many builders had a hand in working on it during its long and turbulent history. The essentially Gothic church was rebuilt in Baroque style in the 17th century, the neo-classic facade was tacked on 200 years later and then the damage incurred during World War II had to be repaired. The chapel dedicated

to St. Ladislas, among the oldest parts of the cathedral, contains a reli-
quary considered a masterpiece of Hungarian goldsmith's art. Other works
of gold, silver, copper and enamel, as well as ceremonial vestments, can
be seen in the cathedral's treasury. All around the cathedral, the winding
streets of the area called Chapter Hill are full of historic atmosphere as
well as the medieval keep, the Bishop's Palace, and remains of the town
bastions along the river. The Carmelite Church is a Baroque achievement
of the early 18th century. The gracefully proportioned facade has statues
in three niches and a large window in the organ loft. Rich paintings and
carvings decorate the interior. The Town Hall is a U-shaped, neobaroque,
turreted building. Strolling through the historic heart of the town today, in
pedestrians-only streets plentifully supplied with outdoor cafes, you might
well agree with the local promoters that it all turned out rather nicely in
the end.

List of Participants

ALBIN, Patrik Department of Mathematics, Chalmers University of Technology and Gothenburg University, Gothenburg, S-41296 Sweden
e-mail: palbin@math.chalmers.se phone: 46-31-7723512 fax: 46-31-7723508

ARATÓ, Miklós Dept. of Probability Theory and Statistics, Eötvös Loránd University, Múzeum krt. 6-8. H-1088 Budapest Hungary
e-mail: aratonm@ludens.elte.hu phone: 361-266-9833 fax: 361-266-3556

BENTKUS, Vidmantas Fakultät für Mathematik, Bielefeld University, Postfach 100131 , Bielefeld 33501 Germany
e-mail: bentkus@mathematik.uni-bielefeld phone: 49-521-1064974

BOKOR, Rózsa University of Szeged, Aradi Vértanúk tere 1., H-6720 Szeged, Hungary
e-mail: bokor@jate.math.hu phone: 62-326-246 fax: 62-326-246

BOSHNAKOV, Georgi Institute of Mathematics and Informatics, Sofia 1113 Bulgaria
e-mail: boshnakov@bgearn.acad.bg phone: 359-2-713-3857 fax: 359-2-971-3649

BRIGO, Damiano Department of Mathematics, University of Padova c/o Prof. G.B. Di Masi, via Belzoni 7, Padova 35121, Italy
brigo@pdmat1.unipd.it phone: 39-49-8254981 fax: 39-49-875-8596

CABALLERO, Emilia María Instituto de Matemáticas, University of Mexico UNAMI, Circuito Exterior c.v. Mexico, DF 4510, Mexico
e-mail: emilia@servidor.unam.mx phone: 616-0348

CHIANG, Tzuu-Shuh Institute of Mathematics, Academia Sinica, Taipei, Taiwan
e-mail: matsch@ccvax.sinica.edu.tw phone: 886-2-785-1211/417 fax: 886-2-782-7432

CHUPRUNOV, Alexey Kazan State University, Cheboratev Research Inst. of Mathematics and Mechanics, Universitetskaia str. 17., 420008 Kazan Russia
e-mail: alexey.chuprunov@ksu.ras.ru phone: 8432-387525 fax: 8432-387321

List of Participants

CRUZEIRO, Ana Bela Grupo de Física-Matemática, Univ. Lisboa, Av. prof. Gama Pinto 2., Lisboa, 1699 Codex Portugal
e-mail: abc@gfm.cii.fc.ul.pt phone: 351-1-7904853 fax: 351-1-7954288

CSISZÁR, Imre chairman, Mathematical Institute of the Hungarian Academy of Sciences, Reáltanoda u. 13-15., Budapest 1053, Hungary
e-mail: csiszar@math-inst.hu phone: 361-1182875

DALANG, Robert Departement de Mathematiques, Ecole Polytechnique Federale de Lausanne, DMA-EPFL, Lausanne 1015, Switzerland
e-mail: dalang@math.epfl.ch phone: 41-21-693-2551 fax: 41-21-693-4303

DUNCAN, Tyrone University of Cansas, Mathematics Department, Lawrence, Kansas, 66045 USA,
e-mail: duncan@math.ukans.edu phone: 1-913-864-3032 fax: 1-913-864-5255

FAZEKAS, István KLTE, Institute of Psychology, Egyetem tér 1, Debrecen, 4010 Hungary
e-mail: fazekas@math.klte.hu phone: 36-52-316666 fax: 36-52-431216

FRITZ, József Eötvös Loránd University, Department of Probability Theory and Statistics, Múzeum krt. 6-8., Budapest H-1088, Hungary
e-mail: jofri@math.elte.hu phone: 361-266-9833 fax: 361-266-3556

FUKUSHIMA, Masatoshi Department of Mathematical Science, Faculty of Engineering Science, Osaka University, Toyonaka, Osaka, Japan
e-mail: fuku@sigmath.es.osaka-u.ac.jp phone: 06-850-6460 fax: 06-850-6496

FUNAKI, Tadahisa Department of Mathematical Sciences, University of Tokyo, 3-8-1 Komaba, Meguro, Tokyo, 153 Japan
e-mail: funaki@ms.u-tokyo.ac.jp phone: 81-3-5465-7033 fax: 81-3-5465-7012

GERENCSÉR, László Computer and Automation Research Institute of the Hungarian Academy of Sciences, Kende u. 13-17, Budapest 1111, Hungary e-mail: laszlo@decst.scl.sztaki.hu phone: 361-1667-483 fax: 361-1667-503

GIELERAK, Roman Institute of Theoretical Physics, University of Wroclaw, p. Maxa Borna 8, Wroclaw 50204, Poland
e-mail: gielerak@proton.ift.uni.wroc.pl phone: 48-71-201438

GYÖNGY, István Department of Probability Theory and Statistics, Eötvös Loránd University, Múzeum krt. 6-8., Budapest 1088, Hungary
e-mail: gyongy@math.elte.hu phone: 361-266-9833 fax: 361-266-3556

HÖGNAS, Göran Department of Mathematics, Abo Akademi University, Abo, FIN-20500 Finland
e-mail: ghognas@abo.fi phone: 358-21-2654224 fax: 358-21-2654865

HANZON, Bernard Department Econometrics, Fewec Vrye Universiteit Amsterdam, De Boelelaan 1105, Amsterdam, 1081 HV Holland
e-mail: bhanzon@econ.vu.nl phone: 31-20-4446017 fax: 31-20-4446020

HEBER, Gerd GMD-FIRST, Institute for Computer Architecture and Software Technology, Rudower Chaussee 5, GEB. 1310 Berlin, 12489 Germany
e-mail: heber@first.gmd.de phone: 30-6392-1908

IMKELLER, Peter Humboldt Universitaät zu Berlin, Institut für Mathematik, Unter den Linden 6, Berlin 10099, Germany
e-mail: imkeller@mathematik.hu-berlin.de phone: 49-30-209-35860

ISPÁNY, Márton Kossuth Lajos University, Institute of Mathematics and Infarmatics, Egyetem tér 1. Pf. 12, Debrecen 4010, Hungary
e-mail: ispany@math.klte.hu phone: 36-52-666-2804 fax: 36-52-416-857

KHASMINSKII, Rafail Department of Mathematics, Wayne State University, Detroit, MI 48202 USA
e-mail: rafail@math.wayne.edu phone: 313-5773182 fax: 313-5777596

KOHATSU-HIGA, Arturo Department of Mathematics, Universitat Pompeu Fabra and Department of Mathematics, Faculty of Science, Kyoto, Kyoto, 606-01 Japan
e-mail: kohatsu@kusm.kyoto-u.ac.jp phone: 81-75-7538698 fax: 81-75-7533711

KOLODII, Alexander Department of Mathematics, Volgograd State University, Vtoraya Prodol'naya, 30, Volgograd 400062, Russia
e-mail: kolodii@math.vgu.tsaritsyn.su phone: 8442-431-426 fax: 8442-433786

KRÁMLI, András University of Szeged, Aradi Vértanúk tere 1., Szeged, 6720 Hungary
e-mail: kramli@ilab.sztaki.hu phone: 36-62-45-4097 fax: 36-62-326-246

KUBO, Izumi Department of Mathematics, Faculty of Science, Hiroshima University, Higashi-Hiroshima, 739 Japan
e-mail: kubo@math.sci.hiroshima-u.ac.jp phone: 81-824-247342 fax: 81-824-240710

KUNITA, Hiroshi Graduate School of Mathematics, Kyushu University, Hakozaki, Fukuoka, 812 Japan
e-mail: kunita@math.kyushu-u.ac.jp phone: 92-642-2790 fax: 92-642-2789

LINDQUIST, Anders Royal Institute of Technology, Optimization and Systems Theory, Stockholm, 10044 Sweden
e-mail: alq@math.kth.se phone: 46-8-790-7311 fax: 46-8-225-320

MÁRKUS, László Conference Secretary, Eötvös Loránd University, Department of Probability Theory and Statistics, Múzeum krt. 6-8., Budapest, 1088 Hungary
e-mail: markus@ludens.elte.hu phone: 361-266-9833/2713 fax: 361-266-3556

MÁTH, János Institute of Psychology, Kossuth Lajos University, Egyetem tér 1., Debrecen, 4010 Hungary
e-mail: jmath@tigris.klte.hu phone: 36-52-316666/2537 fax: 36-52-431-216

MÓRI, Tamás Eötvös Loránd University, Department of Probability Theory and Statistics, Múzeum krt. 6-8., Budapest, 1088 Hungary
e-mail: moritamas@ludens.elte.hu phone: 361-266-9833/2304 fax: 361-266-3556

MAJOR, Péter Mathematical Research Institute of the Hungarian Academy of Sciences, 1364 Pf. 127, Reáltanoda u. 13-15, 1053 Budapest, Hungary
e-mail: major@math-inst.hu phone: 36-1-118-2875

MICHALETZKY, György chairman Eötvös Loránd University, Department of Probability Theory and Statistics, Múzeum krt. 6-8., Budapest, 1088 Hungary
e-mail: michgy@ludens.elte.hu phone: 361-266-9833/2711 fax: 361-266-3556

MILLET, Annie Université Paris X et Université Paris VI, Laboratorie de Probabilités, Tour 56, 3 étage 4, place Jussieu, Paris, 75252 Cedex 05, France
e-mail: amil@ccr.jussieu.fr phone: 33-144277042 fax: 33-144277223

NUALART, David Facultat de Matematiques, University of Barcelona, Gran Via 585, Barcelona, 8007 Spain
e-mail: nualart@cerber.mat.ub.es phone: 343-4021656 fax: 343-4021601

OBER, Raimund Center for Engineering Mathematics EC 35, University of Texas at Dallas, Richardson, TX 75083 USA
e-mail: ober@utdallas.edu phone: 214-8832158 fax: 214-6481259

PAP, Gyula Kossuth Lajos University, Institute of Mathematics and Informatics, Pf. 12., Debrecen, 4010 Hungary
e-mail: papgy@math.klte.hu phone: 36-52-416-857 fax: 36-52-310-936

PAVON, Michele Dipartimento Di Electronica E Informatica, Universita di Padova, v. Gradenigo 6/A, Padova, 35131 Italy
e-mail: pavon@ladseb.pd.cnr.it phone: 39-49-8277604 fax: 39-49-8277699

PICCI, Giorgio University of Padova, DEI, Via Gradenigo 6/a, Padova, 35131 Italy
e-mail: picci@dei.unipd.it phone: 49-827-7832 fax: 49-827-7826

PRŐHLE, Tamás Eötvös Loránd University, Department of Probability Theory and Statistics, Múzeum krt. 6-8., Budapest, 1088 Hungary
e-mail: prohlet@math.elte.hu phone: 361-266-9833 fax: 361-266-3556

SPECTORSKY, Igor National Technical University of Ukraine, Kiev Politechnic Institute, 2 Zalomov str. apt.76, Kiev, 252069 Ukraine
e-mail: dalet@stoch.freenet.kiev.ua phone: 380-44-277-424 fax: 380-44-274-3987

SPREIJ, Peter Department of Econometrics, Vrije Universiteit, De Boelelaan 1105, Amsterdam, 1081 HV The Netherlands
e-mail: pspreij@econ.vu.nl phone: 3120-4446017 fax: 3120-4446020

STEBLOVSKAYA, Victoria Romualdovna National Technical University of Ukraine, Kiev Politechnic Institute, 53-v Scherbakov str. apt. 18, Kiev, 252111 Ukraine
e-mail: vrs@email.polytech.kiev.ua phone: 380-44-449-3925 fax: 380-44-274-3987

STRAMER, Osnat Department of Statistics and Actuarial Science, University of Iowa, Iowa-City, 52242-1419 USA
e-mail: stramer@stat.uiowa.edu phone: 319-335-3182 fax: 319-335-3017

SUBBA RAO, Tata Department of Mathematics, UMIST, Manchester, M60 1QD U.K.
e-mail: tata.subbarao@umist.ac.uk

SZABADOS, Tamás Technical University of Budapest, Department of Mathematics, Egry J. u. 20-22., Budapest, 1521 Hungary
e-mail: szabados@inf.bme.hu phone: 361-463-2986 fax: 361-463-3147

TÓTH, Bálint Mathematical Institute of the Hungarian Academy of Sciences, Reáltanoda u. 13.-15., Budapest, 1053 Hungary
e-mail: balint@math-inst.hu phone: 361-118-2399/3015 fax: 361-117-7166

TERDIK, György Center for Informatics and Computing, Kossuth University of Debrecen, Pf 58., Debrecen, 4011 Hungary
e-mail: terdik@kltesrv.klte.hu phone: 36-52-416-784 fax: 36-52-418-801

TSIKALENKO, Tatiana Forschungszentrum BIBOS, Universität Bielefeld, Postfach 100131, Bielefeld, 33501 Germany
e-mail: tsikalenko@physik.uni-bielefeld phone: 49-521-1065292 fax: 49-521-1062961

TWARDOWSKA, Krystyna Institute of Mathematics Warsaw University of Technology, Plac Politechniki 1, Warsaw, 00-661 Poland
e-mail: tward@alpha.im.pw.edu.pl phone: 22-660-7488 fax: 22-621-9312

VÁGÓ, Zsuzsanna Computer and Automation Research Institute of the Hungarian Academy of Sciences, Kende u. 13-17., Budapest, 1011 Hungary
e-mail: zsuzsa@decst.scl.sztaki.hu phone: 361-166-7483 fax: 361-166-7502

van ZUIJLEN, Martien Department of Mathematics, University of Nijmegen,
Toernooiveld 1, 6525 ED Nijmegen, The Netherlands
e-mail: zuijlen@sci.kun.nl

VERETENNIKOV, Alexander Institute for Information Transmission Problems, 19 Bolshoy Karetnii, Moscow, 101447 Russia
e-mail: ayu@sci.lpi.ac.ru phone: 795-2999415 fax: 795-2090579

WATANABE, Hisao Okayama University of Science, Department of Applied Mathematics, Ridai-chō, Okayama, 700 Japan
e-mail: whsx@dam.ous.ac.jp phone: 81-86 2523161 fax: 81-86-2550238

XIONG, Jie Department of Mathematics, University of Tennessee, Knoxville, TN 37996-1300 USA
e-mail: jxiong@math.utk.edu phone: 423-9744308 fax: 423-974 6576

ZAKAI, Moshe Faculty of Electrical Engineering, Technion, Haifa, 32000 Israel
e-mail: zakai@ee.technion.ac.il phone: 972-48-234402

ZEMPLÉNI, András Eötvös Loránd University, Department of Probability Theory and Statistics, Múzeum krt. 6-8., Budapest, 1088 Hungary
e-mail: zempleni@ludens.elte.hu phone: 361-266-9833 fax: 361-266-3556

ZHAO, Xuelei Shantou University, Institute of Mathematics, Shantou, 515063 P.R. China
e-mail: xlzhao@mailserv.stu.edu.cn phone: 86754-8510000/33131 fax: 86754-8712379

Periodically Correlated Solutions to a Class of Stochastic Difference Equations

Georgi N. Boshnakov

1. Introduction.

The class of the periodically correlated processes sets up one of the possible frameworks for description and modeling of time series having pseudo-periodic behaviour. The mean and the autocovariance functions of the processes from this class are periodic. Many of the concepts of the stationary theory admit generalization to the periodic case. There is a duality between the multivariate stationary processes and the periodically correlated processes which makes the investigation of these two classes theoretically equivalent. A survey on these questions (mainly from an algorithmic point of view) and a lot of references may be found in Boshnakov [Bo].

We investigate the stochastic difference equation which is used to define the periodic autoregression model. We determine conditions for existence, uniqueness and causality of the periodically correlated solutions to that equation and investigate some properties of the corresponding autocovariance function. Our approach is based on a non-stationary Markovian representation of the model which in some respects gives a more intuitive description of the properties of the periodic autoregression than its familiar representation as stationary multivariate autoregression. Proofs are generally omitted. They may be found in [Bo].

2. Deterministic and causal processes

Definition 2.1 *A process $\{X_t\}$ is said to be causal with respect to another process $\{Z_t\}$ if for every t there exists an absolutely summable sequence of constants $\{\psi_{ti}\}$, such that*

$$X_t = \sum_{i=0}^{\infty} \psi_{ti} Z_{t-i}.$$

The set of these sequences $\{\psi_{ti}\}$ is called a linear filter (with time-varying coefficients) and we say that $\{X_t\}$ is obtained by filtering $\{Z_t\}$. The following proposition shows that in typical situations the process $\{X_t\}$ obtained in this way is well defined both in mean square and almost sure sense. It is a straightforward generalization of the corresponding result for time invariant filters, for which $\psi_{ti} = \psi_{si}$ for all t, s, and i.

Proposition 2.2 *Let the time series $\{X_t\}$ be such that $\sup_t E|X_t| < \infty$, and let $\{\psi_{ti}\}$ be absolutely summable sequences. Then*

1. *for each t the series $\sum_{i=-\infty}^{\infty} \psi_{ti} X_{t-i}$ converges absolutely with probability one;*

2. *if $\sup_t E|X_t|^2 < \infty$ then the mean square limit of $\sum_{i=-\infty}^{\infty} \psi_{ti} X_{t-i}$ exists and is the same as in 1;*

3. *$P(A) = 0$, where A is the event $A = \{\omega \in \Omega : $ There exists at least one t, such that the series $\sum_{i=-\infty}^{\infty} \psi_{ti} X_{t-i}$ is not absolutely convergent$\}$;*

4. *if $\sup_t E|X_t|^2 < \infty$ and the series $\sum_i |\psi_{ti}| < \infty$ are uniformly convergent then both limits are uniform.*

The assertion *3* shows that almost all trajectories of the process $\{Y_t\}$, where $Y_t = \sum_{i=-\infty}^{\infty} \psi_{ti} X_{t-i}$, are well-defined.

3. Periodically correlated processes

It is convenient to say occassionally that a function f is d-periodic, instead of the longer expression "f is periodic with period d". A d-periodic function of two arguments is one for which $f(x + d, y + d) = f(x, y)$. Similarly a sequence of sequences $\{\psi_{ti}\}$ is said to be d-periodic if $\psi_{t+d,i} = \psi_{t,i}$ for each t and i.

Definition 3.1 *A process $\{X_t\}$ is said to be periodically correlated (d-periodically correlated) if its mean $\mu_t = EX_t$ and autocovariance function $R(t, s) = E(X_t - \mu_t)(X_s - \mu_s)$ are finite and there exists a positive integer d such that both μ_t and $R(t, s)$ are periodic with period d (i.e. $\mu_{t+d} = \mu_t$, and $R(t + d, s + d) = R(t, s)$ for all t, s).*

The number d is called the period of the process. Usually (but not always) the smallest possible d is selected.

Stationary white noise is the simplest stationary process and it is used as a building block to generate other stationary processes. Periodic white noise plays the same role in the periodic case. It reduces to stationary white noise when σ_t^2 (see below) is constant (or equivalently, $d = 1$).

Definition 3.2 *The process $\{\varepsilon_t\}$ is said to be periodic white noise $PWN(0, \sigma_t^2, d)$ if and only if it has the following properties (1) $E\varepsilon_t = 0$; (2) $\sigma_t^2 = E\varepsilon_t^2$ is d-periodic; (3) $E\varepsilon_t\varepsilon_s = 0$ for $t \neq s$.*

4. The periodic autoregression model

The periodic autoregression models are introduced by Jones and Brelsford [JB]. The basic properties of these models, including the asymptotic theory of autocorrelation based estimators, are established by Pagano [Pa]. He has fully exploited the dual process, giving the definition of a covariance stationary periodic autoregression in terms of it and expressing the properties of a periodic autoregressive process entirely through the dual process.

To avoid confusion we should point out that in Pagano's paper the notion " covariance stationary multivariate autoregression" means "covariance stationary and causal multivariate autoregression." Similarly, "covariance stationary periodic autoregression" is what we call here "causal periodic autoregression."

Our treatment of periodic autoregression is slightly different from that of Pagano. First, we give a definition in terms of the properties of the process itself (plus a difference equation), without referring to the multivariate representation. On the other hand, we define it from the beginning to be periodically correlated. This is in accordance with the definition of the stationary autoregression which requires the process to be stationary and to satisfy a difference equation (see e.g., [BD, Chapter 3]).

Definition 4.1 *The process $\{X_t\}$ is said to be periodic autoregression $PAR(p_1, \ldots, p_d)$ with period $d > 0$, and orders $p = (p_1, \ldots, p_d)$ if it is periodically correlated and if for every t,*

$$X_t - \sum_{i=1}^{p_t} \phi_{ti} X_{t-i} = \varepsilon_t, \tag{4.1}$$

where $\{\varepsilon_t\}$ is periodic white noise $PWN(0, \sigma_t^2, d)$ and all parameters are d-periodic too, i.e. $p_{t+d} = p_t$, $\phi_{t+di} = \phi_{ti}$, $\sigma_{t+d}^2 = \sigma_t^2$.

It is common to write equation (4.1) in operator form by using the backward shift operator B $(BX_t = X_{t-1})$:

$$\phi_t(B)X_t = \varepsilon_t, \qquad \text{where} \qquad \phi_t(z) = 1 - \sum_{i=1}^{p_t} \phi_{ti}z^i. \qquad (4.2)$$

The constants in equation (4.1) are called parameters, the process $\{\varepsilon_t\}$ — innovation series (or simply innovations). One can think of the process $\{X_t\}$ as being a solution of equation (4.1) (or (4.2)). Given a periodic autoregressive model $(\phi_t(z), \{\varepsilon_t\})$ it does not necessarily follow that there exists a periodic autoregressive process following this model. We will see that there exists at most one periodically correlated process corresponding to a given periodic autoregression model. On the other hand, if a time series is periodic autoregression then, in general, it can be represented by several periodic autoregressive models. The most important among them is the causal one.

Definition 4.2 *A PAR($p_1, \ldots, p_d$) model for the time series $\{X_t\}$ given by the equation $\phi_t(B)X_t = \varepsilon_t$ is said to be causal if and only if $\{X_t\}$ is causal with respect to $\{\varepsilon_t\}$.*

The following theorems give relations between the autocovariances and the crosscorrelations between X_t and ε_s for the causal periodic autoregression.

Theorem 4.3 (Periodic Yule-Walker equations, ([Pa])) *Let the periodic autoregression process $\{X_t\}$ have the causal representation*

$$X_t - \sum_{i=1}^{p_t} \phi_{ti}X_{t-i} = \varepsilon_t.$$

Then for every t the autocovariance function of $\{X_t\}$ satisfies the periodic Yule-Walker equations

$$R(t, s) - \sum_{i=1}^{p_t} \phi_{ti}R(t - i, s) = \sigma_t^2 \delta_{t,s}$$

for every $s \leq t$, where $\delta_{ts} = 1$ if $t = s$ and $\delta_{ts} = 0$ otherwise.

Proposition 4.4 *If the PAR($p_1, \ldots, p_d$) model $\phi_t(B)X_t = \varepsilon_t$ is causal then for every t the crosscorrelation $EX_t\varepsilon_{t-k}$ between X_t and ε_{t-k} equals 0, σ_t^2, or $\sum_{i=1}^{\min(p_t,k)} \phi_{ti}E(X_{t-i}\varepsilon_{t-k})$, when $k < 0$, $k = 0$ and $k > 0$, respectively.*

The following lemma holds.

Lemma 4.5 *Let the PAR$(p_1, \ldots, p_d)$ model $\phi_t(B)X_t = \varepsilon_t$ for the time series $\{X_t\}$ be given. Suppose that $\sigma_t^2 > 0, t = 1, \ldots, d$. The model is causal if and only if the sequences $\{\psi_{ti}\}, t = 1, \ldots, d$ are absolutely summable.*

4.1   The Markovian dual model

Let us introduce the notations: $m = \max_{1 \leq i \leq d} p_i$, $E_t = (\varepsilon_t, 0, \ldots, 0)'$,

$$A_t = \begin{pmatrix} -\phi_{t1} & -\phi_{t2} & \cdots & -\phi_{tm-1} & -\phi_{tm} \\ 1 & 0 & \cdots & 0 & 0 \\ 0 & 1 & \cdots & 0 & 0 \\ \vdots & \vdots & \ddots & \ddots & \vdots \\ 0 & 0 & \cdots & 1 & 0 \end{pmatrix},$$

$$\alpha_t = A_t A_{t-1} \ldots A_{t-d+1}, \qquad t = 1, \ldots, d.$$

Definition 4.6 *Let $\{X_t\}$ be e periodic autoregressive process. Its Markovian dual process $\{Z_t\}$ is defined to be the vector of m consecutive X-es*

$$Z_t = (X_t, X_{t-1}, \ldots, X_{t-m+1})'.$$

Clearly $\{Z_t\}$ is nonstationary. Its properties are collected in the following proposition.

Proposition 4.7 *The Markovian dual process $\{Z_t\}$ of a causal periodic autoregressive process $\{X_t\}$ has the following properties:*

1. *$Z_t = A_t Z_{t-1} + E_t$, where E_t is orthogonal to Z_s for $s < t$.*

2. *$M(Z_t | Z_s, s < t) = A_t Z_{t-1}$. (linear Markovian property of $\{Z_t\}$)*

3. *$Z_t = \alpha_t Z_{t-d} + V_t$, where $\alpha_t = A_t A_{t-1} \ldots A_{t-d+1}$, V_t is orthogonal to Z_s, $s \leq t - d$, and is given by the formula*

$$V_t = (A_t A_{t-1} \ldots A_{t-d+2}) E_{t-d+1} + \ldots + A_t E_{t-1} + E_t.$$

4. *For every i the process $\{z_t^{(i)}\}$, obtained by taking every d-th Z_t, $z_t^{(i)} = Z_{i+td}$, is stationary multivariate AR(1) process.*

From *4* it follows, in particular, that for $k > 0$,

$$R_z(t, t - kd) = \alpha_t R_z(t - d, t - kd) = \alpha_t R_z(t, t - (k - 1)d). \qquad (4.3)$$

The autocovariance function of the dual Markovian process is closely related to that of the process $\{X_t\}$ and its properties are studied at the end of this section.

Let $S_t = E(E_t E_t^T)$. Using standard arguments we obtain that for every t the following equations hold for the autocovariance function $R_z(t, s)$ of $\{Z_t\}$

$$
\begin{array}{rll}
R_z(t, t-1) & = & A_t R_z(t-1, t-1) \\
R_z(t, t) & = & A_t R_z(t-1, t) + S_t
\end{array}
\qquad t = 1, \ldots, d \qquad (4.4)
$$

If the model is given, the solution of the system (4.4) gives the autocovariances of the process $\{X_t\}$. It shows also that for every t

$$
R_z(t, t) = A_t R_z(t-1, t-1) A_t^T + S_t \qquad (4.5)
$$

Proposition 4.8 $R_z(t, t)$ *is a solution of the equation*

$$
R_z(t, t) = \alpha_t R_z(t, t) \alpha_t^T + B_t \qquad (4.6)
$$

where

$$
B_t = \sum_{i=1}^{d-1} \alpha_{ti} S_{t-i} \alpha_{ti}^T, \qquad \alpha_{ti} = \prod_{j=0}^{i-1} A_{t-j}, \qquad \alpha_t \equiv \alpha_{td-1}.
$$

Note that $R_z(t, t)$ consists of autocovariances of the process $\{X_t\}$. Moreover, if $R_z(t, t)$ is known for some t then all other autocovariances of $\{X_t\}$ can be generated simply by the use of the Yule-Walker equations. Equation (4.6) can be rewritten in the form,

$$
(I - \alpha_t \otimes \alpha_t) \mathrm{vec} R_z(t, t) = \mathrm{vec} B_t, \qquad (4.7)
$$

where the operator "vec" stacks the columns of a matrix one over another.

The second part of the following proposition generalizes the corresponding result for stationary univariate and multivariate models (see [LM]).

Proposition 4.9   1. *The eigenvalues of the matrices α_t, $t = 1, \ldots, d$, are the same.*

2. *If for each pair of eigenvalues λ_i, λ_j of α_t the condition $|\lambda_i \lambda_j| \neq 1$ holds, then equation (4.6) has unique solution for every right-hand matrix B_t, and it is given by the formula*

$$
\mathrm{vec} R_z(t, t) = (I - \alpha_t \otimes \alpha_t)^{-1} \mathrm{vec} B_t.
$$

If the eigenvalues of A lie inside the unit circle the solution can be expressed in the form

$$\text{vec}R_z(t,t) = \sum_{k=0}^{\infty} \alpha_t^k B {\alpha_t'}^k = \sum_{i=0}^{\infty}(\alpha_t \otimes \alpha_t)^i \text{vec}B_t.$$

From equation (4.3) we can see that $R_z(t, t - kd) = \alpha_t^k R_z(t - d, t - d)$. Therefore the following proposition holds.

Proposition 4.10 $R_z(t, t - kd) \to 0$ *when* $k \to \infty$ *if and only if all eigenvalues of* α_t *have modulus less than one.*

Since the α's are products of companion matrices it follows that the α's are nonsingular if and only if $p_i = m$ for $i = 1, \ldots, d$. It is natural to call the nonzero eigenvalues of α_t characteristic or eigennumbers of the periodic autoregression model. There are at most m of them.

5. Solutions of the PAR($p_1, \ldots, p_d$) equation

Using the results and notations from the previous sections we can establish the following theorem.

Theorem 5.1 *Let a PAR($p_1, \ldots, p_d$) model be given. If the eigenvalues of* α_t *are less than one in absolute value then the equation*

$$R_z(t,t) = \alpha_t R_z(t,t)\alpha_t' + B_t,$$

has unique solution. This solution gives the autocovariance function of a periodic autoregressive process.

Corollary 5.2 *If the eigenvalues of* α_t *are inside the unit circle then the equation* $\phi_t(B)X_t = \varepsilon_t$, *has tje unique solution and it is causal.*

Theorem 5.3 *The periodic autoregression model* $(\phi_t(B), \{\varepsilon_t\})$ *is causal if and only if the moduli of all eigenvalues of* α_t *are less than one.*

Theorem 5.4 *The equation* $\phi_t(B)X_t = \varepsilon_t$ *has a periodically correlated solution if and only if the moduli of all eigenvalues of* α_t *are different from one. When such a solution exists, it is unique.*

Proof. This result follows from the corresponding result for multivariate processes. But since this seems not to be easily available we give here a proof, which in addition throws more light on the required solution.

Consider the Markovian dual process of $\{X_t\}$, as defined above

$$Z_t = \alpha_t Z_{t-d} + V_t.$$

Let the process $\{z_t\}$ be obtained by setting $z_t = Z_{dt}$. Similarly, $v_t = V_{dt}$. The process z_t is a multivariate AR(1) process. Below we will omit the index of α_t.

Let UJU^{-1} be the Jordan decomposition of α. Consider the equations

$$z_t = \alpha z_{t-1} + v_t, \qquad \text{and} \qquad y_t = J y_{t-1} + w_t,$$

where $y_t = U^{-1} z_t$, $w_t = U^{-1} v_t$. If $\{z_t\}$ is a stationary solution of the first equation then $\{y_t\}$ is a stationary solution of the second one. For every eigenvalue λ of J there is a row, say i, in the last equation such that $y_{ti} = \lambda y_{t-1,i} + w_{ti}$, i.e. y_{ti} is (possibly complex) univariate autoregression. There exists a stationary solution for y_{ti} if and only if $|\lambda| \neq 1$. Since the components of a multivariate stationary process are themselves stationary, the same condition applies to z. If the size of the Jordan cell for λ is $k > 1$, we must check that the condition is sufficient. For such a cell we have $y_{ti} = \lambda y_{t-1,i} + w_{ti}$, and

$$y_{t,i-j} = \lambda y_{t-1,i-j} + y_{t-1,i-j+1} + w_{t,i-j}$$

for $j = 0, \ldots, k-1$. From these equations it is easy to see that $y_{t,i-j}$, $j = 0, 1, \ldots, k-1$, are ARMA(j+1,j) processes with autoregressive parts $(1 - \lambda B)^{j+1}$, in which the root λ cannot be cancelled by the moving average parts. Therefore the condition $|\lambda| < 1$ is necessary and sufficient for the existence of the required solution. ∎

The above proof is a constructive one. Namely, we have for each Jordan cell $y_{ti} = \sum_{j=0}^{\infty} \lambda^j w_{t-j,i}$ for $|\lambda| < 1$, $y_{ti} = \sum_{j=0}^{\infty} \lambda^{-j} w_{t+j,i}$ for $|\lambda| > 1$, and if the size of the Jordan cell is $k > 1$, the remaining $y_{t,i-l}$, $l = 1, \ldots, k-1$ can be obtained recurrently from the formulas $y_{t,i-l} = \sum_{j=0}^{\infty} \lambda^j (y_{t-l+j,i-l+1} + w_{t-j,i-l})$ for $|\lambda| < 1$, $y_{t,i-l} = \sum_{j=0}^{\infty} \lambda^{-j} (y_{t-l+j,i-l+1} + w_{t+j,i-l})$ for $|\lambda| > 1$.

Acknowledgments

I would like to thank the Organizing Committee for partial financial support. Many thanks also to Dr Sahib Esa for interesting discussions on the subject of the paper. The partial support of the Ministry of Science and Education (contract No. MM 421/94) is gratefully acknowledged.

References

[Bo]   G. N. Boshnakov. Periodically correlated sequences: Some properties and recursions. Research Report 1, Division of Quality Technology and Statistics, Luleo University, Sweden, Mar 1994.

[BD]   P.J. Brockwell and R.A. Davis. *Time series: theory and methods.* Springer Series in Statistics. Springer, second edition, 1991.

[JB]   R.H. Jones and W.M. Brelsford. Time series with periodic structure. *Biometrika* **54** (1967), 403–408.

[LM]   H. Lütkepohl and E.O. Maschke. Bemerkung zur Lösung der Yule-Walker-Gleihungen. *Metrika*, **35/5** (1988), 287–289.

[Pa]   M. Pagano. On periodic and multiple autoregression. *Ann.Statist.*, **6** (1978), 1310–1317.

Institute of Mathematics,
Bulgarian Academy of Sciences
Sofia 1110 Bulgaria

Received September 11, 1996

On Nonlinear SDE'S whose Densities Evolve in a Finite–Dimensional Family

Damiano Brigo

1. Introduction

This paper moves from the differential geometric approach to nonlinear filtering as developed by Brigo, Hanzon and LeGland [BHL] and Brigo [BR2]. We consider the projection in Fisher metric of the density evolution of a diffusion process onto an exponential manifold. We examine the projected evolution and interpret it as the density evolution of a different diffusion process. This result leads to the following consequence: Given an arbitrary diffusion coefficient and an arbitrary exponential family, one can define a drift such that the density of the resulting diffusion process evolves in the prescribed (for example Gaussian) exponential family. We construct also nonlinear SDEs with prescribed stationary exponential density. We shortly introduce a possible financial interpretation of this result. Furthermore we present some hints on how, for particular models, convergence of the original density towards an invariant distribution implies existence of a finite dimensional exponential family for which the projected density converges to the same distribution. Finally, we use the results proven on diffusion processes to derive existence results for filtering problems. Given a prescribed (possibly nonlinear) diffusion coefficient for the state equation, a prescribed (possibly nonlinear) observation function and a partially prescribed exponential family, one can define a drift for the state equation such that the resulting nonlinear filtering problem has a solution which is finite dimensional and which stays on the prescribed exponential family.

2. Projection of a diffusion density evolution

On the complete probability space $(\Omega, \mathcal{F}, P)$ consider a scalar Itô diffusion

$$dX_t = f_t(X_t)dt + \sigma_t(X_t)dW_t, \quad X_0,$$

where $\{W_t, t \geq 0\}$ is a standard Brownian motion independent of the initial condition X_0. Consider the following assumption.

(A) The initial state X_0 and the coefficients f_t, $a_t := \sigma_t^2$ are such that there exists a unique positive $C^{1,2}$ density $p(t,x) = p_{X_t}(x)$ which solves the Fokker–Planck equation:

$$\frac{\partial p(t,x)}{\partial t} = \mathcal{L}_t^* p(t,x), \qquad \mathcal{L}_t = f_t \frac{\partial}{\partial x} + \tfrac{1}{2} a_t \frac{\partial^2}{\partial x^2},$$

$$\mathcal{L}_t^* p = -\frac{\partial}{\partial x}(f_t p) + \tfrac{1}{2} \frac{\partial^2}{\partial x^2}(a_t p).$$

In the following, we shall use the abbreviation $p_t(\cdot) = p(t, \cdot)$.

More details are available in Brigo [BR1, BR2]. At this point we introduce the geometric structure which permits to project the Fokker–Planck equation onto a finite dimensional manifold of densities. We rewrite the Fokker–Planck equation as an equation in L_2:

$$\frac{\partial \sqrt{p_t}}{\partial t} = \frac{\mathcal{L}_t^* p_t}{2\sqrt{p_t}}. \qquad (2.1)$$

Next, we select a finite dimensional manifold of square roots of exponential densities to approximate $\sqrt{p_t}$. We formalize this idea via the following

Definition 2.1 *Let* $\{c_1, \cdots, c_m\}$ *be scalar functions defined on* $\mathbf{R}$*, having at most polynomial growth and twice continuously differentiable, and such that* $\{1, c_1, \cdots, c_m\}$ *are* linearly independent. *Assume that the convex set*

$$\Theta_0 := \{\theta \in \mathbf{R}^m : \psi(\theta) = \log \int \exp[\theta^T c(x)] \, dx < \infty\},$$

has non–empty interior. *Then*

$$EM(c) = \{p(\cdot, \theta), \, \theta \in \Theta\}, \qquad p(x, \theta) := \exp[\theta^T c(x) - \psi(\theta)],$$

where $\Theta \subseteq \Theta_0$ *is open, is called an exponential family of probability densities.*

Remark 2.2 *Given the linearly independent scalar functions* $\{c_1, \cdots, c_m\}$ *it may happen that the densities* $\exp[\theta^T c(x)]$ *are not integrable. However, it is always possible to extend the family so as to deal with integrable densities only. See Brigo [BR1, BR2] for the details.*

We shall denote by $EM(c)^{1/2}$ the subset of L_2 given by square roots of densities of $EM(c)$. The map $\sqrt{p(\cdot, \theta)} \mapsto \theta$ can be seen as a coordinate system that gives a manifold structure to $EM(c)^{1/2}$ (see Brigo, Hanzon and Le Gland [BHL] and Brigo [BR2] for more details). Now consider a generic differentiable curve $t \mapsto \sqrt{p(\cdot, \theta_t)}$ on L_2. Its tangent vector in θ_t is given according to the chain rule:

$$\frac{d}{dt} \sqrt{p(\cdot, \theta_t)} = \sum_{i=1}^{m} \frac{\partial \sqrt{p(\cdot, \theta_t)}}{\partial \theta_i} \dot{\theta}_t^{\ i} = \sqrt{p(\cdot, \theta_t)} \sum_{i=1}^{m} \tfrac{1}{2} [c_i - E_{\theta_t} c_i] \dot{\theta}_t^{\ i}, \quad (2.2)$$

from which we see that tangent vectors in $\theta_t = \theta$ to all curves lie in the linear (tangent) space

$$\text{span} \{ \tfrac{1}{2} \sqrt{p(\cdot, \theta)} \, [c_1(\cdot) - E_\theta c_1], \ldots, \tfrac{1}{2} \sqrt{p(\cdot, \theta)} \, [c_m(\cdot) - E_\theta c_m] \}. \quad (2.3)$$

Define the following quantity

$$g(\theta)_{ij} := 4 \langle \frac{\partial \sqrt{p(\cdot, \theta)}}{\partial \theta_i}, \frac{\partial \sqrt{p(\cdot, \theta)}}{\partial \theta_j} \rangle = E_\theta [c_i \, c_j] - E_\theta c_i \, E_\theta c_j, \quad i, j = 1, \ldots, m,$$

where $\langle \cdot, \cdot \rangle$ is the inner product of L_2. Note that $g(\theta)$ is the Fisher information matrix. Consider for all $\theta \in \Theta$ the orthogonal projection Π_θ from L_2 onto (2.3):

$$\Pi_\theta [v] := \sum_{i=1}^{m} [\sum_{j=1}^{m} 4 g^{ij}(\theta) \, \langle v, \tfrac{1}{2} \sqrt{p(\cdot, \theta)} \, (c_j - E_\theta c_j) \rangle] \, (c_i - E_\theta c_i) \tfrac{1}{2} \sqrt{p(\cdot, \theta)}.$$

At this point we project the Fokker–Planck equation for $\sqrt{p_t}$, and we obtain the following (m–dimensional) ODE on the manifold $EM(c)^{1/2}$:

$$\frac{\partial}{\partial t} \sqrt{p(\cdot, \theta_t)} = \Pi_{\theta_t} [\frac{\mathcal{L}_t^* p(\cdot, \theta_t)}{2 \sqrt{p(\cdot, \theta_t)}}] . \quad (2.4)$$

This equation is well defined and admits locally a unique solution if

$$(B) \qquad \frac{\mathcal{L}_t^* p(\cdot, \theta)}{2 \sqrt{p(\cdot, \theta)}} \in L_2 \ \ \forall \theta \in \Theta \iff E_\theta \{ \alpha_{t,\theta}^2 \} < \infty \ \ \forall \theta \in \Theta,$$

$$\alpha_{t,\theta} := \frac{\mathcal{L}_t^* p(\cdot, \theta)}{p(\cdot, \theta)} = -f_t \frac{\partial}{\partial x} (\theta^T c) - \frac{\partial f_t}{\partial x} + \tfrac{1}{2} [a_t \frac{\partial^2}{\partial x^2} (\theta^T c) +$$

$$+ a_t \, (\frac{\partial}{\partial x} (\theta^T c))^2 + 2 \frac{\partial a_t}{\partial x} \frac{\partial}{\partial x} (\theta^T c) + \frac{\partial^2 a_t}{\partial x^2}] ,$$

(see Brigo [BR1, BR2] for the details and for the following case: (B) holds when the functions f_t, $\partial_x f_t$, a_t, $\partial_x a_t$, $\partial_{xx}^2 a_t$, $\partial_x c$, $\partial_{xx}^2 c$ have at most polynomial growth). Writing the projection map explicitly, and by using standard results on exponential families and duality $\mathcal{L}^* / \mathcal{L}$ one obtains the following

Theorem 2.3 (projected evolution of the density of an Itô diffusion) *Assumptions (A) and (B) on the initial value X_0 and coefficients f, a of the Itô diffusion X and on the exponent functions c of the exponential family $EM(c)$ in force. Then the projection of Fokker–Planck equation describing the local evolution of $p_t = p_{X_t}$ onto $EM(c)^{1/2}$ reads, in local coordinates:*

$$\frac{\partial}{\partial t}\sqrt{p(\cdot, \theta_t)} = E_{\theta_t}\{\mathcal{L}_t\, c\}^T\, g^{-1}(\theta_t)\, [c(\cdot) - E_{\theta_t}c]\, \frac{\sqrt{p(\cdot, \theta_t)}}{2}, \qquad (2.5)$$

and the differential equation describing the local evolution of the parameters for the projected evolution of the density is

$$\dot{\theta}_t = g^{-1}(\theta_t)\, E_{\theta_t}\{\mathcal{L}_t\, c\}. \qquad (2.6)$$

3. An interpretation of the projected density evolution

Problem 3.1 *Let be given a diffusion coefficient $a_t(\cdot) := \sigma_t^2(\cdot)$, $t \geq 0$, a drift $f_t(\cdot)$, $t \geq 0$, and an initial condition X_0 satisfying assumption (A). Consider the SDE*

$$dX_t = f_t(X_t)dt + \sigma_t(X_t)dW_t, \quad X_0.$$

Let be given an exponential family $EM(c)$. Characterize the SDE's whose initial condition is X_0, whose diffusion coefficient is a, and whose density evolutions coincide with the projected density evolution of X onto $EM(c)$ (Theorem 2.3).

This is just a particular case of a more general problem, as explained in Brigo (1996) [BR2]. In order to solve this problem, define a diffusion

$$dY_t = u_t(Y_t)dt + \sigma_t(Y_t)dW_t, \quad Y_0 = X_0, \qquad (3.1)$$

with the same diffusion coefficient as X. We shall try to define the drift u in such a way that the density of Y_t coincides with $p(\cdot, \theta_t)$. Let $\mathcal{T}_t$ denote the backward differential operator of Y_t (and $\mathcal{T}_t^*$ its adjoint operator):

$$\mathcal{T}_t = u_t\frac{\partial}{\partial x} + \tfrac{1}{2}a_t\frac{\partial^2}{\partial x^2}.$$

Clearly, the density of Y_t coincides with $p(\cdot, \theta_t)$ if

$$\frac{\mathcal{T}_t^* p(\cdot, \theta_t)}{2\sqrt{p(\cdot, \theta_t)}} = \Pi_{\theta_t}[\frac{\mathcal{L}_t^* p(\cdot, \theta_t)}{2\sqrt{p(\cdot, \theta_t)}}],$$

for all $t \geq 0$, which we can rewrite as

$$T_t^* p(\cdot, \theta_t) = E_{\theta_t}\{\mathcal{L}_t c\}^T g^{-1}(\theta_t) [c(\cdot) - E_{\theta_t} c] \, p(\cdot, \theta_t) \, ,$$

for all $t \geq 0$. By simple calculations one can rewrite this equation as a differential equation for u, whose solution u^* yields (details in Brigo [BR1, BR2]).

Theorem 3.2 (Solution of problem 3.1) *Assumptions (A) and (B) on the initial value X_0 and the coefficients f, a of the Itô diffusion X in force. Let $p(\cdot, \theta_t)$ be the projected density evolution, according to Theorem 2.3. Define*

$$
\begin{aligned}
dY_t &= u_t^*(Y_t)dt + \sigma_t(Y_t)dW_t, \\
u_t^*(x) &= \tfrac{1}{2}\frac{\partial a_t}{\partial x}(x) + \tfrac{1}{2}a_t(x)\theta_t^T\frac{\partial c}{\partial x}(x) + \\
&\quad - E_{\theta_t}\{\mathcal{L}_t c\}^T g^{-1}(\theta_t) \int_{-\infty}^{x} (c(y) - E_{\theta_t}c) \, \exp[\theta_t^T(c(y) - c(x))]dy \, .
\end{aligned} \tag{3.2}
$$

Then Y is an Itô diffusion whose density evolves according to the projected evolution $p(\cdot, \theta_t)$ of X onto $EM(c)$.

Example 3.3 *An arbitrage–theory interpretation of the solution of Problem 3.1*

Given a price process $\{B_t, t \in [0 \ T]\}$ for a risk–free asset and a price process $\{S_t, t \in [0 \ T]\}$ for a stock,

$$
\begin{aligned}
dB_t &= r_t B_t dt, \quad B_0, \\
dS_t &= S_t \, \phi_t(S_t)dt + S_t \, v_t(S_t)dW_t, \quad S_0,
\end{aligned}
$$

and a simple contingent claim $Z = \phi(S_T)$, one can replace ϕ obtaining that the density of S stays in a prescribed exponential family $EM(c)$ (via Theorem 3.2), without changing the pricing of Z. See Brigo [BR1, BR2] for the details.

4. A simple convergence result

In the particular case where we select a *constant* diffusion coefficient, convergence of the density of the original process X implies existence of at least an exponential family such that the projected density $p(\cdot, \theta_t)$ converges towards the same stationary distribution, no matter how we choose θ_0. Assume then $\sigma_t(x) = 1$ for all x, t. We have the following

Theorem 4.1 (Global stability of the projected evolution) *Assume that the diffusion process*

$$dX_t = f(X_t)dt + dW_t, \quad X_0$$

satisfies assumption (A), with f having at most polynomial growth and nonzero in a set with positive Lebesgue measure. Let F be a primitive of f and assume that X admits the stationary density $\bar{p} \propto \exp[2F]$. Assume

$$\int f(x)^4 \exp(2F(x))dx < \infty, \quad \int (\partial_x f)^2(x) \exp(2F(x))dx < \infty.$$

Then there exists at least an exponential family $EM(c)$ such that $\bar{p} \in EM(c)$ and the projected density $p(\cdot, \theta_t)$ given by Theorem 2.3 converges pointwise towards $\bar{p}$ for all possible initial $\theta_0 \in \Theta$.

For the proof, see Brigo [BR1, BR2]. We remark that one uses the exponential family $EM(\bar{c})$, where $\bar{c}(x) := 2\int_0^x f(z)dz = 2F(x)$.

5. Examples

In this section we consider some simple applications of the above theory which lead to nonlinear SDEs whose solutions have exponential densities. We shall be concerned mainly with nonlinear SDEs with Gaussian densities, but we shall present also an example of nonlinear SDEs with densities in $EM(x^4)$ and of nonlinear SDEs with stationary density in an arbitrary one–dimensional family $EM(c)$. More examples are available in Brigo [BR1, BR2].

5.1 The case of a one dimensional unit-variance Gaussian Manifold

Consider the one-dimensional exponential manifold $EM(x)$ given by

$$p(\cdot, \theta) = \sqrt{\frac{1}{2\pi}} \exp[\theta x - \tfrac{1}{2}x^2 - \tfrac{1}{2}\theta^2] \sim \mathcal{N}(\theta, 1) , \quad \theta \in \mathbf{R}.$$

Notice that the term $-\tfrac{1}{2}x^2$ in the exponent defines a reference measure different from the Lebesgue measure. There are no modifications for the projected Fokker–Planck equation. Notice that $\mathcal{L}_t x = f_t$. Moreover, the inverse of the Fisher metric is $g^{-1}(\theta) = 1$ and does not depend on θ. The projected density is described by the solution of the ODE

$$\dot{\theta}_t = E_{\theta_t}\{f_t\}.$$

By applying the previously found formula (3.2) with slight modifications due to the term $-\frac{1}{2}x^2$ we obtain

$$u_t^*(x) := \frac{1}{2}\frac{\partial a_t}{\partial x}(x) + \frac{1}{2}a_t(x)[\theta_t - x] + E_{\theta_t}\{f_t\}.$$

Now, according to the choice of f_t, one can obtain different results. By choosing respectively $f_t(x) = 0$, $f_t(x) = k$, $f_t(x) = x$ one has the following result. Let

$$u_t^1(x) := \frac{1}{2}\frac{\partial a_t}{\partial x}(x) + \frac{1}{2}a_t(x)[\theta_0 - x],$$

$$u_t^2(x) := \frac{1}{2}\frac{\partial a_t}{\partial x}(x) + \frac{1}{2}a_t(x)[\theta_0 + kt - x] + k,$$

$$u_t^3(x) := \frac{1}{2}\frac{\partial a_t}{\partial x}(x) + \frac{1}{2}a_t(x)[\theta_0 \exp(t) - x] + \theta_0 \exp(t).$$

The SDEs $\quad dY_t^i = u_t^i(Y_t^i)dt + \sigma_t(Y_t^i)dW_t \quad Y_0^i \sim \mathcal{N}(\theta_0, 1), \quad i = 1, 2, 3,$ have densities $Y_t^1 \sim \mathcal{N}(\theta_0, 1)$, $Y_t^2 \sim \mathcal{N}(\theta_0 + kt, 1)$, and $Y_t^3 \sim \mathcal{N}(\theta_0 \exp(t), 1)$ for all possible σ_t.

5.2 SDEs with densities evolving in $EM(x^4)$

Consider the exponential family $EM(x^4)$. For this family, we have

$$p(x, \theta) = \rho\,\theta^{1/4}\,\exp(-\theta x^4), \quad \theta > 0, \quad \rho := \frac{2}{\Gamma(1/4)}.$$

In this case $\psi(\theta) = -(\log\theta)/4 - \log\rho$, $E_\theta[x^4] = -1/(4\theta)$, and $g(\theta) = 1/(4\theta^2)$. Let us consider an arbitrary diffusion coefficient a, and take a drift f defined ad hoc according to

$$f_t(x) := -\frac{3\,a_t(x)}{2x} - \frac{x^5}{5}$$

such that $E_\theta\{\mathcal{L}_t x^4\} = 4\theta^2$, which causes equation (2.6) to become $\dot\theta_t = 1$. As a consequence, $\theta_t = \theta_0 + t$ and, according to Theorem 3.2, if we set

$$u_t^*(x) \quad := \quad \frac{1}{2}\frac{\partial a_t(x)}{\partial x} - 2a(x)\,(\theta_0 + t)\,x^3 - \exp[(\theta_0 + t)x^4]$$

$$\int_{-\infty}^{x}(-y^4 + \frac{1}{4(\theta_0 + t)})\exp[-(\theta_0 + t)y^4]dy\ ,$$

we obtain that the SDE

$$dY_t = u_t^*(Y_t)dt + \sigma_t(Y_t)dW_t, \quad Y_0 \sim \rho\,\theta_0^{1/4}\,\exp(-\theta_0 x^4),$$

has density $Y_t \sim \rho\,(\theta_0 + t)^{1/4}\,\exp[-(\theta_0 + t)x^4]$ for all possible σ_t.

5.3 SDEs with prescribed diffusion coefficient and with prescribed stationary exponential density

In Brigo [BR1, BR2] we have proved in detail the following result. Under suitable regularity and growth assumptions on the functions c and σ_t, the SDE

$$dY_t = \tfrac{1}{2}\frac{\partial a_t}{\partial x}(Y_t)dt + \tfrac{1}{2}a_t(Y_t)\theta_0\frac{\partial c}{\partial x}(Y_t)dt + \sigma_t(Y_t)dW_t,$$
$$p_{Y_0}(x) = \exp[\theta_0 c(x) - \psi(\theta_0)],$$

has stationary density $p_{Y_t}(x) = p_{Y_0}(x), \quad t \geq 0$, for all possible σ_t.

6. Application to filtering

Consider the filtering problem with continuous time state X and discrete time observations Z (for references and details, see Brigo [BR1, BR2]):

$$
\begin{aligned}
dX_t &= f_t(X_t, Z^t)dt + \sigma_t(X_t)\,dW_t, &(6.1)\\
Z_n &= h(X_{t_n}) + V_n.
\end{aligned}
$$

In this model only discrete–time observations are available, at time instants $0 = t_0 < t_1 < \cdots < t_n < \cdots$, and $\{V_n, n \geq 0\}$ is a Gaussian white noise sequence independent of $\{X_t, t \geq 0\}$. $Z^t := [Z_1, \ldots, Z_k]$, where $t_k \leq t < t_{k+1}$. It is possible, under rather general conditions, to solve the following problem:

Problem 6.1 (Existence of finite–dimensional filters for nonlinear filtering problems) *Given nonlinear and arbitrary (regular enough) h, σ and $c_1, \ldots, c_m$, find f such that the filtering problem for the system (6.1) admits a finite–dimensional filter expressed by a conditional density $p_{X_t|Z^t}$ evolving in $EM([h, \ h^2, \ c_1, \ldots, c_m]^T)$ when $p_{X_0} \in EM([h, \ h^2, \ c_1, \ldots, c_m]^T)$.*

The required f can be constructed easily by using the previous results, in particular Theorem 3.2. For the details, see Brigo [BR1, BR2].

Acknowledgments

While working on this article the author was supported by the *Istituto Nazionale di Alta Matematica 'F. Severi'*. Participation to the conference on SDDE's was supported by the Commission of the European Communities, under an individual TMR fellowship, contract number ERBFMBICT-960791.

References

[BR1]   D. Brigo, On diffusions with prescribed diffusion coefficients whose densities evolve in prescribed exponential families, *Internal Report CNR–LADSEB* **02/96** (1996), CNR–LADSEB, Italy, March 1996.

[BR2]   D. Brigo, Filtering by Projection on the Manifold of Exponential Densities, PhD Thesis, Free University of Amsterdam, forthcoming.

[BHL]   D. Brigo, B. Hanzon, F. Le Gland, A differential geometric approach to nonlinear filtering: the projection filter, *Publication Interne IRISA* **914**, IRISA, France, (1995) (available on the internet at the URL address: ftp://ftp.irisa.fr/techreports/1995/PI-914.ps.Z).

c/o Prof. G.B. Di Masi, Dipartimento di Matematica Pura e Applicata
Università di Padova, Via Belzoni 7, I-35131 Padova, Italy
E-mail: brigo@pdmat1.unipd.it

Received September 12, 1996

Composition of Skeletons and Support Theorems

María Emilia Caballero, Begoña Fernández, and David Nualart

Abstract

In this paper we introduce a notion of composition of skeletons for Wiener functionals and, as an application, we characterize the topological support of the law of the solution to a stochastic differential equation with random initial condition.

1. Introduction

The support theorem characterizes the topological support of the law of a diffusion process as the closure of the set of deterministic paths obtained replacing the driving white noise by smooth functions. This type of result was first proved by Stroock and Varadhan in [SW] using the notion of approximative continuity. A more direct approach to establish support theorems was introduced by Mackevičius in [Ma], and it was later developed by Gyöngy [Gy] and Millet and Sanz-Solé [MS2]. In these papers the main idea is to establish the support theorem by a direct approximation argument. A general notion of skeleton for a Wiener functional which leads to a characterization of the support of the law of this functional was presented by Millet and Sanz-Solé in [MS2]. In this note we introduce a notion of composition of skeletons, and deduce a support theorem for the composition of a continuous random field with a Wiener functional. As an application we deduce the support theorem for the solution to stochastic differential equations with random initial condition, generalizing the results established in [MN] by the method of approximative continuity.

2. Composition of skeletons

Let us first introduce some preliminaries and notation. Suppose that H is a real separable Hilbert space whose norm and inner product are denoted by $\|\cdot\|_H$ and $\langle\cdot,\cdot\rangle_H$, respectively. Consider a Gaussian and centered family of random variables $W = \{W(h), h \in H\}$ such that

$$E(W(h)W(g)) = \langle h, g \rangle_H,$$

for all $h, g \in H$. The family W is defined in some complete probability space $(\Omega, \mathcal{F}, P)$. We will assume that $\mathcal{F}$ is generated by W.

In the sequel we will assume that $\{\Pi^N, N \geq 1\}$ is a sequence of linear maps in H of finite-dimensional range converging strongly to the identity. Set $(\Pi^N W)(h) = W(\Pi^N h)$. Then $\Pi^N W$ is an element of $L^2(\Omega; H)$ such that $\langle \Pi^N W, h \rangle_H$ converges in $L^2(\Omega)$ to $W(h)$ for each $h \in H$, as N tends to infinity. On the other hand, for all $h \in H$, we fix a sequence of measurable transformations $T_N^h : \Omega \to \Omega$ such that $P \circ (T_N^h)^{-1}$ is absolutely continuous with respect to P. We will first introduce the definition of skeleton:

Definition 2.1 *Let $G : \Omega \to S$ be a random variable taking values on a Polish space space (S, d). We will say that a measurable function $\Psi : H \to S$ is a skeleton of G if the following two conditions are satisfied:*

(i) For all $\epsilon > 0$ we have

$$\lim_N P \left\{ d(\Psi(\Pi^N W), G) > \epsilon \right\} = 0.$$

(ii) For all $h \in H$ and for every $\epsilon > 0$

$$\lim_N P \left\{ d(G \circ T_N^h, \Psi(h)) > \epsilon \right\} = 0.$$

This notion of skeleton allows to characterize the topological support of the law of the random variable G (see Millet and Sanz-Solé [MS2]):

Proposition 2.2 *Suppose that Ψ is a skeleton of G in the sense of Definition 2.1. Then the support of the law of G is the closure in S of the set $\{\Psi(h), h \in H\}$.*

Consider now the following situation. Let $F(\omega, x)$ be a continuous random field indexed by a parameter $x \in \mathbb{R}^m$. Suppose that ξ is an m-dimensional random variable. We can form the random variable $G(\omega) = F(\omega, \xi(\omega))$. The following question arises: Given a skeleton $\Phi(h, x)$ for $F(x)$, for each $x \in \mathbb{R}^m$, and a skeleton $\phi(h)$ for ξ, under what conditions is the composition $\Phi(h, \phi(h))$ a skeleton for the random variable G? We need the following notion of uniform skeleton:

Definition 2.3 *Let $F : \Omega \times \mathbb{R}^m \to S$ be a random variable taking values on a Polish space space (S, d), such that $F(\omega, x)$ is continuous in x, for each ω. We will say that a mapping $\Phi : H \times \mathbb{R}^m \to S$ such that $\Phi(h, x)$ is continuous in x for each h is a uniform skeleton of F if:*

(i) For all $\epsilon > 0$ and $K > 0$ we have

$$\lim_N P \left\{ \sup_{|x| \leq K} d(\Phi(\Pi^N W, x) = F(x)) > \epsilon \right\} = 0.$$

(ii) For all $h \in H$, $\epsilon > 0$, and $K > 0$ we have

$$\lim_N P \left\{ \sup_{|x| \leq K} d(F(T_N^h(\omega), x), \Phi(h, x)) > \epsilon \right\} = 0.$$

The next proposition assures that the composition $\Phi(h, \phi(h))$ is a skeleton for the composition $F(\omega, \xi(\omega))$:

Proposition 2.4 *Let $F : \Omega \times \mathbb{R}^m \to S$ be a random variable taking values on a Polish space space (S, d), such that $F(\omega, x)$ is continuous in x, for each ω. Suppose that $\Phi : H \times \mathbb{R}^m \to S$ is a uniform skeleton of F. Assume that ξ is an m-dimensional random variable with a skeleton ϕ. Then $\Psi(h) := \Phi(h, \phi(h))$ a skeleton for $G(\omega) := F(\omega, \xi(\omega))$.*

Proof. We have to show that Ψ and G satisfy the conditions (i) and (ii) of Definition 2.1. In order to check (i) we write

$$P \left\{ d(\Phi(\Pi^N W, \phi(\Pi^N W)), F(\xi)) > \epsilon \right\}$$

$$\leq P \left\{ d(\Phi(\Pi^N W, \phi(\Pi^N W)), F(\phi(\Pi^N W))) > \frac{\epsilon}{2} \right\}$$

$$+ P \left\{ d(F(\phi(\Pi^N W)), F(\xi)) > \frac{\epsilon}{2} \right\}$$

$$\leq P \{ |\phi(\Pi^N W)| > K \} + P \left\{ \sup_{|x| \leq K} d(\Phi(\Pi^N W, x), F(x)) > \frac{\epsilon}{2} \right\}$$

$$+ P \left\{ d(F(\phi(\Pi^N W)), F(\xi)) > \frac{\epsilon}{2} \right\} = a_1 + a_2 + a_3.$$

The term a_1 verifies

$$\lim_{K \uparrow \infty} \limsup_{N \to \infty} a_1 = 0.$$

The term a_2 tends to zero as N tends to infinity, for each fixed K, due to condition (i). Finally, the term a_3 tends to zero as N tends to infinity due to the fact that ϕ is a skeleton for ξ and $F(x)$ is continuous in x.

To show condition (ii) we write

$$P \left\{ d(F(T_N^h(\omega), \xi(T_N^h(\omega))), \Phi(h, \phi(h))) > \epsilon \right\}$$

$$\leq P \left\{ d(F(T_N^h(\omega), \xi(T_N^h(\omega))), \Phi(h, \xi(T_N^h(\omega)))) > \frac{\epsilon}{2} \right\}$$

$$+ P \left\{ d(\Phi(h, \xi(T_N^h(\omega))), \Phi(h, \phi(h))) > \frac{\epsilon}{2} \right\}$$

$$\leq P \{ |\xi(T_N^h(\omega))| > K \} + P \left\{ \sup_{\{|x| \leq K\}} d(F(T_N^h(\omega), x), \Phi(h, x)) > \frac{\epsilon}{2} \right\}$$

$$+ P \left\{ d(\Phi(h, \xi(T_N^h(\omega))), \Phi(h, \phi(h))) > \frac{\epsilon}{2} \right\} = b_1 + b_2 + b_3.$$

The term b_1 verifies

$$\lim_{K \uparrow \infty} \lim_{N \to \infty} \sup b_1 = 0.$$

The term b_2 tends to zero as N tends to infinity, for each fixed K, due to condition (ii). Finally, the term b_3 tends to zero as N tends to infinity due to the fact that ϕ is a skeleton for ξ and $\Phi(h, x)$ is continuous in x. ■

3. Application to diffusion processes

Let us now describe the application of the above results to the case of a diffusion process. In the sequel we will assume that the underlying Gaussian process is a d-dimensional Brownian motion $\{W(t), t \in [0, 1]\}$, defined in the canonical probability space $\Omega = C_0([0, 1], \mathbb{R}^d)$. Then $H = L^2([0, 1]; \mathbb{R}^d)$ and $W(h) = \sum_{i=1}^{d} \int_0^1 h^i(t) dW_t^i$.

For each positive integer N, we set

$$(\Pi^N h)_t = \sum_{i=1}^{2^N - 1} 2^N \int_{(i-1)2^{-N}}^{i2^{-N}} h_s ds \mathbf{1}_{(i2^{-N}, (i+1)2^{-N}]}(t).$$

With this definition we have

$$(\Pi^N W)_t = \sum_{i=1}^{2^N - 1} 2^N (W(i2^{-N}) - W((i-1)2^{-N}) \mathbf{1}_{(i2^{-N}, (i+1)2^{-N}]}(t).$$

That is, $\Pi^N W(\omega) = \dot{\omega}^N$, where for any continuous function $\omega \in \Omega$ we denote by ω^N the element of Ω such that on each interval $(i2^{-N}, (i + 1)2^{-N})$, $1 \le i \le 2^N - 1$, has a constant derivative equal to $2^N(\omega(i2^{-N}) - \omega((i-1)2^{-N}))$. More precisely, for each $N \ge 1$ ω^N is the adapted linear interpolation of ω given by

$$\omega_t^N = \omega_{\underline{t}_N} + 2^N (t - \tilde{t}_N)[\omega_{\tilde{t}_N} - \omega_{\underline{t}_N}]$$

where for $\frac{k}{2^N} \le t < \frac{k+1}{2^N}$:

$$\tilde{t}_N = \frac{k}{2^N}, \quad \underline{t}_N = \frac{k-1}{2^N} \vee 0.$$

We will denote by H^1 the subspace of Ω formed by the functions of the form $\varphi(t) = \int_0^t \dot{\varphi}(s) ds$, $\dot{\varphi} \in H = L^2([0, 1]; \mathbb{R}^d)$. H^1 is called the Cameron-Martin space and we will denote by i the isometry between H and H^1, that is, for each $h \in H$ we have $i(h)(t) = \int_0^t h(s) ds$.

For each $h \in H$, let T_N^h be the sequence of absolutely continuous transformations given by

$$T_N^h(\omega) = \omega - \omega^N + i(h).$$

In what follows we will assume the following condition:

H: Let $\sigma : I\!\!R^m \to I\!\!R^m \otimes I\!\!R^d$ and $b : I\!\!R^m \to I\!\!R^m$ be functions such that σ is of class C_b^3 (i.e., the partial derivatives up to order 3 are continuous and bounded), and $b \in C_b^2$.

Consider the following stochastic differential equation of Stratonovich type on $I\!\!R^m$:

$$X_t(x) = x + \int_0^t b(X_s(x))ds + \int_0^t \sigma(X_s(x)) \circ dW_s. \qquad (3.1)$$

We recall that $X(x)$ can be expressed as the solution of the following Itô equation:

$$
\begin{aligned}
X_t(x) &= x + \int_0^t [b(X_s(x)) + \frac{1}{2}\nabla\sigma(X_s(x))\sigma(X_s(x))]ds \\
&+ \int_0^t \sigma(X_s(x))dW_s.
\end{aligned}
$$

For any $h \in H$ let $\Phi(h, x)$ be the solution of

$$\Phi(h, x)_t = x + \int_0^t b(\Phi(h, x)_s)ds + \int_0^t \sigma(\Phi(h, x)_s)h_s ds. \qquad (3.2)$$

For each $N \geq 1$, and $h \in H$ let $X^N(x) = X(x, T_N^h(\omega)) = X(x, \omega - \omega^N + i(h))$. Clearly, $X^N(x)$ satisfies:

$$
\begin{aligned}
X_t^N(x) &= x + \int_0^t [b(X_s^N(x)) + \frac{1}{2}\nabla\sigma(X_s^N(x))\sigma(X_s^N(x))]ds \\
&+ \int_0^t \sigma(X_s^N(x))h_s ds + \int_0^t \sigma(X_s^N(x))dW_s \\
&- \int_0^t \sigma(X_s^N(x))\dot\omega_s^N ds. \qquad (3.3)
\end{aligned}
$$

In order to prove the main result we will need the following Lemma:

Lemma 3.1 *Let σ and b be bounded functions that satisfy hypothesis* **H.** *Then for each $p \geq 1$ and for every compact set K there exists a constant C*

such that for $s, t \in [0, 1]$ we have

$$\sup_n \sup_{x \in K} (|X_t^N(x) - X_s^N(x)|^{2p}) \leq C|t - s|^p,$$

$$\sup_n \sup_{x \in K} (|\nabla(X_t^N(x) - X_s^N(x))|^{2p}) \leq C|t - s|^p.$$

The first inequality has been proved, for instance, in [MS2, Proposition 3.1], without the supremum in x, and it is easy to see that this estimates holds uniformly for $x \in K$. The second inequality can be proved by the same arguments.

The main result of this section is the following:

Theorem 3.2 *The mapping $\Phi : H \times \mathbb{R}^m \to C([0, 1]; \mathbb{R}^m)$ is a uniform skeleton of $X(x)$ in the sense of Definition 2.3.*

Proof. The proof will follow from the following convergences:

$$\lim_N P \left(\sup_{t \in [0,1], |x| \leq K} |\Phi(\Pi^N W)_t - X_t| > \epsilon \right) = 0, \tag{3.4}$$

and

$$\lim_N P \left(\sup_{t \in [0,1], |x| \leq K} |X_t \circ T_N^h - \Phi(h)_t| > \epsilon \right) = 0. \tag{3.5}$$

The proof will be done in two steps.

Step 1. We assume in addition to condition **H** that σ and b are bounded functions. Condition (3.4) was proved by Bismut ([Bi, Theorem 1.2, p. 39]) when σ and b are bounded C^∞-functions with bounded partial derivatives. In fact the proof is valid under less regularity conditions on the coefficients.

To prove Condition (3.5) we use Sobolev's embedding theorem to obtain for each compact set $K \subset R^d$:

$$E[\sup_{x \in K} \sup_{t \in [0,1]} |X_t^N(x) - \Phi(h, x)_t|^2]$$

$$\leq M \sup_{x \in K} E[\sup_{t \in [0,1]} |X_t^N(x) - \Phi(h, x)_t|^2]$$

$$+ M \sup_{x \in K} E[\sup_{t \in [0,1]} |\nabla(X_t^N(x) - \Phi(h, x)_t)|^2]. \tag{3.6}$$

We will give a sketch of the proof of the convergence to zero for the second summand in (3.6). The proof of the convergence for the first summand is similar and in fact is contained in [MS2, Theorem 3.5], since the estimations given there can be taken uniformly in $x \in K$. For each multi-index $\alpha = (\alpha_1, \cdots, \alpha_d)$ such that $|\alpha| = 1$, we have:

$$E[\sup_{s \leq t} |\nabla^\alpha (X_s^N(x) - \Phi(h, x)_s)|^2]$$

$$\leq K \Big\{ E[\sup_{s \leq t} | \int_0^s (\nabla b(X_u^N(x)) \nabla^\alpha X_u^N(x) - \nabla b(\Phi(h, x)_u) \nabla^\alpha \Phi(h, x)_u) du|^2]$$

$$+ E[\sup_{s \leq t} | \int_0^s (\nabla \sigma(X_u^N(x)) \nabla^\alpha X_u^N(x) - \nabla \sigma(\Phi(h, x)_u) \nabla^\alpha \Phi(h, x)_u) h_u du|^2]$$

$$+ E[\sup_{s \leq t} | \int_0^s (\nabla \sigma(X_u^N(x))) \nabla^\alpha X_u^N(x) dW_u$$

$$- - \int_0^s \nabla \sigma(X_u^N(x)) \nabla^\alpha X_u^N(x) \dot{\omega}_u^N du$$

$$+ \frac{1}{2} \int_0^s [\nabla(\nabla\sigma)(X_u^N(x))\sigma(X_u^N(x)) + (\nabla\sigma)(\nabla\sigma)(X_u^N(x))\nabla^\alpha X_u^N(x) du|^2 \Big\}$$

From the Lipschitz and boundedness conditions on $b, \nabla b, \sigma, \nabla \sigma$, the fact that $|\Phi(h, x)_u| + |\nabla\Phi(h, x)_u| \leq M$ for $x \in K$ and Schwartz inequality, we obtain:

$$E[\sup_{s \leq t} |\nabla^\alpha(X_s^N(x) - \Phi(h, x)_s)|^2]$$

$$\leq C \Big\{ \int_0^t E[\sup_{s \leq u} |X_s^N(x) - \Phi(h, x)_s|^2](|h_u|du + du)$$

$$+ \int_0^t E[\sup_{s \leq u} |\nabla^\alpha X_s^N(x) - \nabla^\alpha \Phi(h, x)_s|^2](|h_u|du + du)$$

$$+ E[\sup_{s \leq t} | \int_0^s \nabla\sigma(X_u^N(x))) \nabla^\alpha X_u^N(x) dW_u$$

$$- \int_0^s \nabla\sigma(X_u^N(x)) \nabla^\alpha X_u^N(x) \dot{\omega}_u^N du$$

$$+ \frac{1}{2} \int_0^s [\nabla(\nabla\sigma)(X_u^N(x))\sigma(X_u^N(x)) + (\nabla\sigma)(\nabla\sigma)(X_u^N(x))\nabla^\alpha X_u^N(x)]dx|^2 \Big\}$$

$$= A_1^N(x) + A_2^N(x) + A_3^N(x),$$

where

$$A_2^N = \int_0^t E[\sup_{s \leq u} |\nabla^\alpha X_s^N(x) - \nabla^\alpha \Phi(h, x)_s|^2](|h_u|du + du).$$

In order to apply Gronwall's inequality, we need to show that

$$\sup_{x \in K} \lim_{N \to \infty} A_i^N(x) = 0, \; i = 1, 3.$$

We know that $\sup_{x \in K} \lim_{N \to \infty} A_1^N(x) = 0$. From Burkholder's inequality, we have

$$
\begin{aligned}
A_3^N(x) \;\leq\; C\Big\{ &\int_0^t E[|\nabla^\alpha X_{\underline{u}_N}^N(x) - \nabla^\alpha X_u^N(x)|^2]\,du \\
&+\int_0^t E[|\nabla^\alpha X_u^N(x||X_{\underline{u}_N}^N(x) - X_u^N(x)|]^2\,du \\
&+E[\sup_{s \leq t} |\int_0^s \nabla\sigma(X_{\underline{u}_N}^N(x))(\nabla^\alpha X_{\underline{u}_N}^N(x))(dW_u - \dot{w}_u^N du)|^2] \\
&+E[\sup_{s \leq t} [\int_0^s \nabla\sigma(X_{\underline{u}_N}^N(x))(\nabla^\alpha X_u^N(x) - \nabla^\alpha X_{\underline{u}_N}^N(x)]\dot{w}_u^N\,du \\
&-\frac{1}{2}\int_0^s (\nabla\sigma)\nabla\sigma)(X_u^N(x))\nabla^\alpha X_u^N(x)du]^2 \\
&+E[\sup_{s \leq t}[\int_0^s \nabla^\alpha X_u^N(\nabla\sigma(X_u^N(x)) - \nabla\sigma(X_{\underline{u}_n}^N(x))\dot{w}_u^N\,du \\
&-\frac{1}{2}\int_0^s \nabla(\nabla\sigma)\sigma(X_u^N(x))\nabla^\alpha X_u^N(x)du]^2] \\
=\; &B_1^N(x) + B_2^N(x) + B_3^N(x) + B_4^N(x) + B_5^N(x).
\end{aligned}
$$

From Lemma 3.1 and Schwartz inequality $\sup_{x \in K} \lim_{N \to \infty}(B_1^N(x) + B_2^N(x)) = 0$. Following the same lines as Millet and Sanz-Solé [MS2, Lemma 3.2], we obtain $\sup_{x \in K} \lim_{N \to \infty} B_3^N(x) = 0$. For the term $B_4^N(x)$, we consider the Taylor's expansion for $\nabla\sigma$ around $X_{\underline{u}_N}^N(x)$, that is,

$$
\nabla\sigma(X_u^N(x)) - \nabla(X_{\underline{u}_N}^N(x)) = \nabla(\nabla\sigma)(X_{\underline{u}_N}^N(x))(X_u^N(x) - X_{\underline{u}_N}^N(x)) + R,
$$

where $|R| \leq C|X_u^N(x) - X_{\underline{u}_N}^N(x)|^2$ and decompose the term $X_u^N(x) - X_{\underline{u}_N}^N(x)$ in the following way:

$$
\begin{aligned}
X_u^N(x) - X_{\underline{u}_N}^N(x) \;=\; &\int_{\underline{u}_N}^u (\sigma(X_s^N(x)) - \sigma(X_{\underline{s}_N}^N(x)))dW_s \\
&-\int_{\underline{u}_N}^u (\sigma(X_s^N(x)) + \sigma(X_{\underline{s}_N}^N(x)))\dot{w}_s^N\,ds \\
&+\int_{\underline{u}_N}^u \sigma(X_s^N(x))h_s\,ds \\
&+\int_{\underline{u}_N}^u [b(X_s^n(x)) + \frac{1}{2}\nabla\sigma\sigma(X_s^N(x))]ds \\
&+\int_{\underline{u}_n}^u \sigma(X_{\underline{s}_N}^N(x))(dW_s - \dot{w}_s^N\,ds).
\end{aligned}
$$

Finally, for the term $B_5^N(x)$ we follow a similar procedure.

Step 2. The boundedness hypothesis on the coefficients can be removed by the localization argument used, for instance, in [GNS]. For each $R > 0$, let $b_R(x)$ and $\sigma_R(x)$ be functions satisfying Hypothesis **H** and such that

$$b_R(x) = \begin{cases} b(x), & \text{if } |x| \le R - 1 \\ 0, & \text{if } |x| > R \end{cases}$$

and

$$\sigma_R(x) = \begin{cases} \sigma(x), & \text{if } |x| \le R - 1 \\ 0, & \text{if } |x| > R \end{cases}$$

We will denote by X^R the solution to

$$X_t^R(x) = x + \int_0^t b_R(X_s^R(x))ds + \int_0^t \sigma_R(X_s^R(x)) \circ dW_s, \qquad (3.7)$$

and $\Phi^R : H \times R^m \to C([0,1]; I\!\!R^m)$ the uniform skeleton of X^R, whose existence is guaranteed by Step 1.

In order to prove Condition (3.4), we introduce the following stopping times for $\varepsilon > 0$, $R > 0$ and $N \in I\!\!N$

$$\begin{aligned}
\tau_N^\varepsilon &:= \inf\{t : \sup_{x \in K} |\Phi(\Pi^N W)_t - X_t(x)| > \varepsilon\}, \\
\gamma^R &:= \inf\{t : \sup_{x \in K} |X_t(x)| \ge R - 1\} \wedge 1, \\
\tau_N^{\varepsilon,R} &:= \tau_N^\varepsilon \wedge \gamma^R.
\end{aligned}$$

Then we have

$$\begin{aligned}
&P[\sup_{t \in [0,1], x \in K} |\Phi(\Pi^N W, x)_t - X_t(x)| \ge \varepsilon] \\
&= P[\sup_{t \in [0, \tau_N^\varepsilon]}]|\Phi(\Pi^N W, x)_t - X_t(x)| \ge \varepsilon] \\
&\le P[\sup_{t \in [0, \tau_N^\varepsilon]} |\Phi(\Pi^N W, x)_t - X_t(x)| \ge \varepsilon, \ \gamma^R = 1] \\
&\quad + P[\sup_{t \in [0, \tau_N^\varepsilon]} |\Phi(\Pi^N W, x)_t - X_t(x) \ge \varepsilon, \ \gamma^R < 1] \\
&\le P[\sup_{t \in [0, \tau_N^\varepsilon]} |\Phi(\Pi^N W, x)_t - X_t(x) \ge \varepsilon, \gamma^R = 1] + P[\gamma^R < 1] \\
&= P[\sup_{t \in [0, \tau_N^\varepsilon]} |\Phi^R(\Pi^N W, x)_t - X_t^R(x) \ge \varepsilon] + P[\gamma^R < 1] \\
&\le P[\sup_{t \in [0,1], x \in K} |\Phi^R(\Pi^N W, x)_t - X_t^R(x) \ge \varepsilon] + P[\gamma^R < 1].
\end{aligned}$$

So we get:

$$\limsup_{N\to\infty} P[\sup_{t\in[0,1],x\in K} |\Phi(\Pi^N W, x)_t - X_t(x)| \geq \varepsilon]$$

$$\leq \limsup_{N\to\infty} \lim_{R\to\infty} P[\sup_{t\in[0,1],x\in K} |\Phi^R(\Pi^N W, x)_t - X_t^R(x)| \geq \varepsilon]$$

$$+ \lim_{R\to\infty} P[\gamma^R < 1] = 0.$$

A similar argument yields Condition (3.5).   ■

4. Examples

Suppose that $d = 1$, that is, $W = \{W_t, t \in [0,1]\}$ is a one-dimensional Brownian motion and $H = L^2([0,1])$. It is easy to check that an element X_0 of the Wiener chaos of order one has always a skeleton. More precisely, if $X_0 = \int_0^1 g_s dW_s$, with $g \in L^2([0,1])$, then $\phi(h) = \int_0^1 g_s h_s ds$ is a skeleton for X_0 in the sense of Definition 2.1. In fact the convergence conditions (i) and (ii) of this definition are satisfied in $L^2(\Omega)$. Actually in this case the approximative continuity property holds (see Nualart and Millet [MN]).

For elements of the Wiener chaos of order greater than one, that is, of the form $X_0 = I_n(g)$, where $g \in L^2([0,1]^n)$ is a symmetric function, and I_n denotes the multiple stochastic integral of order n, the existence of a skeleton in the sense of Definition 2.1 has been proved by Millet and Sanz-Solé in [MS1], assuming that g is continuous. By Theorem 3.2 we deduce the support theorem for an anticipating stochastic differential equation, when the coefficients satisfy hypotheses **H** and the initial condition $X_0 = \sum_{k=0}^n I_k(g_k)$ is a finite sum of multiple stochastic integrals such that the kernels g_k are continuous if $k \geq 2$.

In what follows we will prove the existence of a skeleton for a Wiener chaos of order two without assuming that the kernel is continuous.

Let g be a symmetric function in $L^2([0,1]^2)$ and set

$$X_0 = I_2(g) = 2 \int_0^1 \int_0^t g(s,t) dW_s dW_t.$$

Observe that for $h \in H$, we have:

$$X_0(\omega - \omega^N + h) = X_0(\omega) + 2 \int_0^1 \int_0^1 g(s,t) h_s ds dW_t$$

$$-2 \int_0^1 \int_0^1 g(s,t) h_s ds \dot\omega_t^N dt - 2 \int_0^1 \int_0^t g(s,t) \dot\omega_s^N ds dW_t$$

$$-2\int_0^1\int_0^t g(s,t)dW_s\dot\omega_s^N dt + \int_0^1\int_0^1 g(s,t)\dot\omega_s^N\dot\omega_t^N dsdt$$

$$+\int_0^1\int_0^1 g(s,t)h_s h_t dsdt.$$

Then we obtain the following Proposition:

Proposition 4.1 *Let g be a symmetric function in $L^2([0,1]^2)$ such that*

$$\lim_{n\to\infty}\sum_{i=0}^{2^N-1} 2^N\left(\int_{i/2^N}^{(i+1)/2^N}\int_{i/2^N}^{(i+1)/2^N} g(s,t)dsdt\right) = L(g).$$

Then $\phi : H \to \mathbb{R}$ given by

$$\phi(h) = \int_0^1\int_0^1 g(s,t)h_s h_t dsdt - L(g),$$

is a skeleton for X_0.

Proof. In order to show conditions (i) and (ii) it suffices to show that each one of the following terms converge to zero in $L^2(\Omega)$:

$$\int_0^1\int_0^1 g(s,t)h_s dsdW_t - \int_0^1\int_0^1 g(s,t)h_s ds\dot\omega_t^N dt, \qquad (4.1)$$

$$X_0 - \int_0^1\int_0^1 g(s,t)\dot\omega_s^N\dot\omega_t^N dsdt + L(g), \qquad (4.2)$$

$$X_0 - \int_0^1\left(\int_0^1 g(s,t)dW_s\right)\dot\omega_t^N dt + L(g). \qquad (4.3)$$

The convergence of the first term is immediate, since it can be considered as a Wiener chaos of order one with $f(t) = \int_0^1 g(s,t)h_s ds$. Let us check the convergence to zero of the second term. For simplicity we will write $t_i = i/2^N$. We have

$$X_0 - \int_0^1\int_0^1 g(s,t)\dot\omega_s^N\dot\omega_t^N dsdt + L(g)$$

$$=\sum_{i\neq j}\left\{\int_{t_{i\vee j}}^{t_{i\vee j+1}}\int_{t_{i\wedge j}}^{t_{i\wedge j+1}} g(s,t)dW_s dW_t\right.$$

$$-2^{2N}\left(\int_{t_{j+1}}^{t_{j+2}}\int_{t_{i+1}}^{t_{i+2}} g(s,t)dsdt\right)(W_{t_{j+1}} - W_{t_j})(W_{t_{i+1}} - W_{t_i})\Big\}$$

$$+\sum_j\left\{\int_{t_j}^{t_{j+1}}\int_{t_j}^{t_{j+1}} g(s,t)dW_s dW_t\right.$$

$$-2^{2N}\left(\int_{t_{j+1}}^{t_{j+2}}\int_{t_{j+1}}^{t_{j+2}}g(s,t)dsdt\right)(W_{t_{j+1}}-W_{t_j})^2$$

$$+2^N\left(\int_{t_{j+1}}^{t_{j+2}}\int_{t_{j+1}}^{t_{j+2}}g(s,t)dsdt\right)\Big\}$$

$$+L(g)-\sum_j 2^N\left(\int_{t_{j+1}}^{t_{j+2}}\int_{t_{j+1}}^{t_{j+2}}g(s,t)dsdt\right)$$

$$=A_N^{(1)}+A_N^{(2)}+A_N^{(3)}+A_N^{(4)}+A_N^{(5)}+A_N^{(6)}.$$

The term $A_N^{(2)}$ can be written as a double stochastic integral of a suitable approximation of the kernel g. Taking into account the isometry property of multiple stochastic integrals we deduce that $A_N^{(1)}+A_N^{(2)}$ tends to zero an N tends to infinity. The term $A_N^{(3)}+A_N^{(4)}+A_N^{(5)}$ can be written as

$$I_2\left(\sum_j \mathbf{1}_{(t_j,t_{j+1}]\times(t_j,t_{j+1}]}\left(g-2^{2N}\int_{t_{j-1}}^{t_j}\int_{t_{j-1}}^{t_j}g(s,t)dsdt\right)\right),$$

which tends to zero an N tends to infinity. Finally the term $A_N^{(6)}$ converges to zero due to the definition of $L(g)$. The term (4.3) can be handled in a similar way.

Acknowledgments

We would like to thank Francis Hirsch for stimulating discussions on the subject.

References

[Bi] J.M. Bismut, Mécanique Aléatoire. Lecture Notes in Mathematics **866**, Springer Verlag, 1981.

[Gy] I. Gyöngy, On the support of the solutions of stochastic differential equations. *Theory Probab. Appl.* (1989), 649–653.

[GNS] I. Gyöngy, D. Nualart and M. Sanz-Solé, Approximation and support theorems in modulus spaces, *Probab. Theory Rel. Fields* **101** (1995), 495-509.

[Ma] V. Mackevičius, On the support of the solution of stochastic differential equations, *Lietuvos Matmatikos Rinkinys* **XXXVI** (1) (1986), 91–98.

[MN] A. Millet and D. Nualart, Support theorems for a class of anticipating stochastic differential equations. *Stochastics and Stochastics Reports* **39** (1992), 1-24.

[MS1] A. Millet and M. Sanz-Solé. On the support of a Skorohod anticipating stochastic differential equation, in: *Proceedings of the Barcelona Seminar on Stochastic Analysis*, Progress in Probability **32** (1993), 103-131.

[MS2] A. Millet and M. Sanz-Solé, A simple proof of the support theorem for diffusion processes, in: *Seminaire de Probabilités XXVIII, Lecture Notes in Math.* **1583** (1994), 36–48.

[SW] D. Stroock and S. R. S. Varadhan, On the support of diffusion processes with applications to the strong maximum principle. *Proccedings of the Sixth Berkeley Symp. on Mathematical Statistics and Probability*, Vol. III, (1972), 333–360.

María Emilia Caballero
Instituto de Matemàticas UNAM
Ciudad Universitaria, 04510, México D.F.

Begoña Fernández
Facultad de Ciencias UNAM
Ciudad Universitaria, 04510, México D.F

David Nualart
Facultat de Matemàtiques
Universitat de Barcelona
Gran Via 585, 08007–Barcelona, Spain

October 3, 1996

Invariant Measure for a Wave Equation
on a Riemannian Manifold

A.B. Cruzeiro and Z. Haba

Abstract

We consider a wave equation on a compact Riemannian manifold in the Hamiltonian form. The measure obtained as the product of Wiener measure on path space and the one of the (flat) white noise is infinitesimally invariant under the action of the corresponding flow.

1. Introduction

Wave equations on manifolds appear in Einstein gravity theories, where they describe scalar field self-interactions. There are also cosmological models (Dicke model) where such scalar fields in gravitational (curved) background appear. Invariant measures are then relevant to describe the corresponding statistical mechanics. The measure can be important to construct the natural Hilbert space of functionals on fields in quantum theory of gravity.

Recently the non linear wave equation in a flat space and with a cubical potential has been studied ([F] and [McV]). The equation being taken in the Hamiltonian form, a measure on the phase space was constructed and shown to be invariant under the action of the flow. This measure is essentially given by the product of the white noise measure and the law of Brownian motion, together with a density (a Feynman-Kac) term which accounts for the potential.

We propose here a similar construction for the wave equation on a compact Riemannian manifold, without potential. More precisely, we propose a wave equation which is infinitesimally invariant with respect to a similar product measure in the curved setting. The actual existence of the corresponding flow is not treated here. For the construction we shall use the Wiener measure on the path space associated to the Laplace-Beltrami operator, and a basic result, the Bismut integration by parts formula [B]. We shall follow closely results as well as notations of [FM] (cf. also the first paragraphs of [CM]).

We refer to [DS] for some results on invariant measures for wave type equations although their approach differs from ours in the sense that they do not take into account the momenta as variables the invariant measure depends upon.

2. The differential structure on the path space

We shall denote by M a d-dimensional compact Riemmanian manifold. On M we consider the Levi-Civita connection and $\Gamma_{k,l}^{j}$ will denote the corresponding Christofell symbols. In the local chart the Laplace -Beltrami operator has the following expression:

$$\Delta_M f = g^{i,j} \frac{\partial^2 f}{\partial m^i \partial m^j} - g^{i,j} \Gamma_{i,j}^k \frac{\partial f}{\partial m^k}$$

The Brownian motion $\varphi(\sigma)$, $\sigma \in [0,1]$, associated to this Laplacian satisfies the following Itô stochastic differential equation (cf. for example [IW]):

$$d_\sigma \varphi^i = a^{i,k}(\varphi).dx_k - c^i(\varphi)d\sigma$$

where x is a $\mathbb{R}^d$-valued Brownian motion, a the square root of the metric g and $c^i = \frac{1}{2} \sum_{j,k} g^{j,k} \Gamma_{j,k}^i \frac{\partial a^{i,k}}{\partial m^j}$.

We shall follow the frame bundle point of view for the theory of Riemannian Brownian motion ([M], exposed in [IW]).

A frame above m_0 is an Euclidean isometry r_0 of $\mathbb{R}^d$ onto the tangent space $T_{m_0}(M)$. $O(M)$ will denote the collection of all frames of M and π the projection of $O(M)$ into M, $\pi(r_0) = m_0$. Let $(\theta = \omega)$ be the canonical differential forms realizing for every $r \in O(M)$ an isomorphism between $T_r(O(M))$ and $\mathbb{R}^d \times so(d)$.

Then the horizontal Laplacian on $O(M)$, satisfying the relation

$$\Delta_{0(M)}(f \circ \pi) = (\Delta_M f) \circ \pi ,$$

is equal to $\sum_{k=1}^d L_{A_k}^2$, where A_k denote the horizontal vector fields, defined by

$$\begin{aligned} <\theta, A_k> &= e_k \\ <\omega, A_k> &= 0 \end{aligned}$$

(e_k the canonic basis of $\mathbb{R}^d$). The following Stratonovich stochastic differential equation

$$dr_x = \sum_{k=1}^d A_k(r_x) o dx^k \tag{2.1}$$

$$r_x(0) \quad = \quad r_0$$

defines a flow of diffeomorphisms on $O(M)$, the lifting to $O(M)$ of the Itô parallel transport along Brownian motion. In local coordinates, we have

$$d(r_x)^j_l(\sigma) \quad = \quad -\Gamma^j_{k,i}[d\varphi^k(\sigma)](r_x)^i_l(\sigma) \qquad (2.2)$$
$$r_x(0) \quad = \quad r_0$$

We shall write

$$t^\varphi_{\sigma \leftarrow 0} r_0 = r_x(\sigma)$$

Let us denote by X the Wiener space of the Brownian motion on $\mathbb{R}^d$ and by $P_{m_0}(M)$ the space of continuous paths on M starting at m_0 at time zero.

The Itô map ([M]) $\mathcal{I} : X \to P_{m_0}(M)$ is defined by

$$\mathcal{I}(x)(\sigma) = \pi(r_x(\sigma))$$

and the Wiener measure μ on the path space $P_{m_0}(M)$, namely the law of the Brownian motion on M coincides with the measure induced by this map, i.e., it is equal to the pullback through $\mathcal{I}$ of the Wiener measure μ_0 on X.

We consider a differentiable structure on the probability space. Following [F-M], let Z be an adapted vector field along the Brownian path, i.e., an adapted map $\sigma \in [0,1] \to Z_\sigma \in T_{\varphi(\sigma)}(M)$. The stochastic Itô integral of Z is defined by

$$\int_0^1 Z.d\varphi = \int_0^1 z^k.dx_k$$

where $x = \mathcal{I}^{-1}(\varphi)$ and $z_\varphi(\sigma) = t^\varphi_{\sigma \leftarrow 0} Z_\varphi(\sigma)$.

We have (c.f. [FM])

$$x(\sigma) = \int_0^\sigma r_x^{-1}(\xi).d\varphi(\xi) \qquad (2.3)$$

Finally let us recall the definition of derivative of a functional defined on the path space. When f is a cylindrical functional, namely

$$f(\varphi) = F(\varphi(\sigma_1), \ ... \ , \varphi(\sigma_m))$$

for some m and some smooth F, the partial functional derivatives are defined by

$$[D_{\tau,k}f](\varphi) = \sum_i 1(\tau < \sigma_i)(t^\varphi_{0 \leftarrow \sigma_i}(\partial_{\sigma_i} F) \mid e_k)$$

for $\tau \in [0,1]$ and $k = 1,...,d$. With respect to the norms $\|Df\|^q_{L^q} = E_\mu(\|Df\|^q)$, where

$$\|Df\|(\varphi) = \left[\sum_k \int_0^1 [D_{\tau,k}f]^2 \, d\tau \right]^{\frac{1}{2}}$$

the operator D is closable in L^q. Its domain is the Sobolev space $D_1^q(P_{m_0})$. For a vector field Z such that its parallel transported to the origin along the Brownian curve, z, has a square integrable (in $[0,1]$) time derivative (z belonging to the Cameron-Martin space), one defines

$$D_Z f = \sum_k \int_0^1 \dot{z}_k(\tau) D_{\tau,k} f \, d\tau \ .$$

Notice that this operation can be also be written as $\int_0^1 \sum_k \nabla^\varphi Z(\tau) D_{\tau,k} f \, d\tau$, where $\nabla^\varphi Z(\tau) = \lim_{\epsilon \to 0} \frac{1}{\epsilon}[t^\varphi_{\tau \leftarrow \tau + \epsilon} Z_\varphi(\tau + \epsilon) - Z_\varphi(\tau)]$. The two following theorems will be used in our construction:

Theorem 2.1 *([CM]) A functional f defined on the path space $P_{m_0}(M)$ is differentiable along the vector field $Z \in D_1^q(P_{m_0})$ if and only if $\tilde{f} = f o \mathcal{I}$ is a Wiener functional differentiable along the process z^**

$$dz^*(\sigma) = (\int_0^\sigma \Omega(odx, z))odx(\sigma) + \dot{z}d\sigma$$

where Ω denotes the curvature tensor in M and o Stratonovich integration. Moreover we have

$$[D_z f]o\mathcal{I} = D_{z^*}(f o \mathcal{I})$$

Derivation with respect to vector fields on the path space correspond to derivation on the path space taken along directions which do not belong to the Cameron-Martin space. Nevertheless these directions have an anti-symmetric diffusion coefficient and correspond therefore to Wiener measure preserving transformations. Transferring the differentiable structure of the path space to the flat case leads to an extended differential calculus on the Wiener space (the differential calculus with respect to "tangent processes", cf. [CM]).

Theorem 2.2 (Integration by parts formula [B] and [D]) *If f is a functional in the path space such that $f \in L_\mu^p$ and Z an adapted vector field in $D_1^q(P_{m_0})$, we have*

$$E_{\mu_0}(D_z f) = E_{\mu_0}(f \int_0^1 [\dot{z} + \frac{1}{2}R^M z]\, dx)$$

where

$$R_\tau^M = t_{0 \leftarrow \sigma}^\varphi \circ Ricci_{\varphi(\sigma)} \circ t_{\sigma \leftarrow 0}^\varphi$$

and Ricci denotes the Ricci tensor on M.

The proof of this theorem can be found in [FM].

3. The invariant measure

Let us denote by γ_0 the measure on the space of distributions $\mathcal{S}'([0,1])$ starting at time zero at a fixed point v_0 defined by

$$E_{\gamma_0} exp\, i < h, \pi > = exp(-\frac{1}{2} \int_0^1 h^2(\sigma)d\sigma)$$

for every square integrable function in the interval $[0,1]$.

A vector field on this space will be the data of a pair $Z = (Z^1, Z^2)$, where Z^1 is a vector field along the Brownian path, $Z_\sigma^1 \in T_{\varphi(\sigma)}(M)$ and $Z_\sigma^2 \in L^2[0,1]$. For a functional defined on $P_{m_0}(M) \times \mathcal{S}'(\mathbb{R})$ we define $D_z f = (D_{Z^1} f, D_{Z^2} f)$ with

$$D_{Z^2} f(\pi) = lim_{\epsilon \to 0} \frac{1}{\epsilon}[f(\pi + \epsilon Z^2) - f(\pi)]$$

the limit being taken γ_0 a.e. in π.

As a consequence of Theorem 2.2, the following integration by parts formula with respect to the product measure $\mu \times \gamma_0$ holds:

Theorem 3.1 *If f is a functional defined on $P_{m_0}(M) \times \mathcal{S}'([0,1])$ such that $f \in L_{\mu \times \gamma_0}^p$ and $Z = (Z^1, Z^2)$ an adapted vector field such that $Z^1 \in D_1^q(P_{m_0})$, $Z^2 \in L^q(\gamma_0)$, we have*

$$E_{\mu_0 \times \gamma_0}(D_z f) = E_{\mu_0 \times \gamma_0}(f(\int_0^1 [\dot{z}^1 + \frac{1}{2}R^M z^1]\, dx + \int_0^1 \pi Z^2 d\sigma))$$

4. The wave equation

Let us denote $t_\sigma^\varphi = t_{\sigma \leftarrow 0}^\varphi$ and consider the following wave equation on the Riemannian manifold M

$$\partial_t((t_\sigma^\varphi)^{-1}\partial_t\varphi_t(\sigma)) = \partial_\sigma((t_\sigma^\varphi)^{-1}\partial_\sigma\varphi_t(\sigma)) + \frac{1}{2}(t_\sigma^\varphi)^{-1}Ricci_\varphi o\partial_\sigma\varphi_t(\sigma) \quad (4.1)$$

subject to the initial conditions $\varphi_t(0) = m_0$ and $\partial_t\varphi_t(0) = v_0$. Here $(\Gamma(\partial_\sigma\varphi))_j^i = \Gamma_{k,j}^i[\partial_\sigma\varphi^k]$.

This equation can be written in Hamiltonian form:

$$
\begin{aligned}
\partial_t\varphi_t(\sigma) &= t_\sigma^\varphi \pi_t(\sigma) \\
\partial_t\pi_t(\sigma) &= \partial_\sigma((t_\sigma^\varphi)^{-1}\partial_\sigma\varphi_t(\sigma)) + \tfrac{1}{2}(t_\sigma^\varphi)^{-1}Ricci_\varphi(\partial_\sigma\varphi_t(\sigma))
\end{aligned}
\quad (4.2)
$$

Then *the measure $\mu \times \gamma_0$ is infinitesimally invariant for the wave equation (4.2)*. This means that, if Z denotes the vector field corresponding to equation (4.2), we have:

$$E_{\mu \times \gamma_0}(D_Z f) = 0$$

for sufficiently regular (cylindrical) functionals f.

The rigorous proof of this result is based on its transfer to the Wiener space via the Itô map together with a finite dimensional approximation of the corresponding vector field on the Wiener space. It is a consequence of Theorem 3.1. Formulae (2.2) and (2.3) are used in the proof.

References

[B]   J. M. Bismut, *Large deviations and the Malliavin calculus*, Birkhauser, P. P. bf 45, 1984.

[CM] A. B. Cruzeiro, P. Malliavin, Renormalized differential geometry on path space: structural equation, curvature *J.Funct.Anal.* **139/1** (1996), 119-181.

[DS] Y. L. Daletskii, V. R. Steblovskaya, On absolutely continuous and invariant evolution of smooth measure in Hilbert space, preprint.

[D]   B. K. Driver, A Cameron-Martin type quasi-invariance theorem for Brownian motion on a compact manifold *J.Funct.Anal.* **110/1** (1992), 272-376.

[F]   L. Friedlander, An invariant measure for the equation $u_{tt} - u_{xx} + u^3 = 0$, *Comm. Math. Phys.* **98** (1985), 1-16.

[FM]   S. Fang, P. Malliavin, Stochastic analysis on the path space of a Riemannian manifold, *J.Funct.Anal.* **118** (1993), 249-274.

[IW]   N. Ikeda, S. Watanabe, *Stochastic differential equations and diffusion processes*, North - Holland Kodansha, N. Y. 1981.

[M]   P. Malliavin, Formule de la moyenne, calcul de perturbations et théorème dánnulation pour les formes harmoniques, *J. Funct. Anal.* **1** (1974), 274-291.

[McV] H. P. McKean, K. L. Vaninsky, *Statistical mechanics of nonlinear wave equation*, in Perspectives and Trends in Applied Mathematics, I.Sirovich ed., Springer-Verlag (1994).

A.B. Cruzeiro
Grupo de Física-Matemática
Univ. de Lisboa,
Av. Prof Gama Pinto 2,
1699 Lisboa Codex, Portugal
e-mail: abcgfm.cii.fc.ul.pt

Z. Haba
Institute of Theoretical Physics,
University of Wroclaw,
Poland
e-mail: zhabift.uni.wroc.pl

Received September 30, 1996

Ergodic Distributed Control for Parameter Dependent Stochastic Semilinear Systems[1]

T. E. Duncan, B. Maslowski, and B. Pasik-Duncan

Abstract

A controlled Markov Process in a Hilbert space and an ergodic cost functional are given for a control problem where the process is a solution of a parameter dependent semilinear stochastic differential equation and the control is distributed. The linear term of the semilinear stochastic differential equation arises from the infinitesimal generator of a C_0 semigroup. The noise for the stochastic differential equation is a distributed, cylindrical Wiener process. Some ergodic properties of the controlled Markov process are shown to be uniform in the control and the parameter. It is shown that if a family of measures for the Markov process gives uniformly large measure to balls in the Hilbert space, then the family of measures is tight. An optimal control exists for the ergodic control problem and the optimal cost is shown to depend continuously on the system parameter.

1. Introduction

An ergodic control problem for a controlled Markov process in a Hilbert space is formulated and solved where the process is a solution of a parameter dependent semilinear stochastic differential equation in H. The control and the parameter occur in the distributed drift term of the stochastic differential equation. The noise for the stochastic differential equation is a standard, cylindrical Wiener process. For the ergodic control problem for some other applications it is important to verify that some ergodic properties are uniform in the control and the parameter, so this is done. The existence of an optimal ergodic control can be verified from [11]. It is shown that the optimal cost depends continuously on the system parameter. It is shown that the family of time averaged transition measures for the Markov process is tight if this family uniformly gives large measure to balls in the Hilbert space.

[1]This research was partially supported by the National Science Foundation grants DMS 9305936 and 9623439, the Alexander von Humboldt Foundation and the GAČR Grant No. 201/95/0629.

A brief outline of the paper is given now. In Section 2, the control problem is formulated and the basic assumptions are made. The controlled process is the unique, weak, mild solution of the stochastic differential equation and induces a Markov process in H. Some estimates are made of this process. In Section 3, a uniform version of the strong Feller property is given and the variation norm continuity of the transition measures and the invariant measures with respect to the parameter, that is uniform in the control, is given. In Section 4, some tightness properties are verified. It is shown that a "tightness" on balls in H implies tightness. The "tightness" of balls is verified by a Lyapunov method. The optimal cost for the control problem is shown to be a continuous function of the parameter.

A brief description and a comparison of some previous results on these topics in this paper are given now. There seems to be very few results available on the infinite time horizon control problems in infinite dimensional spaces. For infinite horizon discounted cost functions, the existence of an optimal stationary control is shown in [2] and the stationary Hamilton-Jacobi-Bellman equation is investigated in [4,14]. The only previous work on ergodic distributed control seems to be [11]. There seems to be no previous work on parameter dependent ergodic control problems except in [10] where an ergodic boundary control problem for a parameter dependent system is studied.

2. Preliminaries

Consider a controlled, infinite dimensional process $(X(t), t \geqslant 0)$ that satisfies the stochastic differential equation

$$
\begin{aligned}
dX(t) &= AX(t)dt + (f(\alpha, X(t)) + h(\alpha, X(t), u(X(t))))dt + Q^{1/2}dW(t) \\
X(0) &= x
\end{aligned}
$$

(2.1)

where $X(0), X(t) \in H$, H is a separable, infinite dimensional Hilbert space with inner product $\langle \cdot, \cdot \rangle$ and norm $|\cdot|$, $\alpha \in A \subset {}^d$ is a parameter and A is compact, K is a compact metric space with metric ρ, $A : \mathrm{Dom}(A) \to H$ is the infinitesimal generator of a C_0- semigroup $(S(t), t \geqslant 0)$,

$$
\begin{aligned}
f &: \ A \times H \to H \\
h &: \ A \times H \times K \to H
\end{aligned}
$$

are Borel measurable functions. It is assumed that $Q \in L(H)$ is invertible, $Q^{1/2}, Q^{-1/2} \in L(H)$ and $(W(t), t \geqslant 0)$ is a standard cylindrical Wiener process in H, that is defined on a filtered, complete probability space $(\Omega, F, (F_t),)$. The family of controls, U, is

$$U = \{u : H \to K \mid u \text{ is Borel measurable}\}.$$

The control problem is to minimize, over $u \in U$, the ergodic cost functional

$$J(x, u, \alpha) = \limsup_{T \to \infty} \frac{1}{T} \int_0^T c(X(s), u(X(s))) ds \qquad (2.2)$$

where $c : H \times K \to {}_+$ is bounded and Borel measurable. Define the family of linear operators, $(Q_t, t \geqslant 0)$, as follows

$$Q_t = \int_0^t S(r) Q S^*(r) dr. \qquad (2.3)$$

The following assumptions, (A1)–(A6) are used selectively in this paper.

(A1)

$$\int_0^T |S(r)|^2_{\text{HS}} dr < \infty$$

for some $T > 0$ where $|\cdot|_{\text{HS}}$ is the Hilbert-Schmidt norm on H and S is the semigroup whose infinitesimal generator is A.

(A2) For each $\alpha \in A$ the function $h(\alpha, \cdot, \cdot) : H \times K \to H$ is continuous and $f(\alpha, \cdot) : H \to H$ is Lipschitz continuous on the bounded subsets of H and there are constants k, k_f, k_h and $\tilde{k}(\alpha)$ such that $|f(\alpha, x)| \leqslant k + k_f|x|$, $|h(\alpha, x, u)| \leqslant k + k_h|x|$ and $|h(\alpha, x, u)| \leqslant \tilde{k}(\alpha)$ for all $x \in H$, $u \in K$ and $\alpha \in A$.

(A3) There is a continuous, increasing function $w : {}_+ \to {}_+$ with $w(0) = 0$ such that

$$|f(\alpha, x) - f(\beta, x)| + |h(\alpha, x, u) - h(\beta, x, u)| \leqslant w(|\alpha - \beta|)(1 + |x|)$$

for all $\alpha, \beta \in A$, $x \in H$ and $u \in K$.

(A4) For each $u \in U$ and $\alpha \in A$ there is an invariant measure $\mu(\alpha, u)$ for the process $(X(t), t \geqslant 0)$ that satisfies (2.1) and the family of measures $(\mu(\alpha, u), \alpha \in A, u \in U)$ is tight.

(A5) The function $c : H \times K \to {}_+$ given in (2.2) is bounded and Borel measurable and $c(x, \cdot) : K \to {}_+$ is continuous for each $x \in H$.

(A6) The set $h(\alpha, x, \mathrm{K}) \times c(x, \mathrm{K}) \subset H \times {}_+$ is convex for each $\alpha \in \mathrm{A}$ and $x \in H$.

Consider the following two stochastic differential equations

$$\begin{aligned} \mathrm{d}Z(t) &= AZ(t)\mathrm{d}t + Q^{1/2}\mathrm{d}W(t) \\ Z(0) &= x \end{aligned} \tag{2.4}$$

and

$$\begin{aligned} \mathrm{d}X(t) &= AX(t)\mathrm{d}t + f(\alpha, X(t))\mathrm{d}t + Q^{1/2}\mathrm{d}W(t) \\ X(0) &= x. \end{aligned} \tag{2.5}$$

Under the assumptions (A1) and (A2) it is easy to verify that each of the equations (2.4) and (2.5) has one and only one mild solution on the probability space (Ω, F, P), that is,

$$Z(t) = S(t)x + \int_0^t S(t - r)Q^{1/2}\mathrm{d}W(r) \qquad t \geqslant 0 \tag{2.6}$$

and

$$X(t) = S(t)x + \int_0^t S(t - r)f(\alpha, X(r))\mathrm{d}r + \int_0^t S(t - r)Q^{1/2}\mathrm{d}W(r) \quad t \geqslant 0. \tag{2.7}$$

These solutions are H-valued processes that belong to $C([0, T], L^p(\Omega, H))$ for any $p \geqslant 1$ and $T > 0$ (cf.[Ma1]). Furthermore, $(X(t), t \geqslant 0)$ and $(Z(t), t \geqslant 0)$ have H-continuous versions (cf. [dPZ2, Se1]).

Let $P_\alpha : {}_+ \times H \times \mathrm{B}(H) \to [0, 1]$ be the transition probability function for $(X(t), t \geqslant 0)$ in (2.7), that is,

$$P_\alpha(t, x, \Gamma) = (X(t) \in \Gamma) \tag{2.8}$$

and let $(T(t), t \geqslant 0)$ be the Markov transition semigroup for $(Z(t), t \geqslant 0)$ in (2.6), that is,

$$T_t\varphi(x) = \varphi(Z(t)) \tag{2.9}$$

where $t \geqslant 0$ and $\varphi \in \mathrm{M}(H)$, the bounded, Borel measurable functions on H. It is clear that $T_t 1_\Gamma(x) = N(S(t)x, Q_t)(\Gamma)$ where $t \geqslant 0$, $\Gamma \in \mathrm{B}(H)$, $x \in H$ and Q_t is given by (2.3) so it is self-adjoint, nonnegative and nuclear and $N(S_t x, Q_t)$ is the Gaussian measure on H with mean $S_t x$ and covariance Q_t.

Let $\xi_T^{\alpha,u}$ be the random variable as follows

$$\begin{aligned} \xi_T^{\alpha,u} = \ &\int_0^T \langle Q^{-1/2}h(\alpha, X(t), u(X(t))), \mathrm{d}W(t) \rangle \\ &- \tfrac{1}{2} \int_0^T |Q^{-1/2}h(\alpha, X(t), u(X(t)))|^2 \mathrm{d}t \end{aligned} \tag{2.10}$$

for $\alpha \in A$, $u \in U$ and $T > 0$ where $(X(t), t \in [0,T])$ is the solution of (2.7). A weak solution of (2.1) is constructed by an absolutely continuous change of probability measure (cf. [dPZ1, DV, GG, Ko]). Note that $\exp(\xi_T^{\alpha,u}) = 1$ by (A2). So

$$\mathbb{P}_x^{\alpha,u}(d\omega) = \exp(\xi_T^{\alpha,u})(d\omega), \tag{2.11}$$

the process $(W^*(t), t \geq 0)$ given by

$$W^*(t) = W(t) - \int_0^t Q^{-1/2}h(\alpha, X(s), u(X(s)))ds$$

is a cylindrical Wiener process on H and using $\mathbb{P}_x^{\alpha,u}$ and the solution of (2.7) it follows that

$$
\begin{aligned}
X(t) \quad = \quad & S(t)x + \int_0^t S(t-r)f(\alpha, X(r))dr \\
& + \int_0^t S(t-r)h(\alpha, X(r), u(X(r)))dr \\
& + \int_0^t S(t-r)Q^{1/2}dW^*(r).
\end{aligned}
\tag{2.12}
$$

So there is a weak solution to (2.1) which is weakly unique and induces a Markov process on H whose Markov transition semigroup is denoted as

$$P_t^{\alpha,u}\varphi(x) = \mathbb{E}_x^{\alpha,u}\varphi(X(t)) \tag{2.13}$$

for $t \geq 0$ and $\varphi \in M(H)$ where $\mathbb{E}_x^{\alpha,u}$ is the expectation using the probability measure $\mathbb{P}_x^{\alpha,u}$. Furthermore,

$$P^{\alpha,u}(t,x,\Gamma) = P_t^{\alpha,u}1_\Gamma(x) \tag{2.14}$$

for $t \geq 0$, $\Gamma \in B(H)$ and $x \in H$ denotes the corresponding transition probability function.

In the remainder of this section, three technical lemmas are given that are useful in the sequel. Initially (Proposition 2.2 [Ma1]) is given.

Lemma 2.1 *If* (A1) *and* (A2) *are satisfied,* $p \geq 2$ *and* $x \in H$, *then for each* $T > 0$ *there is a constant* $C = \hat{C}(p)$ *such that*

$$|X(T)|^p \leq C(1 + |x|^p) \tag{2.15}$$

where $(X(t), t \geq 0)$ *is the solution of (2.7).*

Lemma 2.2 *If* (A1) *and* (A2) *are satisfied then for each* $T > 0$ *and* $R > 0$

$$\lim_{N \to \infty} \inf \mathbb{P}_x^{\alpha,u}\left(\sup_{t \in [0,T]} |X(t)| \leq N \right) = 1 \tag{2.16}$$

where the infimum is taken over $\alpha \in A$, $u \in U$, *and* $x \in H$ *with* $|x| \leq R$ *and* $(X(t), t \geq 0)$ *is the solution of (2.7)*

By (A2) it follows that there is a sequence $(f_m, h_m, m \in)$ such that for each $m \in$ $(f_m(\alpha, x), h_m(\alpha, x, u)) = (f(\alpha, x), h(\alpha, x, u))$ for $\alpha \in A$, $u \in K$, $x \in H$ with $|x| \leqslant m$ and $|f_m| + |h_m| \leqslant M_m$ where M_m is a constant depending only on m, $f_m(\alpha, \cdot)$ is Lipschitz continuous and $h(\alpha, \cdot, \cdot)$ is continuous for each $m \in$ and

$$|f(\alpha, x) - f_m(\beta, x)| + |h(\alpha, x, u) - h_m(\beta, x, u)| \leqslant \tilde{\omega}_m(|\alpha - \beta|)$$

for $\alpha, \beta \in A$, $x \in H$ and $u \in K$ where $\tilde{\omega}_m$ has the same properties as ω in (A3) for each $m \in$.

Lemma 2.3 Let $P_m^{\alpha, u} : {}_+ \times H \times B(H) \to [0, 1]$ be the transition probability function for the Markov process that is defined by the solution of (2.1) with f and h replaced by f_m and h_m respectively that are described above. If (A1) and (A2) are satisfied then

$$\lim_{m \to \infty} \|P_m^{\alpha, u}(t, x, \cdot) - P^{\alpha, u}(t, x, \cdot)\| = 0$$

uniformly in $\alpha \in A$, $u \in U$ and x from bounded sets in H where $\| \cdot \|$ is the variation norm.

3. The Continuous Dependence of
Some Measures on a Parameter

In this section, some verifications are made for the continuous dependence of $P^{\alpha, u}(t, x, \cdot)$ on the parameter α, the uniform strong Feller property and, with the tightness condition (A4), the uniform continuity of the invariant measures with respect to the parameter $\alpha \in A$. This last result is used in Section 4 to prove continuity of the optimal cost for the control problem (2.1, 2.2).

Consider the mild Kolmogorov equation of the form

$$v(t, x) = T_t \varphi(x) + \int_0^t T_{t-s}(\langle D_x v(s, \cdot), f(\alpha, \cdot) \rangle$$
$$+ \langle D_x v(s, \cdot), h(\alpha, \cdot, u(\cdot)) \rangle)(x) ds \tag{3.1}$$

for $t \geqslant 0$ where $\varphi \in C_b(H)$ and for notational convenience the dependence of v on α and u is suppressed. The solution $v(t, x)$ of (3.1) is shown to be $P_t^{\alpha, u} \varphi(x)$. Denote by U_c the set of all continuous controls in U, $U_c = C(H; K)$. The following Proposition is given in [dPZ1, ChG].

Proposition 3.1 *If* (A1)–(A2) *are satisfied,* $u \in U_c$, $\varphi \in C_b(H)$ *and* $|f|$ *and* $|h|$ *are bounded independent of* $\alpha \in A$ *and* $u \in U_c$ *then the equation* (3.1) *has one and only one solution* $v(t, x) = P_t^{\alpha, u} \varphi(x)$ *that satisfies*

$$|D_x v(t, x)| \leqslant \frac{\tilde{c}}{t^{1/2}} \sup |\varphi| \qquad (3.2)$$

for $t \in (0, T]$ *where the constant* $\tilde{c}$ *does not depend on* $\varphi, u \in U_c$ *or* $\alpha \in A$.

The following result is a uniform version of the strong Feller property.

Lemma 3.2 *If* (A1)–(A2) *are satisfied then for each* $t > 0$ *there is a continuous, increasing function* $\tilde{\omega} : _+ \to _+$ *with* $\tilde{\omega}(0) = 0$ *such that*

$$\|P^{\alpha, u}(t, x, \cdot) - P^{\alpha, u}(t, y, \cdot)\| \leqslant \tilde{\omega}(|x - y|)$$

for all $\alpha \in A$, $u \in U$ *and* $x, y \in H$.

The proof follows easily from Proposition 3.1, Lemma 2.3 and the pointwise convergence of $(u_n, n \in)$ to $u \in U$ where $u_n \in U_c$ for all n.

Proposition 3.3 *If* (A1)–(A3) *are satisfied then for each* $T > 0$ *the function*

$$P^{\alpha, u}(T, x, \cdot) : A \to P(H) \qquad (3.3)$$

is continuous in the variation norm uniformly in $u \in U$ *and* $x \in K$ *for each compact set* $K \subset H$.

Proof. By Lemma 2.3 it can be assumed that $|f|$ and $|h|$ are bounded and

$$|f(\alpha, x) - f(\beta, x)| + |h(\alpha, x, u) - h(\beta, x, u)| \leqslant \omega(|\alpha - \beta|) \qquad (3.4)$$

for $x \in H$ and $\alpha, \beta \in A$. Initially the uniform continuity of (3.3) is verified for $u \in U_c$. For $v_{\alpha, u}(t, x) = P_t^{\alpha, u} \varphi(x)$ for $x \in H$ and $\varphi \in C_b(H)$ it follows by Proposition 3.1 that

$$v_{\alpha, u}(t, x) = T_t \varphi(x) + \int_0^t T_{t-s}(\psi_{\alpha, u}(D_x v_{\alpha, u}(s, \cdot)))(x) ds \qquad (3.5)$$

for $t \in [0, T]$ where

$$\psi_{\alpha, u}(D_x v_{\alpha, u}(s, \cdot)) = \langle D_x v_{\alpha, u}(s, \cdot), f(\cdot) \rangle + \langle D_x v_{\alpha, u}(s, \cdot), h(\alpha, \cdot, u(\cdot)) \rangle \qquad (3.6)$$

and

$$|D_x v_{\alpha, u}(t, \cdot)| \leqslant ct^{-1/2} \sup |\varphi| \qquad (3.7)$$

for $t \in (0, T]$ where $c > 0$ does not depend on $t \in (0, T]$, $\alpha \in A$, $u \in U_c$ and $\varphi \in C_b(H)$. By Proposition 3.1 modified to apply to the semigroup $(T_t, t \geqslant 0)$ it follows that

$$\sup_x |v_{\alpha,u}(t, x) - v_{\alpha_0,u}(t, x)| + \sup_x |D_x v_{\alpha,u}(t, x) - D_x v_{\alpha_0,u}(t, x)|$$

$$\leqslant \sup_x \int_0^t |T_{t-s}(\psi_{\alpha,u}(D_x v_{\alpha,u}(s, \cdot)) - \psi_{\alpha_0,u}(D_x v_{\alpha_0,u}(s, \cdot)))(x)| ds$$

$$+ \sup_x \int_0^t |D_x T_{t-s}(\psi_{\alpha,u}(D_x v_{\alpha,u}(s, \cdot)) - \psi_{\alpha_0,u}(D_x v_{\alpha,u}(s, \cdot)))(x)| ds. \tag{3.8}$$

By (3.4) and (3.7) it follows that

$$\sup_x |\psi_{\alpha,u}(D_x v_{\alpha,u}(s, x)) - \psi_{\alpha_0,u}(D_x v_{\alpha_0,u}(s, x))|$$

$$\leqslant c_1 \sup_x |D_x v_{\alpha,u}(s, x) - D_x v_{\alpha_0,u}(s, x)|$$

$$\leqslant c_2 \sup_x |D_x v_{\alpha,u}(s, x) - D_x v_{\alpha_0,u}(s, x)| + c_3 s^{-1/2} \omega(|\alpha - \alpha_0|) \tag{3.9}$$

for some constants c_1, c_2 and c_3 depending only on the bounds for $|f|$ and $|h|$. Let $\lambda_{\alpha,u}(\cdot)$ be the left hand side of (3.8). By (3.8) and (3.9) it follows that

$$\lambda_{\alpha,u}(t) \leqslant \int_0^t \frac{k_1}{(t-s)^{1/2}} \lambda_{\alpha,u}(s) \, ds + \omega(|\alpha - \alpha_0|) \int_0^t \frac{k_2}{(t-s)^{1/2} s^{1/2}} ds \tag{3.10}$$

for $t \in (0, T]$ for some constants k_1 and k_2. By the generalized Gronwall Lemma (Theorem 7.1 [He]) it follows that

$$\lambda_{\alpha,u}(t) \leqslant k_3 \omega(|\alpha - \alpha_0|)$$

for $t \in (0, T]$ so

$$\|P^{\alpha,u}(T, x, \cdot) - P^{\alpha_0,u}(T, x, \cdot)\| \leqslant k_4 \omega(|\alpha - \alpha_0|)$$

for some constants k_3 and k_4 that are independent of $x \in H$ and $u \in U_c$. For $u \in U$ there is a sequence $(u_n, n \in)$ from U_c such that $u_n \to u$ pointwise so the same estimate follows easily for all $u \in U$. ∎

Choose and fix $\alpha_1 \in A$ and let $\eta = P_{\alpha_1}(1, 0, \cdot)$ (recall (2.8)). Note that by [Ma2] all of the transition functions $P_\alpha(t, x, \cdot)$ $\alpha \in A$, $t > 0$ and $x \in H$ are equivalent. The following lemma is (Lemma 3, [DPS]).

Lemma 3.4 *Let $\varphi : H \to U$ and $G : H \to U$ be bounded, Borel measurable functions and let $(G_n, n \in)$ be a sequence of bounded, Borel measurable* *functions that converge to G in* $\sigma(L^\infty(H, \eta, H), L^1(H, \eta, H))$. *If* (A1)–(A2) *are satisfied then*

$$\lim_{n \to \infty} \left(\int_0^t \langle \varphi(X(s)), G_n(X(s)) - G(X(s)) \rangle ds \right)^2 = 0 \qquad (3.11)$$

where $(X(t), t \geqslant 0)$ *satisfies (2.5) and* $\alpha \in A$, $x \in H$ *are arbitrary.*

The following result is a modification of a similar result in [DPS].

Proposition 3.5 *Let* $(\alpha_n, n \in \ )$ *be a sequence in* A *that converges to* $\alpha_0 \in A$ *and let* $(h(\alpha_0, \cdot, u_n(\cdot)), n \in \ )$ *be a sequence that converges to* $h(\alpha_0, \cdot, u(\cdot))$ *in* $\sigma(L^\infty(H, \eta, H), L^1(H, \eta, H))$. *If* (A1)–(A2) *are satisfied then*

$$\lim_{n \to \infty} P_t^{\alpha_n, u_n} \varphi(x) = P_t^{\alpha_0, u} \varphi(x) \qquad (3.12)$$

for each $\varphi \in M(H)$, $x \in H$ *and* $t > 0$.

Remark 3.6 *One of the basic assumptions here is the tightness condition (A4). Using a Lyapunov condition, (A4) is verified in Section 4. Furthermore, if (A1)–(A2) are satisfied then for each* $\alpha \in A$ *and* $u \in U$ *the transition probabilities* $(P^{\alpha, u}(t, x, \cdot), t > 0, x \in H)$ *are equivalent which follows from the equivalence of the transition probabilities* $(P_\alpha(t, x, \cdot), t > 0, x \in H)$. *This latter fact is an immediate consequence of the strong Feller property and irreducibility (cf.cite18). From the equivalence of* $(P^{\alpha, u}(t, x, \cdot), t > 0, x \in H)$, *it follows that for each* $\alpha \in A$ *and* $u \in U$ *the invariant measure* $\mu(\alpha, u)$ *is ergodic and unique (Proposition 2.5 [Se2]).*

The following lemma follows basically from Roxin [Ro] (cf. also Appendix [BS]).

Lemma 3.7 *Let* $\alpha \in A$ *be fixed. If* (A2), (A5) *and* (A6) *are satisfied then* $\{(h(\alpha, \cdot, u(\cdot)), c(\cdot, u(\cdot)) : u \in U\} \subset L^\infty(H, \eta, H \times)$ *is compact in the* $\sigma(L^\infty(H, \eta, H \times), L^1(H, \eta, H \times))$ *topology.*

Proposition 3.8 *If* (A1)–(A6) *are satisfied then*

$$\lim_{\alpha \to \alpha_0} \sup_{u \in U} \rho_*(\mu(\alpha, u), \mu(\alpha_0, u)) = 0 \qquad (3.13)$$

where ρ_* *is a metric for weak* convergence of measures,* $\alpha_0 \in A$ *and* $u_0 \in U$.

Using Proposition 3.8 and Proposition 3.3 a strong version of (3.13) is obtained.

Proposition 3.9 *If* (A1)–(A6) *are satisfied then*

$$\lim_{\alpha \to \alpha_0} \sup_{u \in U} \|\mu(\alpha, u) - \mu(\alpha_0, u)\| = 0 \qquad (3.14)$$

where $\| \cdot \|$ *is the variation norm.*

4. The Tightness of Some Measures and the Continuity of the Optimal Cost

In this section some more explicit sufficient conditions for the validity of (A4) are given. Some additional assumptions are
(T1) The semigroup $S(\cdot)$ is compact.

(T2) $$\int_0^T t^{-2\gamma} |S(t)|_{\mathrm{HS}}^2 ds < \infty$$

is satisfied for some $T > 0$ and $\gamma > 0$.

It should be noted that if S is the semigroup constructed from a uniformly elliptic operator on a bounded, one-dimensional, interval with the Dirichlet boundary conditions, then (T1) is satisfied for $0 < \gamma < 1/4$.

Define $\mu_T^{\alpha,u}$ as follows

$$\mu_T^{\alpha,u}(\cdot) = \frac{1}{T} \int_0^T P^{\alpha,u}(t, 0, \cdot) dt \qquad (4.1)$$

for $\alpha \in A$, $u \in U$ and $T > 0$. Since the solution of (2.1) is Feller, to verify (A4) it suffices to show that the family of measures $(\mu_T^{\alpha,u}, \alpha \in A, u \in U, T \geqslant 1)$ is tight. In the following proposition it is shown that the tightness of $(\mu_T^{\alpha,u}, \alpha \in A, u \in U, T \geqslant 1)$ follows from a similar property, where compact sets are replaced by balls (4.2). Note that (4.2) does not guarantee the existence of an invariant measure in general (cf. [Vr]).

Proposition 4.1 *If* (A1), (A2) *and* (T1) *are satisfied and*

$$\lim_{n \to \infty} \mu_T^{\alpha,u}(H \setminus B_n) = 0 \qquad (4.2)$$

where the convergence is uniform in $\alpha \in A$, $u \in U$ *and* $T \geqslant 1$ *and* $B_n = \{x \in H : |x| \leqslant n\}$ *then the family of measures* $(\mu_T^{\alpha,u}, \alpha \in A, u \in U, T \geqslant 1)$ *is tight.*

To verify this proposition the following two lemmas are used.

Lemma 4.2 *For each $T > 0$ and $p \geqslant 0$ there exists a constant $C = \bar{C}(T, p)$ such that*

$$E_x^{\alpha, u} |X(t)|^p \leqslant C(1 + |x|^p) \tag{4.3}$$

holds for $x \in H$, $\alpha \in A$, $u \in U$.

The idea of the proof of the following Lemma 4.3 is based on the method of factorization (cf. [dPGZ, DMP2] for similar results).

Lemma 4.3 *There exists a sequence $(K_n, n \in)$ of compact subsets of H such that for each $p > 1/\gamma$ and $R > 0$ there exists some $L > 0$ such that*

$$P^{\alpha, u}(1, y, H \setminus K_n) \leqslant \frac{L(1 + R^p)}{n}$$

is satisfied for all $x \in H$, $\alpha \in A$, $u \in U$, and $n \in$.

Proof. Define the operator $R_\nu : L^p(0, 1, H) \to L^p(0, 1, H)$ as

$$R_\nu h(t) = \int_0^t (t - s)^{\nu - 1} S(t - s) h(s) ds,$$

where $\nu \in (1/p, 1]$. By (T1) it follows that R_ν is compact as a mapping $L^p(0, 1, H) \to C(0, 1, H)$ (cf. [dPZ1]), so the mapping $J_\nu : L^p(0, T, H) \to H$ given by $J_\nu h := R_\nu h(1)$ is also compact. Define a sequence of compact sets K_n in H,

$$K_n = \{y = S(1)x + J_1 g + J_1 h + J_\gamma k; \ |x|^p + |g|_p^p + |h|_p^p + |k|_p^p \leqslant n\}$$

where $|\cdot|_p$ denotes the norm in $L^p(0, T, H)$. Note that

$$X(1) = S(1)x + J_1(f(\alpha, X(\cdot))) + J_1(h(\alpha, X(\cdot), u(\cdot))) + \frac{\sin \pi \gamma}{\pi} J_\gamma(Y(\cdot)),$$

where

$$Y(s) = \int_0^s (s - u)^{-\gamma} S(s - u) Q^{1/2} dW_u.$$

By Lemma 4.2 and (A2) it follows that

$$E_x^{\alpha, u}(|F(\alpha, X(\cdot))|_p^p + |h(\alpha, X(\cdot), u(X(\cdot)))|_p^p) \leqslant C_1(1 + |x|^p),$$

and (T1) implies that

$$E_x^{\alpha, u} \int_0^1 |Y(s)|^p ds \leqslant C_2 E_x^{\alpha, u} \int_0^1 \left(\int_0^s (s - u)^{-2\gamma} |S(s - u)|_{HS}^2 du \right)^p ds$$

$$< C_3,$$

for some universal constants $C_1 - C_3$. By the Chebyshev inequality it follows for $n > R^p$ that

$$P_x^{\alpha,u}[X(1) \notin K_n] \leq \quad \frac{1}{n}(E_x^{\alpha,u}|F(\alpha, X(\cdot)|_p^p + E_x^{\alpha,u}|h(\alpha, X(\cdot), u(X(\cdot)))|_p^p$$

$$+ \frac{(\sin \gamma \pi)^p}{\pi^p} E_x^{\alpha,u}(|Y(\cdot)|^p) \leq \frac{L(1+|x|^p)}{n}$$

for some universal constant L.   ■

Proof.(Proposition 4.1) Let $(K_n, n \in)$ be given from Lemma 4.3. It follows that

$$\frac{1}{T}\int_1^T P^{\alpha,u}(t, 0, H \setminus K_n)dt$$

$$= \frac{1}{T}\int_1^T \int_H P^{\alpha,u}(1, y, H \setminus K_n)P^{\alpha,u}(t-1, 0, dy)dt \qquad (4.4)$$

$$\leq \mu_{T-1}^{\alpha,u}(H \setminus B_R) + \mu_{T-1}^{\alpha,u}(B_R) \sup_{|y| \leq R} P^{\alpha,u}(1, y, H \setminus K_n)$$

$$\leq \mu_{T-1}^{\alpha,u}(H \setminus B_R) + \frac{1}{n}L(1 + R^p)$$

for each $R > 0$. By (4.2) the right hand side tends to zero as $n \to \infty$ uniformly in $\alpha \in A$, $u \in U$ and $T \geq 1$.   ■

    To verify the "tightness on balls" condition the Lyapunov method is used. It is required that the semigroup is stable, that is,

(T3) $|S(t)|_{L(H)} \leq Me^{-\omega t}$

for some $M > 0$ and $\omega > 0$ and all $t > 0$.

Proposition 4.4 *If* (A2), (T1)–(T3) *and*

$$k_f + k_h < \frac{\omega}{M^2} \qquad (4.5)$$

where M and ω are given (T3) *then the assumption* (A4) *is satisfied.*

Proof. Let V be given by

$$V = \int_0^\infty S(t)S^*(t)dt.$$

By (T3) it follows that V is well defined, self adjoint and nonnegative. By (T2) V is nuclear and $\langle Ax, Vx \rangle = -|x|^2$ for $x \in \text{Dom}(A)$. It follows from a version of the Itô formula that

$$E_x^{\alpha,u}\langle VX(t), X(t)\rangle - \langle Vx, x\rangle$$
$$\leqslant E_x^{\alpha,u} \int_0^t (-|X(s)|^2 + \langle VX(s), f(\alpha, X(s))\rangle) +$$
$$\langle VX(s), h(\alpha, X(s), u(X(s)))\rangle + \frac{1}{2}\mathrm{Tr}\,VQ)ds$$

for all $t \geqslant 0$. By (4.5) for some $r \in (0,1)$ and $c > 0$ it follows that

$$E_x^{\alpha,u}\langle VX(t), X(t)\rangle - \langle Vx, x\rangle \leqslant E_x^{\alpha,u} \int_0^t ((r-1)|X(s)|^2 + c)ds$$

for all $t \geqslant 0$. Since $V \geqslant 0$ it follows that

$$\sup_{t \geqslant 1} \frac{1}{t} \int_0^t E_x^{\alpha,u}|X(s)|^2 ds \leqslant \sup_{t \geqslant 1} \frac{\langle Vx, x\rangle}{t(1-r)} + \frac{c}{1-r} \leqslant c_1. \qquad (4.6)$$

By (4.6) and the Chebyshev inequality it follows that (4.2) is satisfied. ∎

Recall that the control problem is described by the system (ref2.1) and the cost functional (2.2) and the optimal cost is $J^*(\alpha) = \inf_{u \in U} J(\alpha, u)$. If (A1)–(A2), (A4) and (A5) are satisfied then the following equality is satisfied

$$J(\alpha, u) = \int_H c(y, u(y))\mu(\alpha, u)(dy)$$

(cf. Remark 3.6) so the cost $J(\alpha, u)$ does not depend on the initial condition $X(0) = x \in H$.

The following result is verified in [DPS].

Theorem 4.5 *If* (A1)–(A2) *and* (A4)–(A6) *are satisfied for each* $\alpha \in A$ *then there is an optimal control for the control problem given by (2.1) and (2.2).*

The following result shows that the optimal cost is a continuous function of the system parameter.

Theorem 4.6 *If* (A1)–(A6) *are satisfied then* $J^* : A \to$ *is continuous.*

Proof. It follows that

$$\sup_{u \in U} |J(\alpha, u) - J(\alpha_0, u)| \leqslant \sup |c| \sup_{u \in U} \|\mu(u, \alpha) - \mu(u, \alpha_0)\|.$$

By Proposition 3.9 it follows that the right hand side of this inequality tends to zero as $n \to \infty$. Given $\varepsilon > 0$ there is a $\delta > 0$ such that if $|\alpha - \alpha_0| < \delta$ then

$$\sup_{u \in U} |J(\alpha, u) - J(\alpha_0, u)| < \varepsilon.$$

Let $u_\alpha \in U$ be an optimal control for the control problem (2.1) and (2.2) for $\alpha \in A$. Since $J(\alpha, u_{\alpha_0}) \geqslant J(\alpha, u_\alpha)$ it follows that $J(\alpha_0, u_{\alpha_0}) \geqslant J(\alpha, u_\alpha) - \varepsilon$. Since $J(\alpha_0, u_{\alpha_0}) \leqslant J(\alpha_0, u_\alpha)$ it follows that $J(\alpha_0, u_{\alpha_0}) \leqslant J(\alpha, u_\alpha) + \varepsilon$ for $\alpha \in A$ and $|\alpha - \alpha_0| < \delta$.   ∎

References

[BS]   T. Bielecki and L. Stettner, On ergodic control problems for singularly perturbed Markov processes, *Appl. Math. Optim.* **20** (1989), 131–161.

[BG]   V. S. Borkar and T. E. Govindan, Optimal control for semilinear evolution equations, Nonlinear Analysis, *Theory, Methods Appl.* **23** (1994), 15–35.

[ChG]   A. Chojnowska-Michalik and B. Goldys, Existence, uniqueness and invariant measures for stochastic semilinear equations on Hilbert spaces, *Probab. Th. Related Fields* **102** (1995), 331-356.

[ChM]   P. Chow and J. L. Menaldi, Infinite-dimensional Hamilton-Jacobi-Bellman equations in Gauss-Sobolev spaces, preprint.

[dPFZ]   G. Da Prato, M. Fuhrman and J. Zabczyk, Differentiability of Ornstein-Uhlenbeck semigroups, Preprint 1995/26, Scuola Normale Superiore, Pisa.

[dPGZ]   G. Da Prato, D. Gatarek and J. Zabczyk, Invariant measures for semilinear stochastic equations, *Stochastic Anal. Appl.* **10** (1992), 387–408.

[dPZ1]   G. Da Prato and J. Zabczyk, *Stochastic Equations in Infinite Dimensions*, Cambridge Univ. Press, Cambridge (1992).

[dPZ2]   G. Da Prato and J. Zabczyk, Evolution equations with white noise boundary conditions, *Stochastics Stochastic Rep.* **42** (1993), 167–182.

[DMP1]   T. E. Duncan, B. Maslowski and B. Pasik-Duncan, Adaptive boundary and point control of linear stochastic distributed parameter systems, *SIAM J. Control Optim.* **32** (1994), 648–672.

[DMP2]   T. E. Duncan, B. Maslowski and B. Pasik-Duncan, Ergodic control of some semilinear systems in Hilbert spaces, preprint.

[DPS]   T. E. Duncan, B. Pasik-Duncan and L. Stettner, On ergodic control of stochastic evolution equations, to appear in *Stochastic Anal. Appl.*

[DV]   T. E. Duncan and P. Varaiya, On the solutions of a stochastic control system II, *SIAM J. Control* **13** (1975), 1077–1092.

[GG]   D. Gatarek and B. Goldys, On solving stochastic evolutions by the change of drift with application to optimal control, Proceedings SPDE's and Applications, Ed. G. Da Prato and L. Tubaro, Pitman, 1992, 180–190.

[GR]   F. Gozzi and E. Rouy, Regular solutions of second-order stationary Hamilton-Jacobi equations, Preprint 1995/18, Scuola Normale Superiore, Pisa.

[He]   D. Henry, *Geometric Theory of Semilinear Parabolic Equations*, Springer-Verlag, 1981.

[Ko]   S. M. Kozlov, Some questions of stochastic equations with partial derivatives, *Trudy Sem. Petrovsk.* **4** (1978), 147–172.

[Ma1]   B. Maslowski, Stability of semilinear equations with boundary and pointwise noise, *Annali Scuola Normale Superiore Pisa* **22** (1995), Serie IV, 55–93.

[Ma2]   B. Maslowski, On probability distributions of solutions of semilinear stochastic evolution equations, *Stochastics Stochastic Rep.* **45** (1993), 11–44.

[Ro]   E. Roxin, The existence of optimal controls, *Michigan Math. J.* **91** (1962), 109–119.

[Se1]   J. Seidler, Da Prato-Zabczyk's maximal inequality revisited I, *Mathematica Bohemica* **118** (1993), 67–106.

[Se2]   J. Seidler, Ergodic behaviour of stochastic parabolic equations, to appear in *Czechoslovak Math. J.*

[Vr]   I. Vrkoč, A dynamical system in a Hilbert space with a weakly attractive nonstationary point, *Mathematica Bohemica* **118** (1993), 401–423.

T.E. Duncan and B. Pasik-Duncan
Department of Mathematics,
University of Kansas,
Lawrence, KS 66045.

B. Maslowski
Institute of Mathematics,
Czech Academy of Sciences,
Prague, Czech Republic.

Received October 1, 1996

Dirichlet Forms, Caccioppoli Sets and the Skorohod Equation

Masatoshi Fukushima

1. Dirichlet forms and reflecting Brownian motions

Let R^d be the Euclidean d−space and m be the Lebesgue measure on it. We consider a bounded domain $D \subset R^d$ with $m(\partial D) = 0$ and and the Sobolev space

$$H^1(D) = \{u \in L^2(D) : \partial_i u \in L^2(D), \ 1 \le i \le d\} \qquad (1.1)$$

with inner product

$$\mathcal{E}(u,v) = \frac{1}{2} \int_D \nabla u \cdot \nabla v dx, \quad u,v \in H^1(D) \qquad (1.2)$$

$(H^1(D), \mathcal{E})$ is a specific Dirichlet space on $L^2(D)$ but it is not necessarily regular.

In general, let X be a locally compact separable metric space and m be an everywhere dense positive Radon measure on X. A symmetric form $\mathcal{E}$ with a domain $\mathcal{F}$ dense in $L^2(X; m)$ is called a *Dirichlet form* on $L^2(X; m)$ if $\mathcal{F}$ is a real Hilbert space with inner product

$$\mathcal{E}_\alpha(u,v) = \mathcal{E}(u,v) + \alpha(u,v)$$

for each $\alpha > 0$, where (,) denotes the L^2- inner product, and if the *unit contraction operates* on $\mathcal{E}$ in the sense that

$$u \in \mathcal{F} \Rightarrow v = (0 \vee u) \wedge 1 \in \mathcal{F}, \quad \mathcal{E}(v,v) \le \mathcal{E}(u,u).$$

It is called *regular* if $\mathcal{F} \cap C_0(X)$ is dense both in $\mathcal{F}$ and in $C_0(X)$, where $C_0(X)$ denotes the space of continuous functions on X with compact support. $\mathcal{E}$ is called *local* if $\mathcal{E}(u,v) = 0$ whenever $u,v \in \mathcal{F}$ have disjoint compact support. The importance of these notions lies in the general fact proven in [FOT] that, given any regular local Dirichlet form $\mathcal{E}$, there exists a diffusion process (a strong Markov process with continuous sample paths) $\mathbf{M} = (X_t, P_x)$ on X which is associated with the form $\mathcal{E}$ in the sense that the transition semigroup $p_t f(x) = E_x(f(X_t)), x \in X$, is a version of the L^2-semigroup $T_t f$ generated by $\mathcal{E}$ for any non-negative L^2- function f. **M** is unique up to a set of zero capacity. If further

$$1 \in \mathcal{F} \quad \mathcal{E}(1,1) = 0, \tag{1.3}$$

then **M** is conservative: $p_t 1(x) = 1$.

In what follows, we exclusively use the notation m for the d-dimensional Lebesgue measure and $\mathcal{E}$ for the Dirichlet integral defined by (2). Denote by $C_0^1(D)$ the space of continuously differentiable functions on D with compact support and by $C^1(\overline{D})$ the restrictions to $\overline{D}$ of continuously differentiable functions on R^d. Their closures in the space $(H^1(D), \mathcal{E})$ will be designated by $H_0^1(D)$ and $\hat{H}^1(D)$ respectively. Then $(H_0^1(D), \mathcal{E})$ is a regular local Dirichlet form on $L^2(D)$ and $(\hat{H}^1(D), \mathcal{E})$ is a regular local Dirichlet form on $L^2(\overline{D})$ satisfying (3). The diffusion process on D associated with the former is known to be identical with the absorbing Brownian motion on D ([FOT]). From now on, the conservative diffusion process on $\overline{D}$ associated with the latter will be denoted by $\mathbf{M} = (X_t, P_x)$ and called the $\hat{H}^1(D)$-*diffusion* on $\overline{D}$. This is the object of our investigation in this paper.

As will be explained in the next section, we can apply a general decomposition theorem of additive functionals [FOT] to the process **M** to see that its sample path $X_t \in \overline{D}$ admits a decomposition

$$X_t - X_0 = B_t + N_t, \quad t \geq 0, \quad P_x - a.s. \text{ for } q.e.x \in \overline{D}, \tag{1.4}$$

where $B_t, t \geq 0$, is a d-dimensional Brownian motion and each component of $N_t, t \geq 0$, is a continuous additive functional of zero energy, which is known to be of zero P_m-quadratic variation but not necessarily of bounded variation (on each finite time interval). 'q.e.' is an abbreviation of 'quasi-everywhere' meaning 'except for a subset of $\overline{D}$ of zero capacity'. We are interested in finding a condition on the domain D to ensure that the second process N_t in the above decomposition is of bounded variation and admits a Skorohod type expression.

A bounded domain $D \subset R^d$ is said to be a *Caccioppoli set* if

$$|\int_D \partial_i \varphi \, dx| \leq C \|\varphi\|_\infty, \quad \varphi \in C^1(\overline{D}), \quad 1 \leq i \leq d, \tag{1.5}$$

for some positive constant C. Clearly, D is Caccioppoli iff the next *Gauss formula* holds:

$$\int_D div \, \varphi \, dx = \int_{\partial D} \varphi \cdot n \, d\sigma, \quad \varphi \in C^1(\overline{D}; R^d), \tag{1.6}$$

for some finite positive measure σ concentrated on the boundary ∂D and some measurable vector $n(x) = (n_1(x), \cdots, n_d(x))$ such that $|n(x)| = 1$, σ-a.e. $x \in \partial D$. Note that the Caccioppoli set is purely measure theoretic notion and should be considered as an equivalence class of sets coincident m-a.e.

We aim at proving the next theorem:

Theorem 1.1 *The following conditions are equivalent:*
(i) *D is a Caccioppoli set.*
(ii) *N_t admits an expression*

$$N_t = \frac{1}{2} \int_0^t n(X_s) dL_s \quad P_x-a.s. \text{ for } q.e.x, \tag{1.7}$$

for a positive continuous additive functional L_t such that

$$L_t = \int_0^t I_{\partial D}(X_s) dL_s, \quad E_m(L_t) < \infty, \tag{1.8}$$

and for a measurable vector $n(x)$ on ∂D with $|n(x)| = 1$.

The above theorem says that, when D is Caccioppoli, the sample path X_t of the $\hat{H}^1(D)$-diffusion is a solution of the Skorohod type equation

$$X_t - X_0 = B_t + \frac{1}{2} \int_0^t n(X_s) dL_s \quad P_x-a.s. \tag{1.9}$$

for q.e. starting point $x \in \overline{D}$. Here n coincides with the generalized inward unit normal vector on the boundary ∂D appearing in the Gauss formula (6). Therefore we may regard the $\hat{H}^1(D)$-diffusion as a reflecting Brownian motion on $\overline{D}$. However the space $\hat{H}^1(D)$ can be a proper subspace of $H^1(D)$ and the name 'reflecting Brownian motion' has been exclusively attributed to some diffusion associated with the maximum Dirichlet space $H^1(D)$ dominating $H_0^1(D)$ ([F1],[F2]). Indeed a reflecting Brownian motion in this sense was first constructed by the author[F1] in 1967 on a Kuramochi type compactification D^* of an arbitrary bounded domain D. If $\hat{H}^1(D) = H^1(D)$, then D^* can be identified (quasi-everywhere) with the Euclidean closure $\overline{D}$ on account of Proposition 2 stated in §3, so that we may well call the $\hat{H}^1(D)$-diffusion a reflecting Brownian motion on $\overline{D}$. But otherwise the reflecting Brownian motion can hardly be realized on the Euclidean closure $\overline{D}$.

For instance, let D_1 be the open unit disk in R^2, Γ be the segment $[0, 1)$ on the x-axis and put

$$D = D_1 \setminus \Gamma, \tag{1.10}$$

$\hat{H}^1(D) = H^1(D_1)$ is a proper subspace of $H^1(D)$. In this case, the Kuramochi type boundary of D is (up to a homeomorphism) the union of the unit circle and $\Gamma^+ \cup \Gamma^-$, Γ^+ and Γ^- representing the positive and negative sides of Γ, while the $\hat{H}^1(D)$-diffusion is just the reflecting Brownian motion on the closed unit disk. This is an example where ∂D is not equal to ∂D. Even when they are equal, similar situations may occur; see Kolsrud's example in [M].

Nevertheless the reflecting Brownian motion (X_t^*, P_x^*) on the Kuramochi type compactification D^* always induces a stationary process on $\overline{D}$. In fact, there is a mapping Ψ from D^* to $\overline{D}$ such that the process $\Psi(X_t^*)$ is continuous and stationary with respect to P_m. In order to characterize the quasimartingale property (semimartingale property with some integrability condition) of this induced stationary process, Z.Q.Chen,P.J.Fitzsimmons and R.J.Williams [CFW] has introduced the notion of a *strong Caccioppoli set* by requiring the inequality (5) to hold for any $\varphi \in H^1(D) \cap C_b(D)$. Their theorem asserts that the induced stationary process $\Psi(X_t^*)$ is P_m−quasimartingale iff D is strong Caccioppoli. In proving Theorem 1, we follow their method in [CFW], but we need to utilize a uniform bound of the resolvent density associated with the space $\hat{H}^1(D)$ instead of $H^1(D)$ and a result on a capacity preserving quasi-homeomorphism obtainable by invoking author's previous papers [F2] and [F3].

The present treatment of the Skorohod equation is in the q.e. level. Its treatment in the everywhere level, see the recent works [BH2], [FT],[FT2].

2. Decomposition of additive functionals and Gauss formula

Consider the coordinate functions $g_i(x) = x_i$, $1 \leq i \leq d$. Since they are continuous functions in the space $\hat{H}^1(D)$, we can apply the decomposition theorem in [FOT] to $X_t^i = g_i(X_t)$ to get

$$X_t^i - X_0^i = M_t^i + N_t^i, \quad P_x - a.s. \ q.e.x \in \overline{D}.$$

Here $M^i, 1 \leq i \leq d$, are continuous martingale additive functionals and the co-variation of M^i and M^j is related to the co-energy measure of g_i and g_j by the Revuz correspondence. By the expression (2) of the form $\mathcal{E}$, we know that the latter equals $\delta_{ij} \, m$ and accordingly

$$\langle M^i, M^j \rangle_t = \delta_{ij} \, t.$$

Hence we conclude that $(M^1, \cdots, M^d)$ is a d-dimensional Brownian motion. N_t^i is a continuous additive functional of zero energy but not necessarily of bounded variation on each finite time interval. We can see from §5.4 of [FOT] that the next lemma holds.

Lemma 2.1 N_t^i *is of bounded variation on each finite time interval and satisfies the integrability*

$$E_m \left(\int_0^t |dN_t^i| \right) < \infty \qquad (2.1)$$

if and only if the following conditions are satisfied:
(i) *there exists a finite signed measure ν_i such that*

$$\mathcal{E}(g_i, \varphi) = - \int_D \varphi \, d\nu_i \quad \varphi \in C_0^1(\overline{D}). \qquad (2.2)$$

(ii) ν_i *is smooth, namely, it charges no set of zero capacity with respect to the space $(\hat{H}^1(D), \mathcal{E})$.*
  In this case, N_t^i and ν_i are related by the Revuz correspondence.

Evidently condition (i) of the above lemma holds if and only if D is a Caccioppoli set, and in this case, $d\nu_i = n_i \, d\sigma$. For the proof of Theorem 1, it only remains to show that the measure ν_i satisfying (12) is smooth. If ν_i were not a signed measure, then formula (12) immediately implies that it is of finite energy integral and hence smooth. To check the smoothness of a signed measure is a non-trivial task which we would like to carry out in the next section.

3. Smoothness of signed measures

Consider any linear space $\mathcal{F}$ such that

$$H_0^1(D) \subset \mathcal{F} \subset H^1(D), \quad \mathcal{F} \text{ is } \mathcal{E}_1 - closed, \quad 1 \in \mathcal{F}. \qquad (3.1)$$

Then $(\mathcal{F}, \mathcal{E})$ is a Dirichlet space on $L^2(D)$. We say that a compactification D^* of D is $\mathcal{F}-admissible$ if $(\mathcal{F}, \mathcal{E})$ can be regarded as a regular Dirichlet space on $L^2(D^*)$. Here we identify $L(D^*)$ with $L^2(D)$ by setting $m(D^* - D) = 0$.

For instance, $\overline{D}$ is $\hat{H}^1(D)$-admissible. Given $\mathcal{F}$ and admissible D^* as above, there exists an associated conservative diffusion process $\mathbf{M}^* = (X_t^*, P_x^*)$ on D^*. The resolvent of $\mathbf{M}^*$ is denoted by $\{G_\alpha, \alpha > 0\}$. The associated capacity of an open set $E \subset D^*$ is defined by

$$Cap(E) = \inf\{\mathcal{E}_1(u, u) : u \geq 1 \text{ on } E, \ u \in \mathcal{F}\},$$

which is extended to any set as an outer capacity. A finite signed measure on D^* is called *smooth* if it charges no set of zero capacity. "$\mathcal{F}$-quasi every-where" or "$\mathcal{F}$-q.e" will mean "except for a subset of D^* of zero capacity".

Suppose that ν is a finite signed measure on D^*such that there exists a function $u \in \mathcal{F}$ and

$$\mathcal{E}(u, \varphi) = \int_{D^*} \varphi \, d\nu \quad \varphi \in \mathcal{F} \cap C(D^*). \tag{3.2}$$

Denote by $B_0(D)$ the space of bounded Borel functions on D with compact support. The next proposition can be proved in exactly the same way as in [CFW].

Proposition 3.1 *If*

$$G_1 \; : \; B_0(D) \longrightarrow C(D^*), \tag{3.3}$$

then ν is smooth.

The next Proposition is a specification to the present situation of a general theorem from [F3](see also Appendix of [FOT]).

Proposition 3.2 *If D_1^* and D_2^* are $\mathcal{F}-$admissible compactifications, then there exists a capacity preserving quasi-homeomorphism Ψ from D_1^* to D_2^* such that Ψ is the identity map on D.*

We need one more proposition.

Proposition 3.3 *There exists an $\mathcal{F}-$admissible compactification D^* for which (15) is valid.*

Taking the validity of the last proposition for granted, we proceed to the proof of Theorem 1.

Proof of Theorem 1

Assume that D is Caccioppoli and let $\mathcal{F} = \hat{H}^1(D)$. Then the inequality (5) holds for any $\varphi \in \mathcal{F} \cap C_b(D)$ by approximation. Choose D^* of Proposition 3 for the present $\mathcal{F}$. Since (5) holds for any $\varphi \in \mathcal{F} \cap C(D^*)$, there is a finite signed measure ν_i^* concentrated on $D^* - D$ such that

$$\mathcal{E}(g_i, \varphi) = \int_{D^*} \varphi \, d\nu_i^*, \quad \varphi \in \mathcal{F} \cap C(D^*), \quad 1 \leq i \leq d. \tag{3.4}$$

On account of Proposition 1, ν^* charges no set of zero capacity on D^*. One can then derive from (16)

$$\mathcal{E}(g_i, \varphi) = \int_{D^*} \tilde{\varphi} \, d\nu_i^*, \quad \varphi \in \mathcal{F}_b, \tag{3.5}$$

where $\tilde{\varphi}$ denotes any $\mathcal{F}$-quasi continuous version of $\varphi \in \mathcal{F}_b$, because we can find a sequence of uniformly bounded functions in $\mathcal{F} \cap C(D^*)$ converging to $\tilde{\varphi}$ $\mathcal{F}$-q.e. and in $\mathcal{E}_1$-metric as well.

On the other hand, $\overline{D}$ is also $\mathcal{F}$-admissible and there is, by Proposition 2 a capacity preserving quasi-homeomorphism Ψ from D^* to $\overline{D}$ which is the identity on D. Now, for any $\varphi \in C^1(\overline{D})$, $\varphi(\Psi(\cdot))$ is its $\mathcal{F}$-quasi continuous version and the left hand side of (17) equals

$$\int_{D^*} \varphi(\Psi(x)) \nu_i^*(dx) = \int_{\overline{D}} \varphi(y)(\Psi \nu_i^*)(dy),$$

which, combined with (6), leads us to $n_i \sigma = \Psi \nu_i^*$. Accordingly σ charges no set of zero capacity on $\overline{D}$. The proof of Theorem 1 is now complete.

Finally we give

Outline of the proof of Proposition 3

Let $\{G_\alpha, \alpha > 0\}$ be the resolvent on $L^2(D)$ associated with the Dirichlet space $(\mathcal{F}, \mathcal{E})$. By invoking the work [F2], we can conclude that it admits a density function $G_\alpha(x, y)$ such that

$$G_\alpha(x, y) = G_\alpha^0(x, y) + R_\alpha(x, y)$$

where $G_\alpha^0(x, y)$ is the resolvent density of the absorbing Brownian motion on D and $R_\alpha(x, y)$ is strictly positive, symmetric, α-harmonic in each variable and

$$\sup_{x \in D, y \in K} R_\alpha(x, y) < \infty \qquad (3.6)$$

for any compact $K \subset D$. In fact, the kernel $R_\alpha(x, y)$ was constructed in [F2] as

$$R_\alpha(x, y) = \int_M K_\alpha^x(\xi) \tilde{R}^\alpha K_\alpha^y(\xi) \mu(d\xi),$$

where M is the Martin boundary of D, $K_\alpha^x(\xi)$ is the α-order Martin kernel and $\tilde{R}^\alpha$ is a positivity preserving operator on a function space on M with $\tilde{R}^\alpha 1(\xi)$ being bounded. Therefore the Kuramochi type compactification D^* with respect to this resolvent $G_1(x, y)$ as in [F1] provides us with desired property (15).

References

[BH1] R.F.Bass and P.Hsu, The semimartingale structure of reflecting Brownian motion, *Proc.Amer. Math. Soc.*, **108** (1990), 1007-1010.

[BH2] R.F.Bass and P.Hsu, Some potential theory for reflecting Brownian motion in Hölder domains, *Ann. Probab.*, **19** (1991), 486-506.

[CFW] Z.Q.Chen, P.J.Fitzsimmons and R.J.Williams, Quasimartingales and strong Caccioppoli set, *Potential Analysis*, **2** (1993), 281-315.

[F1] M.Fukushima, A construction of reflecting barrier Brownian motions for bounded domains, *Osaka J. Math.*, **4** (1967), 183-215.

[F2] M.Fukushima, On boundary conditions for multi-dimensional Brownian motions with symmetric resolvent densities, *J.Math.Soc.Japan* **21** (1969), 58-93.

[F3] M.Fukushima, Dirichlet spaces and strong Markov processes, *Trans. Amer. Math. Soc.*, **162** (1971), 185-224.

[FOT] M.Fukushima, Y.Oshima and M.Takeda, *Dirichlet forms and symmetric Markov processes*, Walter de Gruyter, Berlin-New York, 1994.

[FT] M.Fukushima and M.Tomisaki, Reflecting diffusions on Lipschitz domains with cusps-Analytic construction and Skorohod representation, *Potential Analysis* **4** (1995), 377-408.

[FT2] M.Fukushima and M.Tomisaki, Construction and decomposition of reflecting diffusions on Lipschitz domains with Hölder cusps, *Probab.Theory Relat.Fields*, to appear.

[G] E.Giusti, *Minimal surfaces and functions of bounded variations*, North-Holland, Amsterdam, 1980.

[M] V.G.Maz'ja, *Sobolev spaces*, Springer-Verlag, Berlin-Heidelberg-New York, 1985.

Department of Mathematical Science, Faculty of Engineering Science
Osaka University

Received September 24, 1996

Rate of Convergence of Moments of Spall's SPSA Method[1]

László Gerencsér

1. Introduction

The aim of this paper is to present a rate of convergence theorem for a class of stochastic approximation processes for function minimization developed by J. Spall in [Spa92]. The main feature of this method is a new way of estimating the gradient using only two measurements at properly selected random parameter values. The main advances of the present paper is that a crucial boundedness hypothesis of [Spa92] is removed by forcing the estimator to stay in a bounded domain and we get the rate of convergence of higher order moments of the estimation error. The rate that we get is identical with the normalization that is used in proving asymptotic normality of the estimator sequence (cf. Proposition 2 in [Spa92]).

The analysis given in [Spa92] is based on the early work of Fabian on the Kiefer-Wolfowitz method (cf. [Fab68]). The present paper uses advanced techniques of the theory of recursive identification. This paper also complements recent results of [CDPD96], where an almost sure convergence rate has been given (Theorem 3 [CDPD96]) for a modified version of Spall's algorithm.

2. The problem formulation

We consider the following problem: minimize the function $L(\theta)$ defined for $\theta \in D$, where $D \subset \mathbb{R}^p$ is an open domain, for which only noise corrupted measurements are available, given in the form $L(\theta)+\varepsilon(n,\omega)$ where $\varepsilon(n,\omega)$ is a zero mean random variable over some probability space $(\Omega, \mathcal{F}, \mathcal{P})$. Note that the measurement noise $\varepsilon(n,\omega)$ does not depend on θ, thus we have what is called a state-independent noise. The extension of the results of this paper to state-dependent noise seems is possible, but a number of

---

[1]This work was supported by a grant from The Johns Hopkins University, Applied Physics Laboratory Independent Research and Development Program and by the National Research Foundation of Hungary (OTKA) under Grant no. 20984

additional technical details have to be clarified. It is assumed that the measured values of $L(\theta) + \varepsilon(n)$ can be obtained for each n, ω via a physical experiment, and if necessary the experiment can be repeated.

Condition 2.1 *The function $L(\theta)$ is three-times continuously differentiable with respect to θ for $\theta \in D$. It is assumed that $L(\theta)$ has a unique local minimum in D, which will be denoted by θ^*.*

Condition 2.2 *The compound noise-process $\varepsilon_1(n, \omega)$ (cf.(3.2)) is assumed to be L-mixing with respect to a pair of families of σ-fields $(\mathcal{F}_n, \mathcal{F}_n^+)$.*

For the definition of L-mixing cf. [Ger89]. Actually we need less than L-mixing, namely it is sufficient that an "improved Hölder-inequality" (cf. Lemma 2.1 of [Ger89]) holds for $\varepsilon_1(n, \omega)$.

3. Gradient estimation

To minimize $L(\theta)$ we need an estimator of its gradient, denoted by $G(\theta) = L_\theta(\theta)$. The conventional finite difference approximation of partial derivatives that requires a large number of function evaluations is replaced by an approximation using simultaneous random perturbations of the components of θ. Let k denote the iteration time for the stochastic gradient algorithm to be developed. At time k we take a random vector $\Delta_k = (\Delta_{k1}, ..., \Delta_{kp})^T$, where Δ_{ki} is a double sequence of i.i.d., symmetrically distributed, bounded random variables such that $E\Delta_{ki}^{-2} < \infty$ (cf. Section III of [Spa92]). A standard perturbation that will be used in the rest of the paper is the double sequence Δ_{ki}

$$P(\Delta_{ki} = +1) = 1/2 \qquad P(\Delta_{ki} = -1) = 1/2.$$

Now let $0 < c_k \leq 1$ be a fixed sequence of positive numbers and let D_0 be a compact, convex domain specified in Condition 4.1 below. For each $\theta \in D_0$ we take two measurements that are denoted by $M_k^+(\theta, \omega) = L(\theta + c_k\Delta_k) + \varepsilon(2k - 1, \omega)$, $M_k^-(\theta, \omega) = L(\theta - c_k\Delta_k) + \varepsilon(2k, \omega)$. Then the i-th component of the gradient at time k for $\theta \in D_0$ is estimated as

$$H_i(k, \theta, \omega) = (M_k^+(\theta, \omega) - M_k^-(\theta, \omega)/2c_k\Delta_{k1}.$$

A convenient representation is obtained if we define the random vector $\Delta_k^{-1} = (\Delta_{k1}^{-1}, ..., \Delta_{kp}^{-1})^T$, namely then the gradient estimator at $\theta \in D_0$ can be written as

$$H(k, \theta, \omega) = (M_k^+(\theta, \omega) - M_k^-(\theta, \omega))(2c_k)^{-1}\Delta_k^{-1}. \qquad (3.1)$$

Let $\varepsilon_1(k, \omega)$ be the compound measurement error defined by

$$\varepsilon_1(k, \omega) = \varepsilon(2k - 1, \omega) - \varepsilon(2k, \omega). \tag{3.2}$$

Then we can write $H(k, \theta, \omega)$ as

$$\left(L(\theta + c_k \Delta_k) - L(\theta - c_k \Delta_k) \right)(2c_k)^{-1}\Delta_k^{-1} + +\varepsilon_1(k, \omega)(2c_k)^{-1}\Delta_k^{-1}. \tag{3.3}$$

A standard choice for c_k is $c_k = c/k^\gamma$ with some $\gamma > 0$.

4. The associated ODE

A standard method in numerical analysis for function minimization is the gradient method. A continuous-time and scaled version of it is described by the ordinary differential equation

$$\dot{y}_t = -\frac{1}{t}G(y_t), \qquad y_s = \xi \tag{4.1}$$

$t \geq s$, which will be called the associated differential equation. $G(y)$ is defined in D and it has continuous partial derivatives up to second order. Under the condition above (4.1) has a unique solution in $[s, \infty)$ which we denote by $y(t, s, \xi)$. It is well-known that $y(t, s, \xi)$ is a continuously differentiable function of (t, s, ξ). Let us introduce the notation $\psi(t, s, \xi) = (\partial/\partial\xi)y(t, s, \xi)$ for $s \leq t$.

Condition 4.1 . *Let $D_0 \subset$ int D denote a compact, convex domain such that $\theta^* \in$ int D_0, and the closure of the neighborhood of D_0 of radius $c > 0$, denoted as $D(c)$, is inside D. It is assumed that for every $\xi \in D_0$, $t > s \geq 1$ $y(t, s, \xi) \in D$ is defined and we have with some $C_0, \alpha > 0$*

$$\|(\partial/\partial\xi)y(t, s, \xi)\| = \|\psi(t, s, \xi)\| \leq C_0(s/t)^\alpha. \tag{4.2}$$

Here $\| \cdot \|$ denotes the operator norm of a matrix. Furthermore we assume that the initial condition $\xi\epsilon$ int $D_{00} \subset$ int D_0, where D_{00} is a compact domain which is invariant for (4.1) and such that for any $t > s \geq 1$

$$y(t, s, D_{00}) = \{y(t, s, x) : x \in D_{00}\} \subset \text{int } D_{00}.$$

The last part of the condition ensures that the solution of the associated deterministic differential equation converges to θ^* for any initial condition $\xi \in D_{00}$.

5. The SPSA method

Let a_k be a fixed sequence of positive numbers with a_k denoting the step size at time k. Then in the original form of Spall's method the sequence of estimated parameters, denoted by $\widehat{\theta}_k$, $k = 1, 2, \dots$ is generated recursively by

$$\widehat{\theta}_{k+1} = \widehat{\theta}_k - a_{k+1} H(k+1, \widehat{\theta}_k, \omega). \qquad (5.1)$$

A standard choice for is $a_k = a/k^{\alpha'}$ with $0 < \alpha' \leq 1$, and $a > 0$.

The almost sure convergence of the estimator process for the case when the noise is a martingale difference process has been established in [Spa92] using results of [MP84]. In the same paper asymptotic normality of a properly scaled estimation error process is established by a nontrivial application of [Fab68]. The scaling is nonstandard compared to classical statistical theory: assuming $a_k = a/k^{\alpha'}$ and $c_k = c/k^\gamma$ a normal limit distribution of $k^{\beta/2}(\widehat{\theta}_k - \theta^*)$ exists with $\beta = \alpha' - 2\gamma > 0$ for $\gamma \geq \alpha'/6$ and $\beta = \alpha' - 2\gamma > 0$. A main advance of the present paper is that the "boundedness conditions" (Conditions *A3* and *A5* of [Spa92]) is removed by the application of a resetting or truncation mechanism. A further advance is that by the application of the methods of [Ger92] we get a tight upper bound for the rate of convergence of the moments of the error process.

Resetting: Let the estimator $\widehat{\theta}_{k+1}$ computed according to (5.1) be re-named $\widehat{\theta}_{k+1}^+$. If this tentative value leaves D_0 then the final update will be $\widehat{\theta}_{k+1} = \widehat{\theta}_0$.

Theorem 5.1 *Let* $\beta = \min(4\gamma, \alpha' - 2\gamma) > 0$. *Assume that the Hessian-matrix of the cost function at the minimizing point has all its eigenvalues to the right of* $\beta/2$. *Then under the conditions above we have*

$$\widehat{\theta}_k - \theta^* = O_M(k^{-\beta/2}).$$

The notation $O_M(.)$ means that the the $L_q(\Omega, \mathcal{F}, \mathcal{P})$-norm of the left hand side decreases with the rate given on the right hand side for any $q \geq 1$.

The value of $\beta = \min(4\gamma, \alpha' - 2\gamma)$ is maximized for $4\gamma = \alpha' - 2\gamma$, from which we get $\gamma = \alpha'/6$ and $\beta = 2\alpha'/3$. The best rate is obtained with $\alpha' = 1$ and $\gamma = 1/6$, which is then $k^{-1/3}$.

The assumption on the Hessian can be explained as follows: we require that the solution trajectories of the associated ODE given by (4.1) converge to θ^* with a rate of at least $k^{-\beta/2+\varepsilon}$ with some positive ε, which is a natural requirement if we expect the noise-corrupted process $\widehat{\theta}_k$ to have rate $k^{-\beta/2}$.

The asymptotic distribution of $k^{-\beta/2}(\widehat{\theta}_k - \theta^*)$ under different conditions have been shown in Proposition 2 of [Spa92] and also stated in Theorem

4 of [CDPD96]. An analogous result under the conditions of the present paper seems likely to be true. The arguments in favor of such a result follow [Ger95].

6. The main steps of the analysis

We present the main steps of the analysis for the case $\alpha' = 1$. Then γ is restricted by the inequality $0 < \gamma < 1/2$.

The first step of the proof is to imbed our discrete-time data and procedure into *continuous-time* data and procedure. Consider the piecewise linear curve $\widehat{\theta}_t^{c+}$ defined for $k \leq t \leq k+1$ as $\widehat{\theta}_t^{c+} = (t-k)\widehat{\theta}_{k+1}^+ + (k+1-t)\widehat{\theta}_k$, where $\widehat{\theta}_{k+1}^+$ is the tentative value of the estimator computed by (5.1). Redefining the resetting mechanism in an appropriate manner and introducing a piecewise-constant, continuous-time extension of $(H(k+1, \theta, \omega)), (a_k), (c_k)$ etc. we get a continuous-time process.

In the second step of the analysis we consider the trajectory $\widehat{\theta}_t^c$ on the interval $[\sigma, q\sigma \wedge \tau(\sigma))$ with some fixed $\sigma \geq 1$, and $q > 1$. Let $\tau(\sigma) \leq q\sigma$ denote the first moment after σ, at which $\widehat{\theta}_t^c$ hits the boundary of D_0. If $\widehat{\theta}_t^c$ does not hit the boundary of D_0 at all, then we set $\tau(\sigma) = q\sigma$. Further, let $\overline{y}_t$ denote the solution of (4.1) with initial condition $\overline{y}_\sigma = \widehat{\theta}_\sigma^c = \theta$ i.e. $\overline{y}_t = y(t, \sigma, \theta)$. A main technical tool used in the proof is the development of a locally uniform upper bound for the increments $|\widehat{\theta}_t^c - \overline{y}_t|$. Let

$$I_\sigma(q) = \sup_{\sigma \leq t \leq q\sigma \wedge \tau(\sigma)} |\widehat{\theta}_t^c - \overline{y}_t|$$

and

$$I_s^*(q) = \sup_{s \leq \sigma \leq qs} I_\sigma(q).$$

Lemma 6.1 . Let $\alpha' = 1, 0 < \gamma < 1/2$ and let $\beta = \min(4\gamma, 1 - 2\gamma)$. Then we have $I_s^*(q) = O_M(s^{-\beta/2})$.

This is the key technical lemma, to proof of which is quite involved. The main steps follow the proof of an analogous lemma in [Ger92] (Lemma 2.1). To sketch the proof we first fix σ. Using an integral representation of the respective differential equations, and the fact that $\widehat{\theta}_\sigma^c = \overline{y}_\sigma = \theta$ we have for $\sigma \leq t \leq q\sigma \wedge \tau(\sigma)$ with some Lipschitz-constant C

$$|\widehat{\theta}_t^c - \overline{y}_t| \leq |\int_\sigma^t \frac{1}{r} J_r dr| + \int_\sigma^t \frac{1}{r} 2L|\widehat{\theta}_r^c - \overline{y}_r| dr, \tag{6.1}$$

where J_r is the cumulative error, which can be written as $J_r^{\Delta\varepsilon} + J_r^c$, where $J_r^{\Delta\varepsilon} = -H^c(r, \widehat{\theta}_r^c, w) + G(\widehat{\theta}_r^c)$, and J_r^c is the error that is due to the continuous-time imbedding. To analyze $J_r^{\Delta\varepsilon}$ substitute $\widehat{\theta}_r^c$ by the free variable θ and define

$$J^{\Delta\varepsilon}(r, \theta, w) = -H^c(r, \theta, w) + G(\theta).$$

Then $J_r^{\Delta\varepsilon} = J^{\Delta\varepsilon}(r, \widehat{\theta}_r^c, w)$. Now we continue to decompose $J^{\Delta\varepsilon}(r, \theta, w)$: by writing it as the sum of the following two terms:

$$
\begin{aligned}
J^{\Delta}(r, \theta, w) &= -((L(\theta + c_r\Delta_r)) - L(\theta - c_r\Delta_r))(2c_r)^{-1}\Delta_r^{-1} + G(\theta) \\
J_r^{\varepsilon} &= -\varepsilon_1(r, w)(2c_r)^{-1}\Delta_r^{-1}.
\end{aligned}
\tag{6.2}
$$

$J^{\Delta}(r, \theta, w)$, will be further decomposed using a third-order Taylor-series expansion. We get $J^{\Delta}(r, \theta, w) = J^{\Delta 1}(r, \theta, w) + J^{\Delta 3}(r, \theta, w)$, where

$$J^{\Delta 1}(r, \theta, w) = -(\Delta_r^{-1}\Delta_r^T - I)G(\theta),\tag{6.3}$$

and $J^{\Delta 3}(r, \theta, w)$ is a residual term. The conditions imposed onto Δ_r imply that

$$E(-\Delta_r^{-1}\Delta_r^T + I) = 0.$$

Since Δ_{ki} is an i.i.d. sequence of random variables with values $+1$ or -1, it follows that $\Delta_r^{-1}\Delta_r^T$ is an L-mixing process, and we conclude that $J^{\Delta 1}(r, \theta, w)$ is a zero-mean L-mixing process. The same holds for its first three derivatives. Moreover it is easy to see that with $\theta = \theta_r$ we have

$$J_r^{\Delta 3} = J^{\Delta 3}(r, \widehat{\theta}_r^c, w) = O(c_r^2) = O(k^{-2\gamma}).\tag{6.4}$$

Thus we have the following decomposition of the error process:

$$J_r = J^{\Delta 1}(r, \widehat{\theta}_r^c, w) + J_r^{\Delta 3} + J_r^{\varepsilon} + J_r^c.\tag{6.5}$$

To estimate $J^{\Delta 1}(r, \widehat{\theta}_r^c, w)$ we proceed as in [Ger92]. We approximate it by $J_r^{\Delta 1} = J^{\Delta 1}(r, \overline{y}_r, w) = J^{\Delta 1}(r, y(r, \sigma, \widehat{\theta}_\sigma^c), w)$, and define the modified error process

$$\overline{J}_r = J_r^{\Delta 1} + J_r^{\Delta 3} + J_r^{\varepsilon} + J_r^c.\tag{6.6}$$

The advantage of this approximation is that the randomness of $\overline{y}_r$ is purely due to the randomness of the initial condition $\widehat{\theta}_\sigma^c$. Taking into account the inequality $|G(\widehat{\theta}_r^c) - G(\overline{y}_r)| \leq C|\widehat{\theta}_r - \overline{y}_r|$, we get

$$|\widehat{\theta}_t^c - \overline{y}_t| \leq |\int_\sigma^t \frac{1}{r}\overline{J}_r dr| + \int_\sigma^t \frac{1}{r}2C|\widehat{\theta}_r^c - \overline{y}_r|dr.\tag{6.7}$$

We will get a tight upper bound for the first term on the right hand side. Define

$$\delta_s^*(J^{\Delta 3}) = \sup_{\substack{\sigma \le t \le q\sigma \wedge \tau(\sigma) \\ s \le \sigma \le qs}} \left| \int_\sigma^t \frac{1}{r} J^{\Delta 3}(r, \widehat{\theta}_r^c, \omega) dr \right|$$

$$\delta_s^*(J^\varepsilon) = \sup_{\substack{\sigma \le t \le q\sigma \wedge \tau(\sigma) \\ s \le \sigma \le qs}} \left| \int_\sigma^t \frac{1}{r} J_r^\varepsilon dr dr \right|$$

The expressions $\delta_s^*(J^{\Delta 1}), \delta_s^*(J^c)$ are defined analogously. Following the arguments of Lemma 3.1 of [Ger92] it is easy to see that $\delta_s^*(J^{\Delta 1}) = O_M(s^{-1/2})$. Moreover $\delta_s^*(J^c) = O(s^{-1})$. For $\delta_s^*(J^{\Delta 3})$ we get using elementary arguments that $\delta_s^*(J^{\Delta 3}) \le C_1 s^{-2\gamma}$ with some $C_1 > 0$. For the estimation of $\delta_s^*(J^\varepsilon)$ we follow the arguments of the proof of Lemma 2.1 of [Ger92]. Adding a suitable extension of Lemma 3.1 of [Ger92] we get $\delta_s^*(J^\varepsilon) = O_M(s^{-1/2+\gamma})$.

Summarizing the above inequalities it is easy to see that the dominant terms are $\delta_s^*(J^{\Delta 3})$, arising from the approximation error in the numerical differentiation scheme, and $\delta_s^*(J_{12})$, arising from the measurement error. Thus we get for the compound error term δ_s^* defined as

$$\delta_s^* = \delta_s^*(J^{\Delta 1}) + \delta_s^*(J^{\Delta 3}) + \delta_s^*(J^\varepsilon) + \delta^*(J^c),$$

the inequality

$$\delta_s^* = O_M(s^{-\beta/2}) \quad \text{with} \quad \beta = \min(4\gamma, 1 - 2\gamma).$$

Thus we get from (6.7) for any $\sigma \le t \le q\sigma \wedge \tau(\sigma)$ with $s \le \sigma \le qs$

$$|\widehat{\theta}_t^c - \overline{y}_t| \le \delta_s^* + \int_\sigma^t \frac{1}{r} 2C |\widehat{\theta}_r^c - \overline{y}_r| dr. \tag{6.8}$$

Now from (6.8) we get using the Bellman-Gronwall-lemma with fixed σ

$$I_\sigma(q) = \sup_{\sigma \le t \le q\sigma \wedge \tau(\sigma)} |\widehat{\theta}_t^c - \overline{y}_t| \le \kappa \delta_s^* \tag{6.9}$$

with some nonrandom κ. Since the right hand side of (6.9) is independent of σ, we can take supremum over σ on the left hand side, and thus Lemma 6.1 is proved.

In the next step we estimate the hitting probabilities. Let C_s denote the event that a resetting takes place in the interval $[s, qs)$. It has been shown in Lemma 2.3 of [Ger92] that Lemma 6.1 implies that $P(C_s) = O(s^{-m})$ for all $m > 0$. It follows that in the whole interval $[s, qs)$ we have

$$\sup_{s \leq t \leq qs} |\widehat{\theta}_t^c - \overline{y}_t| \leq \kappa \delta_s^* + \chi_{C_s} \cdot K = O_M(s^{-\beta/2}),$$

where K is the diameter of D_0.

In the final step we paste together the interval estimates. Let us now take a subdivision of $[1, \infty]$ by the points $s_n = q^n$ with some $q > 1$. Then $\delta_{s_n}^* = O_M(q^{-n\beta/2})$. Following the arguments on p.1208 of [Ger92] and using the stability condition imposed on the associated ODE, it follows that with y_t denoting the solution of the associated ODE starting from $\widehat{\theta}_0$ we have

$$\sup_{q^n \leq t < q^{n+1}} |\widehat{\theta}_t^c - y_t| = O_M(q^{-n\beta/2}).$$

Since $\sup_{q^n \leq t < q^{n+1}} |y_t - \theta^*| = O(q^{-n\beta/2})$, the proposition of the theorem follows.

Acknowledgment

The author expresses his thanks to James C. Spall and John L. Maryak for initiating and cooperating in this research.

References

[CDPD96] H.F. Chen, T.E. Duncan, and B. Pasik-Duncan. A stochastic approximation algorithm with random differences. In J. Gertler, J.B. Cruz Jr, and M. Peshkin, editors, *Preprint of the 13th Triennial IFAC World Congress, San Francisco, USA, June 30/ July 5, 1996*, Volume **H**, (1996), 493–496.

[Fab68]   V. Fabian. On asymptotic normality in stochastic approximation. *Ann. Math. Stat.*, **39** (1968), 1327–1332.

[Ger89]   L. Gerencsér. On a class of mixing processes. *Stochastics*, **26** (1989), 165–191.

[Ger92]   L. Gerencsér. Rate of convergence of recursive estimators. *SIAM J. Control and Optimization*, **30(5)** (1992), 1200–1227.

[Ger95]   L. Gerencsér. A representation theorem for the error of recursive estimators. *SIAM J. Control and Optimization*, page under revision, (1995).

[MP84]   M. Metivier and P. Piouret. Application of a Kushner and Clark lemma to general classes of stochastic algorithms. *IEEE Tr. on Information Theory*, **30** (1984), 140–151.

[Spa92]   J.C. Spall. Multivariate stochastic approximation using a simultaneous perturbation gradient approximation, *IEEE Tr. on Automatic Control*, **37** (1992), 332–341.

Computer and Automation Institute
of the Hungarian Academy of Sciences
H-1518 Budapest, P.O.B. 63, Hungary
Fax: 36-1-1667503, Tel: 36-1-1667483
h2778ger@huella.bitnet

Received October 2, 1996

General Setting for Stochastic Processes Associated with Quantum Fields

Sergio Albeverio, Roman Gielerak and Francesco Russo

Abstract

An axiomatic concept of quantum field theory stochastic processes is introduced and some elementary properties of them are presented. The existence of the trace processes connected to Gibbs random fields obtained in Euclidean Quantum Field Theory is discussed. For particular two-dimensional interactions the Markov property and continuity of paths of the corresponding trace processes have been obtained.

1. Introduction

One of the most challenging open problems in Mathematical Physics is to give an example in space-time dimension $D = 4$ of a quantum field theory fulfilling all Wightman axioms [BLOT, Jo, SW] and describing nontrivial scattering processes. Since the Nelson work [Ne] stochastic analysis methods were successfully applied to solve this problem for $D < 4$ ([AGW, AZ, GJ, Si] and refs therein). A significantly weakened (compared to Nelson's original) system of axioms in terms of random fields has been proposed in [Fr, GJ]. In the present contribution we propose an alternative axiomatic approach based on the notion of distribution valued stochastic processes. The processes fulfilling the proposed set of axioms will be called quantum field theory processes. The system of axioms proposed below is by no means optimal but it is sufficient for the purposes of the present exposition. Comparing to the existing formulation of quantum field theory in terms of random fields our formulation being more restrictive offers however additional opportunities for constructing new examples of quantum field theories using well developed methods of stochastic analysis such as the method of Dirichlet forms [AHK, AR1, AR2, ARZ], the theory of stochastic differential equations [AR1, ARZ], and of partial stochastic (pseudo)-differential equations [AGW, BGL]. The present contribution is devoted to a general introduction to quantum field theory processes whose definition is formulated in the next section and some elementary properties of such processes are presented. Some examples in the case of $D = 2$ of quantum field theory processes are discussed in section 4. The present exposition is based on [AGR1, AGR2, AGR3].

2. Quantum Field Theory Processes

Let us denote by $S(I\!\!R^n)$ the Schwartz space of test functions and let $S'(I\!\!R^n)$ be its topological dual. We denote by $\langle \varphi, f \rangle$ the canonical pairing between S' and S.

A tempered quantum field theory process (tempered qft process) is a stochastic process ξ_t indexed by $I\!\!R^1$ and taking values in $S'(I\!\!R^d)$ fulfilling the following system of axioms:

qft(0)   1. if $f_n \overset{S(I\!\!R^d)}{\underset{n\uparrow\infty}{\longrightarrow}} f$ then for any t: $\langle \xi_t, f_n \rangle \to \langle \xi_t, f \rangle$ in law.

2. all correlation moments of the process ξ_t exist and

$$E \langle \xi_{t_1}, \cdot \rangle \dots \langle \xi_{t_n}, \cdot \rangle \in S'(I\!\!R^{dn})$$

for all $t_1, \dots, t_n \in I\!\!R^1$

3. $E \langle \xi_{\cdot}, f_1 \rangle \dots \langle \xi_{\cdot}, f_n \rangle \in S'(I\!\!R^n)$ for any $n \in N$; $f_i \in S(I\!\!R^d)$

From the nuclear theorem of Schwartz and from qft(0) points 2. and 3. it follows that for any $n \in N$ there exists an unique $S_n \in S'(I\!\!R^{(d+1)n})$ such that for all $g_i \otimes f_i \in S(I\!\!R^1) \otimes_{alg} S(I\!\!R^d)$ we have

$$S_n((g_1 \otimes f_1) \otimes (g_2 \otimes f_2) \dots (g_n \otimes f_n)) = \langle E \langle \xi_{\cdot}, f_1 \rangle \dots \langle \xi_{\cdot}, f_n \rangle, g_1 \otimes \dots \otimes g_n \rangle \tag{2.1}$$

The distributions S_n will be called **Schwinger functions** of the process ξ_t.

qft(1) *for any $F \in C_{pb}(I\!\!R^n)$ (= the space of continuous polynomially bounded functions on $I\!\!R^n$), any $n \in N$, any $t_1, \dots, t_n \in I\!\!R_+ \equiv \{t \in I\!\!R^1 | t \geq 0\}$ and any $f_1, \dots, f_n \in S(I\!\!R^d)$ the following holds*

$$0 \leq E\overline{RF}(\langle \xi_{t_1}, f_1 \rangle, \dots, \langle \xi_{t_n}, f_n \rangle)F(\langle \xi_{t_1}, f_1 \rangle, \dots \langle \xi_{t_n}, f_n \rangle) \tag{2.2}$$

where the reflection R is defined as

$$(RF)(\langle \xi_{t_1}, f_1 \rangle, \dots, \langle \xi_{t_n}, f_n \rangle) = F(\langle \xi_{-t_1}, f_1 \rangle, \dots, \langle \xi_{-t_n}, f_n \rangle) \tag{2.3}$$

A process ξ_t fulfilling the qft(1) axiom will be called **reflection positive** process.

qft(2) *Let $T_{d+1} \triangleleft O(d+1)$ be the Euclidean group of motions (i.e. semidirect product of translations T_{d+1} and the rotations group $O(d+1)$ and let for $f \in S(\mathbb{R}^{d+1})$, $(a, \Lambda) \in T_{d+1} \triangleleft O(d+1)$, $f^{(a,\Lambda)}(x) = f(\Lambda^{-1}(x-a))$. Then the Schwinger functions S_n of the process ξ_t are $T_{d+1} \triangleleft O(d+1)$ invariant i.e., for any $(a, \Lambda) \in T_{d+1} \triangleleft O(d+1)$, $n \in N$, $f_i \in S(\mathbb{R}^{d+1})$*

$$S_n(f_1^{(a,\Lambda)} \otimes \ldots \otimes f_n^{(a,\Lambda)}) \equiv S_n(f_1 \otimes \ldots \otimes f_n) \qquad (2.4)$$

If the Schwinger functions $(S_n)_{n=1,\ldots}$ of the process ξ_t determine uniquely its path space measure μ on the space of maps $\mathbb{R}^1 \ni t \mapsto S'(\mathbb{R}^d)$ then the equivalent formulation of qft(2) axiom is that the measure μ is invariant under the (natural) action of the group $T_{d+1} \triangleleft O(d+1)$ on the space $\mathbb{R}^1 \ni t \mapsto S'(\mathbb{R}^d)$. If the path space measure of the process ξ_t is Euclidean invariant and ξ_t fulfills qft(0) then qft(2) is also fulfilled.

qft(3) *The Schwinger functions of the process ξ_t fulfill the so called Fourier-Laplace property, i.e. (the difference variables) Schwinger functions restricted to the subspace*

$$S_\leq(\mathbb{R}^{(d+1)(n+1)}) \equiv \{f_n \in S(\mathbb{R}^{(d+1)(n+1)}) \,|$$
$$f_{n+1}(t_1, x_1, \ldots, t_{n+1}, x_{n+1}) = 0$$
$$\text{unless } t_1 \leq t_2 \leq \ldots \leq t_{n+1}\}$$

are given by Laplace(in time variables)-Fourier transform of some tempered distributions W_n', the Wightman distributions in the difference variables, (s_i, p_i) being the conjugate variables, supported on

$$\{(s_1, p_1), \ldots, (s_n, p_n)|s_i \geq 0, i = 1, \ldots, n\} \subset \mathbb{R}^{(d+1)n} \, .$$

The Fourier-Laplace property of a (tempered) distribution S_n is characterized in many equivalent ways [BLOT, Si]. The following stronger formulation of qft(3) axiom is more useful in applications.

qft(3)* *there exist a continuous norm $||| \; |||$ on $S(\mathbb{R}^{d+1})$ and a number $\gamma > 0$ such that for any $n \in N$, $f_i \in S(\mathbb{R}^{d+1})$:*

$$S_n(f_1 \otimes \ldots \otimes f_n) \leq (n!)^\gamma \prod_{i=1}^n |||f_i||| \, . \qquad (2.5)$$

It was proven in the paper [OS] that if the system of Euclidean invariant tempered distributions obeys qft(3)* and qft(1) then each S_n has the Fourier-Laplace property. For a recent stronger result see [Zi].

Having a tempered qft process ξ_t it is possible to construct from its Schwinger functions through canonical construction of Osterwalder and Schrader [OS] (see also [GJ, SW]) the corresponding quantum field theory fulfilling all Wightman axioms (with the possible exception of the cluster decomposition property).

In order to become familiar with the qft processes concept the following series of remarks might be useful [AGR3].

Proposition 2.1 Let $\xi_t \in S'(\mathbb{R}^d)$ be a Markov process fulfilling qft(0) and such that the processes ξ_t and $^r\xi_t = \xi_{-t}$ have the same law. Then the process ξ_t fulfills qft(1) i.e. the process ξ_t is reflection positive.

Proposition 2.2 Let $\xi_t \in S'(\mathbb{R}^d)$ be a process fulfilling qft(0) + qft(1) + qft(2). Then the covariance operator $R(t)$ of the process ξ_t is given by

$$R_t(x-y) = \int\limits_0^\infty \int e^{ip(x-y)} \frac{e^{-|t|\sqrt{p^2+m^2}}}{2\sqrt{p^2+m^2}} d\rho(m^2)\, dp + c \cdot 1\,, \qquad (2.6)$$

where ρ is some Borel tempered measure supported on $[0,\infty)$ and such that $\int_0^\infty m^{-1} d\rho(m^2) < \infty$, $c \in \mathbb{R}$, $c > 0$.

Remark In fact weaker conditions on ρ suffice, depending on d, see [Go].

Proposition 2.3 Let $\xi_t \in S'(\mathbb{R}^d)$ be an ergodic process obeying qft(0) + qft(1) + qft(2) and let the measure ρ in (2.6) be supported on a single point $m_0 \in [0,\infty)$. Then ξ_t is Gaussian.

Remark The ergodicity of the process ξ_t is necessary as the following simple example shows. We consider the probability measure whose characteristic functional is of the form $\sum_{n=1}^\infty c_n e^{-\frac{n}{2}||f||^2_{-1}}$, where c_n are non negative constants whose sum is 1; in general this determines the law of some non-Gaussian non-ergodic process with covariance operator proportional to R_t with $d\rho(m^2) = \delta(m^2 - 1)$.

Proposition 2.4 Let $\xi_t \in S'(\mathbb{R}^d)$ be a qft process which is pointwise defined and square integrable. Then ξ_t is a deterministic process, i.e. $\xi_t = c \cdot 1$, $c \in \mathbb{R}$.

For readers who are familiar with basic facts of quantum field theory, we remind that Proposition 2.3 is a stochastic version of the Jost-Schroer theorem [BLOT]. Propositon 2.4 is a stochastic formulation of the fact that a Wigtman field theory in which the (quantum) field theory operator is defined pointwise is a multiple of the identity operator. Proposition 2.1 and Proposition 2.2 are known in the probabilistic context (see i.e. [KL]). In particular Proposition 2.2 is a stochastic version of the Kallen-Lehman representation theorem [BLOT, Jo].

From Proposition 2.2, the following infinite-dimensional extension of the well known theorem characterizing homogeneous Gauss and Markov real valued processes (see i.e. [AGR3, KL]) can be deduced.

Proposition 2.5 *Let $\xi_t \in S'(\mathbb{R}^d)$ be a process fulfilling qft(0) + qft(1) + qft(2). Assume additionally that ξ_t is Gauss and Markov. Then the covariance operator of the process ξ_t is given as*

$$R(t)(x - y) = \int e^{ip(x-y)} \frac{e^{-t\sqrt{p^2+m^2}}}{2\sqrt{p^2 + m^2}} dp. \tag{2.7}$$

3. Gaussian examples

We list below some basic examples of qft-processes which lead however to a trivial (in the sense of involving trivial scattering) quantum field theory models. However certain perturbations of them (discussed briefly in the next section) lead to nontrivial physical examples.

On the space $S(\mathbb{R}^{d+1})$ we define the following characteristc functional:

$$\Gamma_\rho(f) = \exp(-\frac{1}{2}G_0^\rho(f, f)), \tag{3.1}$$

where

$$G_0^\rho(f, f) = \int_0^\infty d\rho(m^2) \int_{\mathbb{R}^{d+1}} dp \frac{|\hat{f}(p)|^2}{p^2 + m^2} \tag{3.2}$$

where $\hat{f}$ is the Fourier transform of f and ρ is some tempered Borel measure on $[0, \infty)$ with the property that $\int_0^\infty m^{-1} d\rho(m^2) < \infty$. By Minlos theorem there exists a unique Borel cylindric measure μ_0^ρ on $S'(\mathbb{R}^{d+1})$ the characteristic functional of which is equal to Γ_ρ. Provided that ρ is a finite measure it is easy to see that for any mollifier $(\chi_\epsilon)_{\epsilon>0}$ (i.e. $\chi_\epsilon \in C_0^\infty(\mathbb{R})$, $\chi_\epsilon \geq 0$ and $\chi_\epsilon \to \delta$ weakly in $S'(\mathbb{R}^{d+1})$ as $\epsilon \downarrow 0$) the limit $\lim_{\epsilon \downarrow 0} \langle \varphi, \chi_\epsilon(-t) \otimes g \rangle$ exists in $L^p(d\mu_0^\rho)$ sense (for all $p \in [1, \infty)$). We denote this limit by $\langle \varphi, \delta_t \otimes g \rangle$. Then we can define a tempered process ξ_t^ρ as (a modification of) the law of the random element (ξ_t^ρ, f) which coincides with the law of $(\varphi, \delta_t \otimes f)$ under the Gaussian distribution μ_0^ρ. The obtained process ξ_t^ρ will be called the coordinate generalized free process. It is well known (see e.g. refs in

[AGR2]) that the process ξ_t^ρ has a version with Hölder continuous paths in the sense that for any $f \in S(\mathbb{R}^d)$ the real valued process $f_{\xi_t^\rho} \equiv \langle \varphi, \delta_t \otimes f \rangle$ has a Hölder continuous version with continuity parameter $\frac{1}{2} - \epsilon$ for any $\epsilon > 0$.

The process ξ_t^ρ is Markov iff the measure $d\rho$ is concentrated on a single point $m_0 \in [0, \infty)$ (and $m_0 > 0$ if $d = 1$) (see e.g. the references in [AZ]).

Another process called following Röckner [Ro1, Ro2] a *trace process* can be associated with μ_0^ρ as a path space measure. For this let $\mathcal{H}_\sigma^{-\alpha}(\mathbb{R}^d)$ be a weighted Sobolev space of a fractional order $\alpha \in \mathbb{R}$ (with the weight $\sigma : \sigma^2 \in L^2(\mathbb{R}^d)$).

Lemma 3.1 *Let $\int_0^\infty d\rho(m^2) < \infty$ and $\operatorname{supp} \rho \subset [m_0, \infty)$, $m_0 > 0$. Then for μ_0^ρ almost every $\varphi \in S'(\mathbb{R}^{d+1})$, any $\alpha > \frac{d-2}{2}$ and any hyperplane $H_t \equiv \{(x_0, x) \in \mathbb{R}^{d+1} | x_0 = t\}$ there exists a trace $T_{\Sigma_t}(\varphi)$ of φ on H_t in the Lions-Magenes sense, which takes values in the space $\mathcal{H}_\sigma^{-\alpha}(\mathbb{R}^d)$.*

Proof: According to the assumptions on $d\rho$ we can easily prove the validity of the following estimate

$$\int \mu_0^\rho(d\varphi)(\varphi, f)^2 \leq C \cdot ||f||_{-1}^2, \tag{3.3}$$

where C is some constant (depending i.e. on ρ) and $|| \cdot ||_{-1}$ is the Sobolev norm of order -1. Using Prop 2.5 of [AGR1] the proof follows. ∎

We define the **trace process** $^0X_t^\rho$ as measurable modification of the process $^0\tilde{X}_t^\rho$ composed from $T_{\Sigma_t}(\varphi)$.

Lemma 3.2 *There exists a version of the process $^0X_t^\rho$ (denoted by the same symbol) with Hölder continuous paths. More precisely for any $n \in N$, $s, t \in [-n, n]$, $|t - s|$ sufficiently small, any $\gamma \in (0, \frac{1}{2})$ there exists an integrable random variable Y_n such that with probability one*

$$||^0X_t^\rho - {}^0X_s^\rho||_{\mathcal{H}_\sigma^{-\alpha}} \leq Y_n \cdot |t - s|^\gamma, \tag{3.4}$$

provided $\alpha > \frac{d-1}{2}$.

The proof and even a more elaborated version of Lemma 3.2 is obtained (as in [Ro1]) by application of version [Ro1] of the Garsia-Preston criterion.

Let us come now to the case $d\rho(m)^2 = \delta(m^2 - m_0^2)dm^2$, $m_0 > 0$. Then the corresponding trace process denoted as X_t^0 can be described by Markovian data [Ro1, Ro2], that is to say an initial law and a transition semigroup. First the traces of a typical realization of $\varphi \in S'(\mathbb{R}^d)$ could be better localized. For this, let us denote by P_t the classical Poisson kernel of the following Dirichlet problem $(-\Delta + 1)f^*(t, x) = 0$ for $t > 0$ and $f^*(t, x) = f(x)$ for a given $f \in C^\infty(\mathbb{R}^d)$. It was proved in [Ro1] that for μ^0 a.e. $\varphi \in S'(\mathbb{R}^{d+1})$ a meaning to $P_t(\varphi)$ could be given. For any $\sigma^2 \in L^2(\mathbb{R}^d)$ fulfilling certain conditions (see [Ro1, Ro2] for details), for any $\alpha > \frac{d-2}{2}$ and for μ_0 a.e. $\varphi \in S'(\mathbb{R}^{d+1})$ the following integral is finite:

$$\pi_\alpha(\varphi)^2 = \int_{\mathbb{R}^d} dx\, \sigma^2(x) \int_0^\infty dt\, t^{2\alpha-1} |P_t(\varphi)|^2(x) < \infty. \tag{3.5}$$

Then we can define the separable real Hilbert space $\mathcal{B}_\sigma^{-\alpha}(\mathbb{R}^d)$ with the norm

$$||\varphi||_{\mathcal{B}_\sigma^{-\alpha}}^2 = ||\varphi||_{\mathcal{H}_\sigma^{-\alpha}}^2 + \pi_\alpha(\varphi)^2,$$

as a space of traces of the random field μ_0. Let r_0 be the Gaussian centered measure on $\mathcal{B}_\sigma^{-\alpha}(\mathbb{R}^d)$ the Fourier transforms Γ_0 of which is given by:

$$\Gamma_0(e^{i(\,\cdot\,, f)}) = \exp -\frac{1}{2}||f||_{-\frac{1}{2}}^2.$$

The following Markov semigroup of Ornstein-Uhlenbeck type on the space $L^2(\mathcal{B}_\sigma^{-\alpha}, dr_0)$ is defined

$$\pi_t^0(\xi; e^{i(\,\cdot\,, f)}) = \exp i(P_t(\xi), f) \exp -\frac{1}{2} G_{\Sigma_0}^0(f \otimes \delta_t \,|\, f \otimes \delta_t),$$

where $G_{\Sigma_0}^0 = (-\Delta_{\Sigma_0} + 1)^{-1}$, $-\Delta_{\Sigma_0}$ being the $d+1$ dimensional Laplacian with zero boundary condition on Σ_0.

Theorem 3.3 *[Ro1, Ro2] The Markov semigroup π_t^0 is Feller. r_0 is the unique stationary measure of π_t^0. The Markov diffusion process $\tilde{X}_t^0$ canonically constructed from (r_0, π_t^0) is equivalent to the trace process X_t^0. The path space measure of the process X_t^0 can be identified with the measure μ_0.*

Due to the essential self-adjointness of the generator of π_t^0 acting on smooth cylinder functions in $L^2(\mu_0)$ one can identify X_t^0 (in law) with the Dirichlet process associated with the classical Dirichlet form given by r_0 [AR1].

Remark Other examples of random fields fulfilling all axioms of Euclidean Quantum Field Theory as formulated for example in [Fr] are provided by white noise. ¿From the point of view of physics they are completely of no interest as they lead to quantum field operators which are multiple of identity. We remark that no reasonnable coordinate and trace processes exist in this case [Ka]. Let us note nevertheless that they can be used as inputs to construct some nontrivial quantum field like structures [AGW, BGL] and refs. therein.

4. Perturbed free processes

Let $\mathcal{U}_\Lambda$ be an additive functional of the free random field μ_0 in the sense of [GJ, Si] and such that $\exp -\mathcal{U}_\Lambda(\varphi) \in \cap_{p\geq 1} L^p(d\mu_0)$. The known example of such functions do exist only for $d+1 = 2$. see, i.e. [GJ, Si]. In order to obtain Euclidean homogeneous random fields, the homogeneous limits $\lim_{\Lambda\uparrow\mathbb{R}^2}$ of $d\mu_\Lambda(\varphi) = (Z_\Lambda)^{-1}\exp -\mathcal{U}_\Lambda(\varphi)d\mu_0(\varphi)$ have to be constructed. A suitable definition of this limiting operation is based on the notion of the Dobrushin-Lanford-Ruelle (DLR) equations which will be formulated now.

For an open set $\Lambda \subset \mathbb{R}^d$ we define $\mathcal{H}_{-1}(\Lambda)$ as metric completion of the space of continuous functions with compact support in the norm $\|\ \|_{-1}$; for closed $\Lambda \subset \mathbb{R}^d$ we define $\mathcal{H}_{-1}(\Lambda) \equiv \cap_{\epsilon>0}\mathcal{H}_{-1}(\Lambda^\epsilon)$ where $\Lambda^\epsilon = \{x \in \mathbb{R}^d \mid \text{dist}(x,\Lambda) < \epsilon\}$. For a given $\Lambda \subset \mathbb{R}^d$ let e_Λ be the orthogonal projector in $\mathcal{H}_{-1}(\mathbb{R}^d)$ onto $\mathcal{H}_{-1}(\Lambda)$. In particular for a region $\Lambda \subset \mathbb{R}^d$ with sufficiently regular boundary $\partial\Lambda$ the action of e_{Λ^c} is described as a solution of the following Dirichlet problem: $(-\Delta+1)e_{\Lambda^c}(f)(x) = 0$ for $x \in \Lambda$ as $e_{\Lambda^c}(f)|_{\Lambda^c} = f|_{\Lambda^c}$. It was proved in [Ro3] that the action of e_{Λ^c} can be extended to a distribution $\varphi \in S'(\mathbb{R}^d)$. We denote this extension by the same symbol. Then the following expression gives rise to regular versions of the conditional expectation values $E_{\mu_\Lambda}\{-|\Sigma(\Lambda^c)\}(\eta)$ of the measure μ_Λ with respect to the σ-algebras $\Sigma(\Lambda^c) = \sigma\{(\varphi,f)|f \in \mathcal{H}_{-1}(\Lambda^c)\} \vee N_{\mu_0}$ where N_{μ_0} is the σ-algebra of μ_0 null sets:

$$E_{\mu_\Lambda}\{e^{i(\varphi,f)}|\Sigma(\Lambda^c)\}(\eta) = e^{i(e_{\Lambda^c}(\eta),f)}\frac{E_{\mu_0}^{\partial\Lambda}(e^{i(\varphi,f)}e^{-\mathcal{U}_\Lambda(\varphi+e_{\Lambda^c}(\eta))})}{E_{\mu_0}^{\partial\Lambda}(e^{-\mathcal{U}_\Lambda(\varphi+e_{\Lambda^c}(\eta))})} \qquad (4.1)$$

$E_{\mu_0}^{\partial\Lambda}$ means the expectation value with respect to a Gaussian measure $\mu_0^{\partial\Lambda}$ defined by $\mu_0^{\partial\Lambda}(\varphi(x)\varphi(y)) = (-\Delta^{\partial\Lambda}+1)^{-1}(x,y)$, $\Delta^{\partial\Lambda}$ being the Laplace operator with the zero boundary condition on $\partial\Lambda$.

Definition 4.1 *Any Borel probability measure μ on $S'(\mathbb{R}^d)$ fulfilling*

$$\mu \circ E_{\mu_{\Lambda^c}}(e^{i(\varphi,f)}) = \mu(e^{i(\varphi,f)})$$

for any bounded region Λ will be called a Gibbs measure corresponding to the interaction $\mathcal{U}_\Lambda$.

The set of all Gibbs measures is denoted as $\mathcal{G}(\mathcal{U}_\Lambda)$. The following subsets of probability measures on $S'(I\!R^d)$ are defined now·

$$2CR \equiv \{\mu|\exists c_{-1}>0 : \mu(\varphi,f)^2 \leq c_{-1}||f||^2_{-1}\};$$

$$2FR \equiv \{\mu|\exists c_{-1},c_1,c_p : \mu(\varphi,f)^2 \leq c_{-1}||f||^2_{-1} + c_1||G_0 * f||_1 + c_p||G_0 * f||_p\};$$

where G_0 is the kernel of $(-\Delta + 1)^{-1}$ and $p > 2$.

Proposition 4.2 1. *Let $\mu \in 2CR$. Then for μ - a.e. $\varphi \in S'(I\!R^d)$ there exists a trace $T_{\Sigma_t}(\varphi)$ (in the Lions-Magenes sense) of φ and takes values in the space $\mathcal{B}_\rho^{-\alpha}(I\!R^{d-1})$ for any $\alpha > \frac{d-2}{2}$ and the weight function ρ as before.*

2. *Let $\mu \in 2FR$ and $d = 2$. Then for μ - a.e. $\varphi \in S'(I\!R^2)$ there exists a trace $T_{\Sigma_t}(\varphi)$ (in the sense as above) of φ and takes values in the space $\mathcal{B}_\rho^{-\alpha,T}(I\!R^1)$ for any $\alpha > \frac{d-2}{2}$, $T < \infty$ where the spaces $\mathcal{B}_\rho^{-\alpha,T}(I\!R^1)$ are defined as*

$$\mathcal{B}_\rho^{-\alpha,T}(I\!R^1) \equiv \{\varphi \in S'(I\!R^1) \,|\, ||\varphi||^2_{\mathcal{H}_\rho^{-\alpha}} + \int_0^T dt \int dx\, \rho^2(x)|\mathcal{P}_t(\varphi)|^2(x) < \infty\}.$$

It follows from this proposition that for any $\mu \in 2FR$ we can define the corresponding trace processes $X_t^\mu \equiv T_{\Sigma_t}(\varphi)$ (or more precisely, their measurable modifications).

A random field μ on $S(I\!R^d)$ is called FR-regular iff $\mu(\prod_{i=1}^n (\varphi,f_i)) \leq c_n \cdot \prod_{i=1}^n |||f|||_p$, where $|||f|||_p = c_{-1}||f||^2_{-1} + c_1||G_0 * f||_1 + c_p||G_0 * f||_p$ with constants c_{-1}, c_1, $c_p \geq 0$, $c_n > 0$ and for any $n = 1, 2, \ldots$. Similarly we say that μ is CR regular iff it is FR regular with $c_1 = c_p = 0$.

Proposition 4.3 *Let μ be an FR regular Euclidean invariant random field. Then there exists a version of the corresponding trace proces X_t^μ (denoted by the same symbol such that for any $\gamma \in (0, \frac{1}{2})$ there exists an integrable random variable $Y(\gamma)$ such that with probability equal to one*

$$|X_t^\mu - X_s^s|_{\mathcal{B}_\rho^{-\alpha,T}} \leq Y(\gamma) \cdot |t - s|^\gamma \tag{4.2}$$

for sufficiently small $|t - s|$ and suitable $\alpha \geq \frac{d-1}{2}$, ρ as above.

We use the following notation below: $\mathcal{G}_{2CR}(\mathcal{U}_\Lambda) \equiv \mathcal{G}(\mathcal{U}_\Lambda) \cap 2CR$, $\mathcal{G}_{2FR}(\mathcal{U}_\Lambda) \equiv \mathcal{G}(\mathcal{U}_\Lambda) \cap 2FR$ and so on.

Proposition 4.4 *Let $\mu \in \mathcal{G}_{FR}(\mathcal{U}_\Lambda)$ be a reflection positive Euclidean invariant Gibbs measure. Then the corresponding trace process X_t^μ fulfills the system of axioms qft(0)–qft(3).*

An important and difficult question is the following: which $\mu \in \mathcal{G}_{2FR}(\mathcal{U}_\Lambda)$ produce a Markovian trace process? This question is related to the preservation of the so called Global Markov Property under the Gibbsian perturbations of the free process X_t^0. A comprehensive summary of the present status of this problem can be found in [AZ]. Here we recapitulate some results on a recent constructive approach to this problem reported in [AGR1].

The idea is to construct the transition kernel π_t^μ for the trace process X_t^μ; then we show that given a Markovian data (r_0, π_t^μ) there is a Markov process whose path space measure can be identified with μ.

Theorem 4.5   1. *Let $\mathcal{U}(x) = \lambda \int d\eta(\alpha) \, e^{i\alpha x}$ where $d\eta$ is some bounded measure supported on $(-2\sqrt{\pi}, 2\sqrt{\pi})$. $|\lambda| < \lambda_0$ being sufficiently small. Then $\mathcal{G}_{2FR}(\mathcal{U}) = \{\mu^\lambda\}$. The corresponding trace process $X_t^{\mu^\lambda} \in \mathcal{B}_\rho^{-\alpha, T}(\mathbb{R})$ (defined for any $|t| < T < \infty$) is a Markovian diffusion. Moreover, the corresponding Markovian data for $X_t^{\mu^\lambda}$ are given by an initial stationary measure r^λ and a transition kernel π_t^λ constructed from the interaction $\mathcal{U}$, explicitly given by a uniformly convergent power series in λ for $|\lambda| < \lambda_0$. Moreover the following holds:*

$$E_{\mu^\lambda}\{e^{i(\,\cdot\,,f\otimes\delta_t)}|\Sigma(t \le 0)\}(\eta) = \pi_t^\lambda(T_{\Sigma_0}(\eta); e^{i(\,\cdot\,,f)})$$

for μ^λ a.e. $\eta \in S'(\mathbb{R}^d)$ and $r^\lambda = \mu^\lambda|_{\Sigma(t=0)}$.

2. *Let $\mathcal{U}(x) = \lambda \int_0^{2\sqrt{\pi}} d\eta(\alpha) \int dx e^{\alpha x}$, $\lambda > 0$, $d\eta$ be a bounded measure supported in $[0, 2\pi)$. Then $\mathcal{G}_{2CR}(\mathcal{U}) = \{\mu^\lambda\}$. The corresponding trace process X^{μ^λ} is a Markovian diffusion.*

Remarks

The explicit formulas for the case 1. are given in [AGR1]. Heuristically one has

– the initial (stationary) measure r^λ:

$$r^\lambda(e^{i(\,\cdot\,,f)}) = \sum_{m\geq0}\frac{1}{m!}\mu_0(e^{i(\varphi,f\otimes\delta_0)};\overbrace{\mathcal{U}_{\mathbb{R}^d}(\phi)\ldots;\mathcal{U}_{\mathbb{R}^d}(\phi)}^{m-\text{times}}),\qquad(4.3)$$

where (informally)

$$\mathcal{U}_{\mathbb{R}^d}(\varphi) = \lambda \int d\eta(\alpha) \int_{\mathbb{R}^d} dx \; : \cos\alpha\varphi : (x);$$

$$: \cos \alpha\varphi : (x) = \exp \frac{\alpha^2}{2} G_0(0) \cos \alpha\varphi(x) \,,$$

$\mu^\lambda(F_1; \ldots F_n)$ are connected (cumulant) moments of μ^λ.

– the transition kernel

$$\pi_t^\lambda(\psi, e^{i(\cdot, f)}) = e^{i(\mathcal{P}_t(\psi), f)} \cdot \sum_{n \geq 0} \frac{1}{n!} \mu_{G_{\tau_0}^0}^0 \overbrace{(e^{i(\cdot, \delta_t \otimes f)}; \mathcal{U}_{\mathbb{R}_+^d}^\psi(\cdot); \ldots; \mathcal{U}_{\mathbb{R}_+^d}^\psi(\cdot))}^{n-\text{times}}$$

$$(4.4)$$

where now (informally):

$$\mathcal{U}_{\mathbb{R}_+^d}^\psi(\varphi) = \lambda \int d\eta(\alpha) \int_{x^0 \geq 0} dx : \cos \alpha(\phi(x) + \mathcal{P}_t(\psi)) : \,.$$

(The explicit expression for the transition kernel corresponding to 2. could be found in [AGR1]).

We expect the processes X^{μ^λ} to be identifiable to ones constructed by the method of Dirichlet forms (see e.g [AR1, AR2]).

Acknowledgments

We are very grateful to H. Gottschalk for discussions. The second (R.G.) and the third author (F.R.) thank BiBoS Research Center for the hospitality while these notes were prepared. The second author (R.G.) gratefully acknowledges the support from Polish National KBN grant 2P03B12211.

References

[AGR1] S. Albeverio, R. Gielerak, F. Russo, *Constructive approach to the global Markov property in Euclidean quantum field theory, I. Construction of transition kernel*, BiBoS preprint bf 721/2/96 Bielefeld, 1996

[AGR2] S. Albeverio, R. Gielerak, F. Russo, On the paths Hölder continuity in the models of Euclidean Quantum Field Theory, Preprint.

[AGR3] S. Albeverio, R. Gielerak, F. Russo, in preparation.

[AGW] S. Albeverio, H. Gottschalk, J. L. Wu, Convoluted generalized white noise, Schwinger functions and their analytic continuation, *Rev. Math. Phys.*, to appear.

[AHK]   S. Albeverio, R. Hoegh-Krohn, Dirichlet forms and diffusion pro-
        cesses on rigged Hilbert spaces, *Z. Wahrsch. verb. Geb.* **40**, (1997),
        1-57.

[AR1]   S. Albeverio, M. Röckner, Stochastic differential equations in in-
        finite dimensions: solutions via Dirichlet forms, *Probab. Th. Rel.
        Fields*, (1991), 347-386.

[AR2]   S. Albeverio, M. Röckner, Classical Dirichlet forms on topologi-
        cal vector spaces: closability and a Cameron-Martin Formula, *J.
        Funct. Anal.* **88** (1990), 355-436.

[ARZ]   S. Albeverio, M. Röckner, T. S. Zhang, Markov uniqueness and its
        applications to martingale problems, stochastic differential equa-
        tions and stochastic quantization, *C.R. Math. Rep. Acad. Sci.*,
        Can. **XV** (1993), 1-6.

[AZ]   S. Albeverio, B. Zegarliński, *Global Markov property in quantum
        field theory and statistical mechanics*, in: S. Albeverio, J.E. Fen-
        stad, H. Holden, T. Lindstrøm, eds. "Ideas and Methods in Quan-
        tum and Statistical Physics", **Vol 2**, Cambridge Univ. Press, 1992.

[BGL]   C. Becker, R. Gielerak, P. Lugiewicz, *Covariant SPDE and Quan-
        tum Field Structures*, SFB 237 Preprint **Nr 331**, Bochum 1996.

[BLOT] V. N. N. Bogoliubov, A. A. Logunov, A. J. Oksak, J. T. Todorov,
        General principles of Quantum Field Theory, Kluwer Academic
        Publishers 1990.

[Fr]   J. Fröhlich, Schwinger functions and their generating functionals,
        I. *Helv. Phys. Acta*, **47** (1974), 256-306.

[GJ]   J. Glimm, A. Jaffe, *Quantum Physics. Functional Integral Point
        of View*, Springer, 1981.

[Go]   H. Gottschalk, Die Momente gefalteten gauss-poissinschen weissen
        Rauschen als Schwingerfunktion. Dissertation and publications in
        preparation, 1996.

[HKPS] T. Hida, H. H. Kuo, J. Potthoff, L. Streit, *White noise: An infinite
        Dimensional Calculus*, Kluwer, Dordrecht (1993)

[Jo]   R. Jost, *The General Theory of Quantized Fields*, Lecture Notes
        in Applied Mathematics, Providence, Rhode Island, 1965.

[Ka]   G. Kallianpur, *Stochastic Filtering Theory*, Springer-Verlag, 1980.

[KL]   A. Klein, L. Landau, Gaussian OS-Positive Processes, *Z. Wahrscheinlichkeit verw. Geb.* **40** (1977), 115-124.

[Ne]   E. Nelson, The Construction of Quantum Field Theory from Markov Fields, *J. Funct. Anal.*, **11** (1973), 97-112.

[OS]   K. Osterwalder, R. Schrader, Axioms for Euclidean Green's Functions, *Comm. Math. Phys.*, **31**, (1973), 83-112.

[RS]   M. Reed, B. Simon, *Methods of Modern Mathematical Physics*, **Vol 2.**, Academic Press, New York, 1972.

[Ro1]   M. Röckner, Traces of harmonic functions and a new path space for the free quantum field, *J. Funct. Anal.*, **79** (1968), 211-246.

[Ro2]   M. Röckner, *On the transition function of the infinite dimensional Ornstein-Uhlenbeck process given by the free quantum field*, in: "Potential Theory", Ed. J. Kral et. al., New York – London, Plenum Press 1988.

[Ro3]   M. Röckner, A Dirichlet problem for distributions and specifications for random fields, *Memoires of the Am. Math. Soc.*, **54** (1985), 324.

[Si]   B. Simon, *The $\mathcal{P}(\phi)_2$ Euclidean (Quantum) Field Theory*, Princeton University Press, 1974.

[SW]   R. F. Streater, A. S. Wightman, *PCT, Spin Statistics and all that*, W.A. Benjamin, New York, 1964.

[Zi]   Yu. M. Zinoview, Equivalence of Euclidean and Wightman field
         theories, *Comm. Math. Phys.*, **174** (1995), 1-27.

Sergio Albeverio,
Fakultät für Mathematik, Ruhr-Univerisität, Bochum,
D-44780 Bochum, Germany

Roman Gielerak,
Institute of Theoretical Physics, University of Wrocław,
PL-50-204 Wrocław, Poland

Francesco Russo,
Département de Mathématiques, Institut Galilée,
Université Paris Nord, F-93430 Villateneuse, France

Received September 30, 1996

On a Class of Semilinear Stochastic Partial Differential Equations

István Gyöngy

1. Introduction

We consider the nonlinear stochastic partial differential equation

$$\frac{\partial}{\partial t}u(t,x) = \frac{\partial^2}{\partial x^2}u(t,x) + f(t,x,u(t,x)) + \frac{\partial}{\partial x}g(t,x,u(t,x))$$
$$+ \quad \sigma(t,x,u(t,x))\frac{\partial^2}{\partial t \partial x}W(t,x) \tag{1.1}$$

with boundary condition

$$u(t,0) = u(t,1) = 0 \qquad t \in [0,T] \qquad u(0,x) = u_0(x) \qquad x \in [0,1], \tag{1.2}$$

where $\frac{\partial^2}{\partial t \partial x}W$ is a space time white noise, and $u_0 \in L^2([0,1])$, the space of square integrable functions. The functions $f = f(t,x,r)$, $g = g(t,x,r)$, $\sigma = \sigma(t,x,r)$ are Borel functions of $(t,x,r) \in \mathbf{R}_+ \times [0,1] \times \mathbf{R}$. The solution of the above problem is an $L^2([0,1])$-valued continuous process which is adapted to a filtration given together with the Brownian sheet W, and which satisfies the above problem in its integral form defined through test functions. (See the rigorous formulation in Section 2.) We refer to this problem as Eq$(u_0; f, g, \sigma)$. If $f = \sigma = 0$ and $g(t,x,r) = \frac{1}{2}r^2$ then the above equation is called Burgers equation. It arose in connection with the study of turbulent fluid motion and it has got significant attention in the literature. (See [Ba], [BP], [Bu1], [Bu2], [Ho] and the references therein.) The Burgers equation perturbed by space-time white noise (i.e., when f=0, $g = \frac{1}{2}r^2$ and $\sigma \neq 0$) has recently become the subject of intensive investigations. (See, e.g. [BCJ], [PDT], [PG] and the references therein.) When $g = 0$ then Eq$(u_0; f, g, \sigma)$ is a stochastic reaction-diffusion equation which has also been studied intensively. (See e.g. [BGy],[Fu],[Gy1],[Ma],[Wa] for existence and uniqueness theorems and [Da],[FR],[Gy2],[GyN] for approximation theorems.)

Our aim is to study the class of equations Eq$(u_0; f, g, \sigma)$. We assume linear growth on f and quadratic growth on g. Thus our class of equations contains both the stochastic Burgers equation and the diffusion-reaction equations as special cases. We prove an existence and uniqueness theorem when f, g and σ are Lipschitz functions. This theorem (Theorem 2.1 below) generalizes the results on existence and uniqueness from [BCJ],[PDT],[PG]. It is based on a simple estimate, Lemma 3.1 below, which shows also the possibility of obtaining a similar result for the equation considered in the whole real line in place of the interval $[0,1]$. We present also a comparison theorem which extends known comparison theorems for stochastic reaction-diffusion equations (see e.g. [DP]) to the case of $Eq(u_0; f, g, \sigma)$. Moreover we show that the space-time white noise has a regularization effect on the above type of equations. Namely, we prove that even if f is only a measurable function $Eq(u_0; f, g, \sigma)$ has a unique solution provided σ is separated from 0. Similar results in the case $g = 0$ are proved earlier by the aid of an estimate on the density $p_{t,x}(r)$ of the law of $u(t, x)$, with respect to the Lebesgue measure, obtained by Malliavin Calculus. (See e.g. [BGy]). In the present paper we show that one does not even need the existence of the density $p_{t,x}(r)$ in order to construct the solution via approximation. This result generalizes the existence and uniqueness theorems from [BGy] and [Gy1]. We remark that an L^p-theory of stochastic partial differential equations is developed in [Kr1] for a broad class of equations in several space dimension, which is applied, in particular, to the stochastic reaction-diffusion equations driven by space-time white noise but it does not cover the above type of equations. Finally we present a theorem from [Gy2] on lattice approximations in the special case $g = 0$. We will study the general case elsewhere.

2. Formulation of the problem and the results

Let $(\Omega, \mathcal{F}, \mathcal{F}_t, P)$ be a stochastic basis carrying an $\mathcal{F}_t$-Brownian sheet $\{W(t, x) : t \geq 0, x \in \mathbf{R}\}$. We recall that W is a continuous centered $\mathcal{F}_t$-adapted Gaussian random field with covariance

$$E(W(s, x)W(t, y)) = (s \wedge t)(x \wedge y),$$

such that $W(s, x) - W(r, x) - W(s, y) + W(r, y)$ and $\mathcal{F}_t$ are independent for every $0 \leq t \leq r \leq s$, $x, y \in [0, 1]$. We use the notation $\mathcal{B}(V)$ for the Borel σ-algebra on V (for a topological space V) and $\mathcal{P}$ for the predictable σ-algebra on $bf R_+ \times \Omega$.

Let τ be an $\mathcal{F}_t$-stopping time. We say that an $L_2([0, 1])$-valued continuous $\mathcal{F}_t$-adapted random field $u = \{u(t, .) : t \in \mathbf{R}_+\}$ is a solution of $Eq(u_0; f, g, \sigma)$ in the interval $[0, \tau)$ if for every $\varphi \in C^2([0, 1])$,

$\varphi(0) = \varphi(1) = 0$

$$\int_0^1 u(t,x)\varphi(x)dx = \int_0^1 u_0(x)\varphi(x)dx + \int_0^t \int_0^1 u(s,x)\frac{\partial^2}{\partial x^2}\varphi(x)\,dxds$$

$$+ \int_0^t \int_0^1 f(s,x,u(s,x))\varphi(x)\,dxds - \int_0^t \int_0^1 g(s,x,u(s,x))\frac{\partial}{\partial x}\varphi(x)\,dxds$$

$$+ \int_0^t \int_0^1 \sigma(s,x,u(s,x,u(s,x))\varphi(x)W(ds,dx) \qquad a.s.$$

for all $t \in [0,\tau)$, where the last integral is a stochastic Ito integral. For finite τ the solution in the closed interval $[0,\tau]$ is defined in the same manner. To formulate our main result we assume the following assumptions:

(1) *The function g is of the form $g(t,x,r) = g_1(t,x,r) + g_2(t,r)$, where g_1 and g_2 are Borel functions of $(t,x,r) \in \mathbf{R}_+ \times [0,1] \times \mathbf{R}$ and of $(t,r) \in \mathbf{R}_+ \times \mathbf{R}$, respectively. The function g_1 satisfies the linear growth and the function g_2 satisfies the quadratic growth conditions, i.e., for every $T \geq 0$ there is a constant K such that*

$$|g_1(t,x,r)| \leq K(1+|r|), \qquad |g_2(t,r)| \leq K(1+|r|^2)$$

for all $t \in [0,T]$, $x \in [0,1]$, $r \in \mathbf{R}$.

(2) *The functions $f = f(t,x,r)$, $\sigma = \sigma(t,x,r)$ are Borel functions, f satisfies the linear growth condition and σ is bounded.*

(3) *σ is globally Lipschitz in r and f,σ are locally Lipschitz functions with linearly growing Lipschitz constant, i.e., for every $T \geq 0$ there exists a constant L such that*

$$|\sigma(t,x,p) - \sigma(t,x,q)| \leq L|p-q|,$$

$$|g(t,x,p) - g(t,x,q)| \leq L(1+|p|+|q|)|p-q|$$

$$|f(t,x,p) - f(t,x,q)| \leq L(1+|p|+|q|)|p-q|$$

for all $t \in [0,T]$, $x \in [0,1]$, $p,q \in \mathbf{R}$.

Theorem 2.1 *Let the initial condition u_0 be an $\mathcal{F}_0$-measurable $L^2([0,1])$-valued random element. Assume (1), (2) and (3). Then $Eq(u_0;f,g,\sigma)$ has a unique solution in the interval $[0,\infty)$.*

If the noise in the equation is nondegenerate then we need not require any smoothness from the nonlinear drift f.

Theorem 2.2 *Let the initial condition u_0 be an $\mathcal{F}_0$-measurable $L^2([0,1])$-valued random element and assume conditions (1) and (2). Suppose that σ is globally Lipschitz in $r \in \mathbf{R}$ and that g is locally Lipschitz in r with linearly growing Lipschitz constant. Then $Eq(u_0; f, g, \sigma)$ has a unique solution in the interval $[0, \infty)$ if σ is separated from 0, i.e., if for every $T \geq 0$ there is constant $\lambda > 0$ such that $|\sigma(t, x, r)| \geq \lambda$ for all $t \in [0, T]$, $x \in [0, 1]$, $r \in \mathbf{R}$.*

Theorem 2.1 will be proved after some preliminary results presented in the next section. In Section 5 a comparison theorem is established, which makes it possible to obtain Theorem 2.2 in Section 6 by generalizing a method from [Kr2].

3. Preliminaries

Let T be a positive number and let $H(s, t; x, y)$ be a real function of $0 \leq s < t \leq T$, $x, y \in \mathbf{R}$. Assume that there are some positive constants K, a, b, c such that for all $0 \leq s < t \leq T$, $x, y \in \mathbf{R}$

(A)
$$|H(s, t; x, y)| \leq K \frac{1}{|t - s|} \exp(-a \frac{|x - y|^2}{|t - s|})$$

(B)
$$|\frac{\partial}{\partial x} H(s, t; x, y)| \leq K \frac{1}{|t - s|^{3/2}} exp(-b \frac{|x - y|^2}{|t - s|})$$

(C)
$$|\frac{\partial}{\partial s} H(s, t; x, y)| \leq K \frac{1}{|t - s|^2} \exp(-c \frac{|x - y|^2}{|t - s|})$$

Define the linear operator J by

$$J(v)(t, x) := \int_0^t \int_0^1 H(r, t; x, y) v(r, y) \, dy dr \qquad t \in [0, T], x \in [0, 1] \quad (3.1)$$

for every $v \in L^\infty([0, T]; L^1([0, 1]))$. In the sequel we use the notation $|h|_p$ for the $L^p([0, 1])$-norm of a function h defined on $[0, 1]$.

Lemma 3.1 *Assume (A), (B) and (C). Then J is a bounded linear operator from $L^\gamma([0,T]; L^1([0,1]))$ into $C([0,T]; L^2([0,1]))$ for $\gamma > 4$. Moreover the following estimates hold.*

(i) For every $T \geq 0$ there is a constant C such that

$$|J(v)(t,\cdot)|_2 \leq C \int_0^t (t-r)^{\frac{-3}{4}} |v(r,\cdot)|_1 \, dr \leq C \frac{4\gamma - 4}{\gamma - 4} t^{\frac{\gamma-4}{4\gamma-4}} \Big(\int_0^t |v(r)|_1^\gamma \, dr \Big)^{\frac{1}{\gamma}}$$
$$(3.2)$$

for all $t \leq T$ and $\gamma > 4$.

(ii) For $T \geq 0$, $0 < \alpha < 1/4$, $0 < \beta < 1/2$, $\gamma > 4/over 1 - 4\alpha$, $\delta > \frac{4}{1-2\beta}$ there is a constant C such that

$$|J(v)(t,\cdot) - J(v)(s,\cdot)|_2 \leq C|t - s|^\alpha \Big(\int_0^t |v(r)|_1^\gamma \, dr \Big)^{1/\gamma} \qquad (3.3)$$

$$|J(v)(t,x) - J(v)(t,\cdot + z)|_2 \leq C|z|^\beta \Big(\int_0^t |v(r)|_1^\delta \, dr \Big)^{1/\delta} \qquad (3.4)$$

for all $s,t \in [0,T]$, $z \in \mathbf{R}$, where we set $v(t,y) := 0$ when $y \notin [0,1]$.

Proof. For the proof we refer to [Gy].

Lemma 3.2 *Let $\zeta_n(t,y)$ be a sequence of random fields on $[0,T] \times [0,1]$ such that $\sup_{t \leq T} |\zeta_n(t,\cdot)|_1 \leq \xi_n$, where ξ_n is a finite random variable for every n. Assume that the sequence ξ_n is bounded in probability, i.e.*

$$\lim_{C \to \infty} \sup_n P(\xi_n \geq C) = 0.$$

Assume moreover (A),(B) and (C). Then the sequence $J(\zeta_n)$ is weakly compact in $V := C([0,T]; L^2([0,1])$.

Proof. By the Arzela-Ascoli theorem one knows that the closure of a set Γ is compact in V if it is uniformly equi-continuous in $t \in [0,T]$ and the closure of $\Gamma(t) := \{v(t,\cdot) : v \in \Gamma\}$ is compact in $L^2([0,1])$ for every $t \in [0,T]$. By a result of Kolmogorov one knows that the closure of a bounded set S in $L^2([0,1])$ is compact if

$$\lim_{z \to 0} \sup \sup\{|h(\cdot) - h(\cdot + z)|_2 : h \in S\} = 0.$$

Therefore by Lemma 3.1 the closure of the set

$$\Gamma_R := \{J(h) : h \in L^\infty([0,T]; L^2([0,1]), \sup_{t \leq T} |h(t)|_1 \leq R\}$$

is compact in V for every positive number R. Hence for every $\varepsilon > 0$

$$1 - \varepsilon \leq \sup_n P(\xi_n \leq R) \leq \sup_n P(J(\zeta_n) \in \Gamma_R)$$

for sufficiently large R, which proves the lemma.   ■

Let $G = G(t, x, y)$ $t \geq 0$, $x, y \in [0, 1]$ denote the Green function for the heat equation with Dirichlet boundary condition. One knows that

$$G(t, x, y) = (4\pi t)^{1/2} \exp\left(-\frac{1}{4}\frac{|x - y|^2}{|t - s|}\right) + L(t, x, y),$$

where L is a smooth function of $(t, x, y) \in \mathbf{R}_+ \times [0, 1] \times [0, 1]$. Hence one can easily see that the kernel H defined by $H(s, t; x, y) := G(t - s, x, y)$ or by $H(s, t; x, y) := \frac{\partial}{\partial y} G(t - s, x, y)$ satisfies conditions (A), (B) and (C). Thus from Lemmas 3.1 and 3.2 we get the following corollary.

Corollary 3.3 *The statements of Lemmas 3.1 and 3.2 hold true if the operator J is defined by $H(s, t; x, y) := G(t - s, x, y)$ or by $H(s, t; x, y) := \frac{\partial}{\partial y} G(t - s, x, y)$ in (3.1).*

Next we reformulate the notion of solution in deterministic intervals $[0, T]$. For the sake of generality, needed in a truncation procedure later, we assume in the following proposition that f, g and σ are $\mathcal{P} \otimes \mathcal{B}(L^2([0, 1]))$-measurable functions mapping $\mathbf{R}_+ \times \Omega \times L^2([0, 1])$ into $L^2([0, 1])$, $L^1([0, 1])$ and $L^2([0, 1])$, respectively.

Proposition 3.4 *Assume that u_0 is an $\mathcal{F}_0$-measurable random variable in $L^2([0, 1])$. Assume moreover that for every $T \geq 0$ there is a constant K such that*

$$|f(t, v)|_2 + |\sigma(t, v)|_2 \leq K|v|_2 \qquad |g(t, v)|_1 \leq |v|_2^2$$

for all $t \in [0, T]$ and $v \in L^2([0, 1])$. Then an L^2-valued $\mathcal{F}_t$-adapted locally bounded stochastic process $\{u(t), t \in [0, T]\}$ has a continuous modification, satisfying

$$\int_0^1 u(t, x)\varphi(x)dx = \int_0^1 u_0(x)\varphi(x)dx + \int_0^t \int_0^1 u(s, x)\frac{\partial^2}{\partial x^2}\varphi(x)\,dxds$$

$$+ \int_0^t \int_0^1 f(s, u(s))(x)\varphi(x)\,dxds - \int_0^t \int_0^1 g(s, u(s))(x)\frac{\partial}{\partial x}\varphi(x)\,dxds$$

$$+ \int_0^t \int_0^1 \sigma(s, u(s))(x)\varphi(x)W(ds, dx) \quad (a.s.) \qquad (3.5)$$

for every test function $\varphi \in C^2([0,1])$, $\varphi(0) = \varphi(1) = 0$ and for all $t \in [0,T]$ if and only if one of the following conditions is met:

(a) *For every test function $\varphi \in C^\infty([0,1])$, $\varphi(0) = \varphi(1) = 0$ and for all $t \in [0,T]$ equation (3.5) holds.*

(b) *For every $t \in [0,T]$ and $\psi \in C^{1,\infty}([0,t] \times [0,1])$, $\psi(s,0) = \psi(s,1) = 0$ $s \in [0,t]$*

$$\int_0^1 u(t,x)\psi(t,x)\,dx =$$

$$\int_0^1 u_0(x)\psi(0,x)\,dx + \int_0^t \int_0^1 u(s,x)[\frac{\partial^2}{\partial x^2}\psi(s,x) + \frac{\partial}{\partial t}\psi(s,x)]\,dxds$$

$$+ \int_0^t \int_0^1 f(s, u(s))(x))\psi(s,x)\,dxds - \int_0^t \int_0^1 g(s, u(s))(x)\frac{\partial}{\partial x}\psi(s,x)\,dxds$$

$$+ \int_0^t \int_0^1 \sigma(s, u(s))(x))\psi(s,x)W(ds, dx) \qquad (3.6)$$

holds almost surely.

(c) *For almost every $\omega \in \Omega$ for all $t \in [0,T]$*

$$u(t,x) = \int_0^t G(t,x,y)u_0(y)\,dy$$

$$+ \int_0^t \int_0^1 G(t-s,x,y)f(s,u(s))(y)\,dyds$$

$$- \int_0^t \int_0^1 G_y(t-s,x,y)g(s,u(s))(y)\,dyds$$

$$+ \int_0^t \int_0^1 G(t-s,x,y)\sigma(s,u(s))(y)\,dW(s,y) \quad (3.7)$$

for dx-almost every $x \in [0,1]$.

Proof. The equivalence of (a) and (b) is obvious. On gets (c) from (b) by taking

$$\psi(s,y) := \int_0^1 G(t-s,z,y)\varphi(z)\,dz \qquad (s,y) \in (0,t] \times [0,1]$$

for functions $\varphi \in C^2([0,1])$. Having expression (3.7) for u in (c) one can directly verify (a). The existence of a continuous L_2-valued modification of u follows from (c) by Corollary 3.3. For the detailed proof we refer to [Gy].

4. Equations with degenerate noise

In order to prove Theorem 2.1 first we prove the uniqueness of the solution and then we construct the solution by approximation using a general approach based on Skorokhod's representation and the following simple observation from [GyK].

Lemma 4.1 *Let $\mathbf{E}$ be a Polish space equipped with the Borel σ-algebra. A sequence of $\mathbf{E}$-valued random elements z_n converges in probability if and only if for every pair of subsequences z_l, z_m there exists a subsequence $w_k := (z_{l(k)}, z_{m(k)})$ converging weakly to a random element w supported on the diagonal $\{(x, y) \in \mathbf{E} \times \mathbf{E} : x = y\}$.*

To carry out our approximation procedure we need some lemmas on the weak compactness and on convergence of stochastic integrals. The proof of these lemmas is an exercise left to reader. (The detailed proofs can be found in [Gy].) Let $\{\phi_k : k \in \Lambda\}$ be an orthonormal subset of $H := L^2([0,1])$. Define

$$W^k(t) := \int_0^1 \int_0^1 \phi_k(x) \, dW(s, x), \quad k \in \Lambda \tag{4.1}$$

where W is an $\mathcal{F}_t$-Brownian sheet. Then W^k, $k \in \Lambda$ are independent $\mathcal{F}_t$-Wiener processes.

Lemma 4.2 *Let $h(t)$ be an $\mathcal{F}_t$-adapted H-valued random process such that*

$$Q := \sup_{t \le T} \sup_{x \in [0,1]} E(|h(s, x)|^{2p}) < \infty.$$

for some numbers $p \ge 1$, $T \ge 0$. Then for the stochastic integral

$$\xi(t, x) := \sum_k \int_0^t \int_0^1 G(t - s, x, y) h(s, y) \phi_k(y) \, dy dW^k(s)$$

the following estimate holds:

$$E(|\xi(t, x) - \xi(s, y)|^{2p}) \le KQ(|t - s|^{\frac{1}{4}} + |x - y|^{\frac{1}{2}})^{2p} \tag{4.2}$$

for all $s, t \in [0, T]$, $x, y \in [0, 1]$, where K is a constant depending only on p and T.

Corollary 4.3 *(cf.[Wa]) Estimate (4.2) holds for the process*

$$\xi(t, x) := \int_0^t \int_0^1 G(t - s, x, y) h(s, y) \, dW(s, y),$$

where h is an H-valued $\mathcal{F}_t$-adapted process satisfying the condition in Lemma 4.1.

Proof. Take an orthonormal basis $\{\phi_k : k \in \Lambda\}$ in Lemma 4.1 and note that in this case

$$\sum_{k\in\Lambda} \int_0^t (\int_0^1 h(s,x)\phi_k(x)\,dx)dW^k(s) = \int_0^t \int_0^1 h(s,x)\,dW(s,x).$$

■

In order to formulate a lemma on convergence of stochastic integrals let $W_n(t,x)$ be an $\mathcal{F}_t^n$-Brownian sheet and let $h_n = h_n(t)$ be an $\mathcal{F}_t^n$-adapted $L^2([0,1])$-valued stochastic process for every integer $n \geq 1$ such that

$$\int_0^T \int_0^1 |h_n(t,x)|^2\,dxds < \infty \text{ (a.s.)}.$$

Let $(\phi_k)_{k\geq 1}$ be an orthonormal basis in $L^2([0,1])$ and define

$$W_n^i : \quad = \quad \int_0^t \int_0^1 \phi_i(x)\,dW_n(s,x)$$

$$\xi_n(t) : \quad = \quad \sum_{i\in\Lambda_n} \int_0^t (\int_0^1 h_n(s,x)\phi_i(x)\,dx)\,dW_n^i(s),$$

where Λ_n is a subset of the positive integers for every $n \geq 1$.

Lemma 4.4 *Assume that for $n \to \infty$*

$$\int_0^T \int_0^1 |h_n(t,x) - h(t,x)|^2\,dxdt \to 0$$

$$\sup_{t\leq T} \sup_{x\in[0,1]} |W_n(t,x) - W(t,x)| \to 0$$

in probability, where $W = W(t,x)$ is an $\mathcal{F}_t$-Brownian sheet and $h = h(t,x)$ is an $\mathcal{F}_t$-adapted random field such that $\int_0^T \int_0^1 |h(t,x)|^2\,dxds < \infty$ (a.s.). Assume moreover that Λ_n is an increasing sequence of sets such that $\cup_{n=1}^\infty \Lambda_n = \{1,2,...\}$. Then for every $\varepsilon > 0$

$$J := P(\sup_{t\leq T} |\xi_n(t) - \int_0^t \int_0^1 h(s,x)\,dW(s,x)| \geq \varepsilon) \to 0$$

Corollary 4.5 *(cf.[Sk]) Let h_n, h, $W_n(s,x)$ and $W(s,x)$ be the same processes, satisfying the same conditions as in Lemma 4.4. Then for every $\varepsilon > 0$*

$$P(\sup_{t \le T} | \int_0^t \int_0^1 h_n(s,x)\, dW_n(s,x) - \int_0^t \int_0^1 h(s,x)\, dW(s,x)| \ge \varepsilon) \to 0$$

for $n \to \infty$.

Lemma 4.6 *Let $f_n = \{f_n(t,x,r) : t \ge 0, x \in [0,1], r \in \mathbf{R}\}$ be a random field for every integer $n \ge 1$ such that for every $R \ge 0$ for $n \to \infty$*

$$\int_0^T \int_0^1 \sup_{|r| \le R} |f_n(t,x,r) - f(t,x,r)|\, dx dt \to 0 \ a.s.,$$

where f is some random field. Assume moreover that for some constant K

$$|f_n(t,x,r)| + |f(t,x,r)| \le K(1 + |r|^2),$$

$$|f_n(t,x,p) - f_n(t,x,r)| \le K(1 + |p| + |r|)|p - r|$$

for all $t \ge 0$, $x \in [0,1]$, $p, r \in \mathbf{R}$ and for all n. Let $u_n = \{u_n(t,x) : t \in [0,T], x \in [0,1]\}$ be a random field for every $n \ge 1$ such that

$$\int_0^T \int_0^1 |u(t,x)|^2\, dx dt < \infty \qquad \int_0^T \int_0^1 |u_n(t,x) - u(t,x)|^2\, dx dt \to 0 \ a.s..$$

Then for $n \to \infty$

$$I := \int_0^T \int_0^1 |f_n(t,x,u_n(t,x)) - f(t,x,u(t,x))|\, dx dt \to 0 \ a.s..$$

Next we prove an existence and uniqueness theorem for the equation obtained from the problem Eq($u_0; f, g, \sigma$) by truncating the operators determined by f and g. Let R be a positive number and consider the equation

$$\frac{\partial}{\partial t} u(t,x) = \frac{\partial^2}{\partial x^2} u(t,x) + \kappa_R(|u(t)|_2) f(t,x,u(t,x))$$

$$+ \frac{\partial}{\partial x} \kappa_R(|u(t)|_2) g(t,x,u(t,x)) + \sigma(t,x,u(t,x)) \frac{\partial^2}{\partial t \partial x} W(t,x) \qquad (4.3)$$

with boundary condition

$$u(t,0) = u(t,1) = 0 \quad t \in [0,T], \qquad u(0,x) = u_0(x) \quad x \in [0,1], \qquad (4.4)$$

where $\kappa_R = \kappa_R(r)$ is a $C^1(\mathbf{R})$ function such that $\kappa_R(r) = 1$ for $r \in [-R, R]$, $\kappa_R(r) = 0$ for $|r| > R + 1$ and $|\frac{d}{dr} \kappa_R(r)| \le 2$ for all $r \in \mathbf{R}$.

Proposition 4.7 *Assume that u_0, f and g satisfy the conditions of Theorem 2.1. Then for every $R > 0$ there is a unique solution of the problem (4.3)-(4.4).*

Proof. First we prove the uniqueness of the solution. Let u and v be two solutions in $[0, T]$. Then by Proposition 3.4 we have

$$u(t, x) - v(t, x) = \zeta_1(t, x) + \zeta_2(t, x) + \zeta_3(t, x) \qquad (4.5)$$

where

$$\zeta_1(t, x) := \int_0^t \int_0^1 G(t - s, x, y)(\kappa_R(|u|_2)f(u)(s, y) - \kappa_R(|v|_2)f(v)(s, y))\, dyds$$

$$\zeta_2(t, x) := -\int_0^t \int_0^1 G_y(t - s, x, y)(\kappa_R(|u|_2)g(u)(s, y) - \kappa_R(|v|_2)g(v)(s, y))\, dyds$$

$$\zeta_3(t, x) := \int_0^t \int_0^1 G(t - s, x, y)(\sigma(u)(s, y) - \sigma(s, y))\, dW(y, s).$$

Clearly there is a constant C such that

$$\int_0^1 E|\zeta_3(t, x)|^2\, dx \leq C \int_0^1 E\left(\int_0^t \int_0^1 G^2(t-s, x, y)|u(s, y) - v(s, y)|^2\right) dsdydx$$

by the Lipschitzness of σ. Hence there is a constant C such that

$$E \int_0^1 \zeta_3^2(t, x)\, dx \leq C \int_0^t (t - r)^{-\frac{1}{2}} E|u(r) - v(r)|_2^2\, dr. \qquad (4.6)$$

By Lemma 3.1 (part (i)) and Corollary 3.3 there is a constant C such that for all $t \in [0, T]$

$$E \int_0^1 \zeta_2^2(t, x)\, dx \leq CE\left(\int_0^t (t-r)^{-\frac{3}{4}}|\eta(r)|_1\, dr\right)^2 \leq CT^{\frac{1}{4}} \int_0^t (t-r)^{-\frac{3}{4}} E(|\eta(r)|_1^2)\, dr, \qquad (4.7)$$

where $\eta(r, y) := |\kappa_R(|u(r)|_2)g(u)(r, y) - \kappa_R(|v(r)|_2)g(v)(r, y)|$. Note that for every $R > 0$ there exists a constant C_R such that

$$D := |\kappa_R(|p|_2)g(p) - \kappa_R(|q|_2)g(q)|_1^2 \leq C_R|p - q|_2^2 \qquad (4.8)$$

for all $p, q \in L^2([0, 1])$ and for all $r \in [0, T]$. (For notational simplicity we omit the variable r in the expression.) It suffices to see this when $|p|_2 \leq |q|_2$. Then by the growth condition and Lipschitz condition on g there is a constant K such that

$$\begin{aligned} D &\leq |\kappa_R(|p|_2) - \kappa_R(|q|_2)||g(p)|_1 + \kappa_R(|q|_2)|g(p) - g(v)|_1 \\ &\leq K|\kappa_R(|p|_2) - \kappa_R(|q|_2)|(1 + |p|_2^2) + \kappa_R(|q|_2)K|p - q|_2(1 + |p|_2 + |q|_2) \\ &\leq 2K|p - q|_2(1 + (R + 1)^2) + K|p - q|_2(2R + 3), \end{aligned}$$

which proves (4.8). Hence by (4.7) there is a constant K such that

$$E \int_0^1 \zeta_2^2(t,x)\, dx \le K \int_0^1 (t-r)^{-\frac{3}{4}} E|u(r) - v(r)|_2^2\, dr \qquad (4.9)$$

for all $t \in [0,T]$. Similarly we can show the existence of a constant C such that

$$E \int_0^1 \zeta_1^2(t,x)\, dx \le C \int_0^t (t-r)^{-\frac{3}{4}} E|u(r) - v(r)|_2^2\, dr \qquad (4.10)$$

for all $t \in [0,T]$. By (4.5), (4.6), (4.9) and (4.10) we have a constant K such that

$$E(|u(t) - v(t)|_2^2) \le K \int_0^t (t-r)^{-\frac{3}{4}} E|u(r) - v(r)|_2^2\, dr$$

for every $t \in [0,T]$. Hence the uniqueness follows by a well-known Gronwall-Bellman type lemma. To prove the existence of the solution let $\mathcal{H}$ denote the Banach space of $L^2([0,1])$-valued $\mathcal{F}_t$-adapted random processes $v(t)$, $t \in [0,T]$, with the norm

$$|v|_{\mathcal{H}} := \sup_{t \le T}\{E(w|v(t)|_2^2)\}^{12} < \infty,$$

where $w := \exp(-|u_0|_2)$. Define the operator $\mathcal{A}$ on $\mathcal{H}$ by $\mathcal{A}(v)(t,x) := \sum_{i=1}^4 A_i(t,x)$ where

$$A_1(t,x) := \int_0^1 G(t,x,y)u_0(y)\, dy$$

$$A_2(t,x) := \int_0^t \int_0^1 G(t-r,x,y)\kappa_R(|v(r)|_2)f(v)(r,y)\, dydr$$

$$A_3(t,x) := -\int_0^t \int_0^1 G_y(t-r,x,y)\kappa_R(|v(r)|_2)g(v)(r,y)\, dydr$$

$$A_4(t,x) := \int_0^t \int_0^1 G(t-r,x,y)\sigma(v)(r,y)\, dW(r,y).$$

Clearly $E(w|\mathcal{A}(v)(t)|_2^2) \le \sum_{i=1}^4 F_i$, where $F_i := E(w|A_i(t)|_2^2)$. By Young's inequality there is a constant C such that

$$|A_1|_{\mathcal{H}}^2 = \sup_{t \le T} F_1 \le CE(w \int_0^1 |u_0(y)|^2\, dy) < \infty$$

By Lemma 3.1 (part (i)) and Corollary 3.3 there is a constant C such that

$$F_3 \leq CE(w(\int_0^t (t-r)^{-\frac{3}{4}} \kappa_R(|v(r)|_2)|g(v)(r)|_1 \, dr)^2).$$

Hence by the growth condition on g we have $|A_3|_{\mathcal{H}}^2 < \infty$. We can easily verify that $|A_2|_{\mathcal{H}}^2 < \infty$ and $|A_4|_{\mathcal{H}}^2 < \infty$. Thus $\mathcal{A}$ is an operator mapping the Banach space $\mathcal{H}$ into itself. Let u, v be from $\mathcal{H}$. Then repeating the same calculation as in the proof of the uniqueness we get a constant K such that

$$\begin{aligned}
E(w|\mathcal{A}(u)(t) - \mathcal{A}(v)(t)|_2^2) &\leq KE(w(\int_0^t (t-r)^{-\frac{3}{4}}|u(r) - v(r)|_2 \, dr)^2) \\
&\leq 4KT^{\frac{1}{2}} \sup_{t \leq T} E(w|u(r) - v(r)|_2^2)
\end{aligned}$$

holds for every $t \in [0,T]$. Hence $\mathcal{A}$ is a contraction on $\mathcal{H}$ if $T < \frac{1}{16K^2}$. Consequently there exists a unique solution in the interval $[0, \frac{1}{32K^2}]$ and hence in every interval $[0,T]$. The proof of the proposition is complete. ∎

Corollary 4.8 *Let τ and ρ be stopping times bounded by some constant T. Let u and v be solutions of (4.3), (4.4), (4.5) in the stochastic intervals $[0,\tau]$ and $[0,\rho]$ respectively. Assume the conditions of Theorem 2.1. Then almost surely $u(r) = v(r)$ in L^2 for for all $t \in [0, \tau \wedge \rho]$.*

Proof of Theorem 2.1. To prove the uniqueness of the solution let us consider two solutions u and v of Eq$(u_0; f, g, \sigma)$ in the interval $[0,T]$. Then by Lemma 3.1 we can easily see that u and v are solutions of the problem (4.3)-(4.4) in the stochastic interval $[0, \rho_R \wedge \tau_R]$ for every positive number R, where

$$\begin{aligned}
\tau_R : &= \inf\{t \geq 0 : |u(t)|_2 \geq R\} \wedge T \\
\rho_R : &= \inf\{t \geq 0 : |v(t)|_2 \geq R\} \wedge T.
\end{aligned}$$

By Corollary 4.8 we have $u(t) = v(t)$ in the stochastic interval $[0, \rho_R \wedge \tau_R]$. Hence we get that almost surely $u(t,x) = v(t,x)$ for almost every $x \in [0,1]$ (with respect to the Lebesgue measure) for all $t \in [0,T]$ since $P(\rho_R \wedge \tau_R < T) \to 0$ as $R \to \infty$. In order to prove the existence of the solution we take sequences of bounded Borel functions $f_n = f_n(t,x,r)$ and $g_n(t,x,r)$ such that they are globally Lipschitz in $r \in \mathbf{R}$ and $f_n = f$, $g_n = g$ for $|r| \leq n$, $f_n = g_n = 0$ for $|r| \geq n+1$. Moreover, f_n and $g_n = g_n^{(1)} + g_n^{(2)}$ satisfy the same Lipschitz conditions and the conditions on growth as f and g

respectively with constants independent of n. Then in the same way as in
the proof of Proposition 4.7 we can prove the existence of a unique solution
u_n of $Eq(u_0; f_n, g_n, \sigma)$. Set $v_n := u_n - \eta_n$, where

$$\eta_n(t, x) := \int_0^t G(t - s, x, y) \sigma(s, y, u_n(s, y)) \, dW(s, y).$$

Since σ is bounded uniformly in n, by Corollary 4.3 for every $p \geq 1$, $T > 0$
there is a constant C such that that for all $n \geq 1$

$$E(|\eta_n(t, x) - \eta_n(s, y)|^{2p}) \leq C(|t - s|^{\frac{1}{4}} + |x - y|^{\frac{1}{2}})^{2p}$$

for all $s, t \in [0, T]$, $x, y \in [0, 1]$. Hence by a well-known lemma of Garsia,
Rodemich and Rumsey [GRR] (or see e.g. in [Wa]) one can easily deduce
that

$$\sup_{n \geq 1} E(\sup_{(t,x) \in [0, T \times [0,1]} |\eta_n(t, x)|^p) < \infty.$$

Consequently, $\eta_n^\star := \sup_{(t,x) \in [0, T \times [0,1]} |\eta_n(t, x)|$ is bounded in probability,
uniformly in n. Next we prove that $\sup_{t \leq T} |u_n(t)|_2$ is also bounded in
probability, uniformly in n. To this end we note first that $v = v_n := u_n - \eta_n$
is a solution of the equation

$$\frac{\partial}{\partial t} v(t, x) = \frac{\partial^2}{\partial x^2} v(t, x) + f_n(t, x, v(t, x) + \eta_n(t)) + \frac{\partial}{\partial x} g_n(t, x, v(t, x) + \eta_n(t))$$
$$(4.11)$$

with boundary condition

$$v(t, 0) = v(t, 1) = 0 \quad t \in [0, T], \quad v(0, x) = u_0(x) \quad x \in [0, 1].$$
$$(4.12)$$

Clearly for every n one has a constant $C = C(n)$ such that

$$|\int_0^1 \psi'(x) \phi'(x) \, dx| \leq |\psi'|_2 |\phi'|_2, \quad |\int_0^1 f_n(\psi)(t, x) \phi(x) \, dx| \leq C |\phi|_2$$
$$(4.13)$$

$$|\int_0^1 g_n(\psi)(t, x) \phi'(x) \, dx| \leq C |\phi'|_2 \qquad (4.14)$$

for every $\psi, \phi \in \mathbf{V} := H^1$, where ϕ' denotes the derivative of ϕ, and
H^1 stands for the Hilbert space of absolutely continuous functions in $[0, 1]$
vanishing at 0 and at 1 with the norm $|\phi|_{\mathbf{V}} = |\phi'|_2$. Thus the operator
$\mathbf{A}(t, \psi)$ defined by

$$< \mathbf{A}(t, \psi), \phi > \quad := \quad -\int_0^1 \psi'(x) \phi'(x) \, dx + \int_0^1 f_n(\psi + \eta_n)(t, x) \phi(x) \, dx$$

$$- \int_0^1 g_n(\psi + \eta_n)(t, x)\phi'(x)\, dx$$

maps $\mathbf{V}$ into its dual space $\mathbf{V}^*$ such that $|\mathbf{A}(t, \psi)|_{\mathbf{V}^*} \le K(1 + |\psi|_V)$ for all $\psi \in \mathbf{V}$, for all $t \in [0, T]$ with a constant K. Thus we can see that the problem (4.11)-(4.12) can be cast in the evolution equation $dv(t) = \mathbf{A}(s, v(s))\, ds$, $u(0) = u_0$ in the triplet $\mathbf{V} \hookrightarrow \mathbf{H} \equiv \mathbf{H}^* \hookrightarrow \mathbf{V}^*$ of spaces based on the Hilbert space $\mathbf{H} := L^2([0, 1])$ identified with its dual $\mathbf{H}$ by the scalar product $(,)$ in $\mathbf{H}$. From (4.13), (4.14) for every n we have a constant K such that

$$< \mathbf{A}(t, \phi), \phi > \le K|\phi|_{\mathbf{H}}^2 - \frac{1}{2}|\phi|_{\mathbf{V}}^2$$

for all $\phi \in \mathbf{V}$. This means the operator $\mathbf{A}$ is coercive. By the Lipschitz condition on f_n and g_n one can easily verify that $\mathbf{A}$ satisfies also the monotonicity and hemi-continuity conditions. Consequently, by a well-known result (see e.g. [Li]) problem (4.11)-(4.12) has a (unique) solution $\bar{v}$ in $C([0, T]; \mathbf{H})$ such that

$$\int_0^T |\bar{v}(t)|_{\mathbf{V}}^2\, dt < \infty \ a.s..$$

Moreover, the energy equality

$$|\bar{v}(t)|_{\mathbf{H}}^2 = |u_0|^2 + 2\int_0^t < \bar{v}, \mathbf{A}(s, \bar{v}(s)) > ds \ a.s. \tag{4.15}$$

holds for for all $t \in [0, T]$. Since v_n solves problem (4.12) through (4.14) and one can easily show that there is no more continuous L^2-valued solution in the interval $[0, T]$ we get $\bar{v} = v_n$. Thus by (4.15)

$$|v_n(t)|_2^2 = |u_0|_2^2 - 2\int_0^t |v_n'(r)|_2^2\, dr + 2A(t) - 2B(t) - 2C(t),$$

where

$$A(t) := \int_0^t (f_n(s, v_n(s) + \eta(s)), v_n(s))\, ds \qquad B(t) := \int_0^t (g_n(s, v_n(s)), v_n'(s))\, ds$$

$$C(t) := \int_0^t (g_n(s, v_n(s) + \eta_n(s)) - g_n(s, v_n(s)), v_n'(s))\, ds.$$

By the linear growth condition on f_n, and the Lipschitz condition on g_n which are uniform in n, we have a constant K such that

$$|A(t)| \le K\int_0^1 (1 + |v_n(s)| + |\eta_n(s)|, |v_n(s)|)\, ds$$

$$\le K \int_0^t (2 + |\eta_n(s)|_2)|v_n(s)|_2^2 \, ds$$

$$|C(t)| \le K \int_0^t (|\eta_n(s)|(1 + |v_n(s)| + |\eta_n(s)|), |v_n'(s)|) \, ds$$

$$\le \int_0^t (\frac{3}{4}|v_n'(s)|_2^2 + 4K^2|\eta_n(s)|_2^2 + 4K^2|\eta_n v_n|_2^2 + 4K^2|\eta_n|_4^4) \, ds$$

using Hölder's inequality. Clearly $B(t) = B_1(t) + B_2(t)$, where

$$B_1(t) := \int_0^t (g_n^{(1)}(s, v_n(s)), v_n'(s)) \, ds \qquad B_2(t) := \int_0^t (g_n^{(2)}(s, v(s)), v_n'(s)) \, ds.$$

By the linear growth condition on $g_n^{(1)}$ we have a constant K such that

$$|B_1(t)| \le K \int_0^t (1 + |v_n(s)|, |v_n'(s)|) \, ds \le \int_0^t (\frac{1}{4}|v_n'(s)|_2^2 + 8K^2 + 8K^2|v_n(s)|_2^2) \, ds.$$

Define the function

$$h_n(t, r) := \int_0^r g_n^{(2)}(t, z) \, dz \qquad t \in [0, T], r \in \mathbf{R}.$$

Then

$$B_2(t) = \int_0^t \int_0^1 \frac{\partial}{\partial y} h_n(s, v_n(s, y)) \, dy \, ds = 0$$

because of the boundary condition. Summing up we get a constant K such that almost surely

$$|v_n(t)|_2^2 \le |u_0|_2^2 + K \int_0^t (1 + |\eta_n(s)|_2^2 + |\eta_n(s)|_4^4 + |v_n(s)|_2^2 + |\eta_n(s)v_n(s)|_2^2) \, ds$$

$$\le K \int_0^t (2 + |\eta_n^\star|^4|v_n(s)|_2^2 + |\eta_n^\star|^2|v(s)|_2^2) \, ds$$

for all $t \in [0, T]$. Hence by the Gronwall-Bellman lemma we have a constant K such that

$$|v_n(t)|_2^2 \le [|u_0|_2^2 + KT(1 + |\eta_n^\star|^4] \exp(K(1 + |\eta_n^\star|^2)t)$$

holds for all $n \ge 1$ and $t \in [0, T]$. Consequently the sequence $\sup_{t \le T} |v_n(t)|_2$ is bounded in probability, since the sequence $\eta_n^\star$ is bounded in probability. Thus the $\sup_{t \le T} |u_n(t)|_2$ is also bounded in probability. Hence by Lemma 3.2 and Corollary 3.3 the sequences of the $L^2([0, 1])$-valued stochastic processes $I_n^{(1)}(t)$, $I_n^{(2)}(t)$ defined by

$$I_n^{(1)}(t) := \int_0^t \int_0^1 G(t - s, x, y) f_n(s, u_n(s)) \, dy \, ds$$

$$I_n^{(2)}(t) := \int_0^t \int_0^1 G_y(t - s, x, y) g_n(s, u_n(s)) \, dy \, ds \qquad t \in [0, T]$$

are weakly compact in $\mathbf{E} := C([0,T]; L^2([0,1]))$. It is well-known that for $u_0 \in L^2([0,1])$ the process

$$I^{(0)}(t) := \int_0^t G(t-s,x,y)u_0(y)\,dy, \qquad t \in [0,T]$$

is in $\mathbf{E}$, and that the sequence of processes

$$I_n^{(3)}(t,x) := \int_0^t \int_0^1 G(t-s,x,y)\sigma(s,u_n(s))\,dW(s,y), \qquad t \in [0,T], x \in [0,1]$$

is weakly compact in $C([0,T] \times [0,1])$. Therefore the sequence of processes

$$u_n(t) = I^{(0)}(t) + I_n^{(1)}(t) - I_n^{(2)}(t) + I_n^{(3)}(t) \qquad t \in [0,T]$$

is weakly compact in $\mathbf{E}$. Thus by Skorokhod's theorem for a given pair of subsequences u_m and u_l there exist subsequences $m(k)$, $l(k)$ of the indices m, l and a sequence of random elements

$$z_k := (\tilde{u}_k, \bar{u}_k, \hat{W}_k), \qquad k := 1,2,3,\ldots$$

in $\mathbf{B} := \mathbf{E} \times \mathbf{E} \times C([0,T] \times [0,1])$ carried by some probability space $(\hat{\Omega}, \hat{\mathcal{F}}, \hat{P})$ such that z_k converges almost surely in $\mathbf{B}$ to a random element $z := (\tilde{u}, \bar{u}, \hat{W})$ for $k \to \infty$ and the distributions of z_k and $(u_{m(k)}, u_{l(k)}, W)$ coincide. The random fields $\hat{W}$ and $\hat{W}_k$ are Brownian fields carried by the stochastic basis $\hat{\Theta} := (\hat{\Omega}, \hat{\mathcal{F}}, (\hat{\mathcal{F}}_t)_{t \geq 0}, \hat{P})$ and $(\hat{\Omega}, \hat{\mathcal{F}}, (\hat{\mathcal{F}}_t^k)_{t \geq 0}, \hat{P})$ respectively, where $\hat{\mathcal{F}}_t$ and $\hat{\mathcal{F}}_t^k$ are the completion of the σ-fields generated by $z(s,x)$ $s \leq t$, $x \in [0,1]$ and by $z_k(s,x)$ $s \leq t$, $x \in [0,1]$ respectively. For every $\phi \in C^2([0,1])$, $\phi(0) = \phi(1) = 0$

$$\int_0^1 u_n(t,x)\phi(x)\,dx = \int_0^1 u_0(x)\phi(x)\,dx + \int_0^t \int_0^1 u_n(s,x)\phi''(x)\,dx\,ds$$

$$+ \int_0^t \int_0^1 f_n(s,x,u_n(s,x))\phi(x)\,dx\,ds - \int_0^t \int_0^1 g_n(s,x,u_n(s,x))\phi'(x)\,dx\,ds$$

$$+ \int_0^t \int_0^1 \sigma(s,x,u_n(s,x))\phi(x)\,dW(s,x) \quad a.s. \qquad (4.16)$$

for all $t \in [0,T]$. Note that this equation holds with $\tilde{u}_k$ and $\hat{W}_k$ in place of u_n and W, where taking the limit $k \to \infty$ by using Corollary 4.5 and Lemma 4.6 we get that $\tilde{u}$ solves $\mathrm{Eq}(\tilde{u}_0; f, g, \sigma)$ on the stochastic basis $\hat{\Theta}$ with the Wiener process $\hat{W}$. In the same way we see that $\bar{u}$ also solves $\mathrm{Eq}(\tilde{u}_0; f, g, \sigma)$ also on $\hat{\Theta}$ and with $\hat{W}$. Hence by the uniqueness of the solution $\tilde{u} = \hat{u}$ and by Lemma 4.1 we conclude that u_n converges in $\mathbf{E}$ in probability to some random element $u \in \mathbf{E}$. Taking now the limit $n \to \infty$ in equation (4.16) we see, using Corollary 4.5 and Lemma 4.6 again, that u is a solution of $\mathrm{Eq}(u_0; f, g, \sigma)$.

5. A comparison theorem

Let u_0, v_0 be $\mathcal{F}_0$-measurable random elements in $L^2([0,1])$ and let f, g, σ be Borel functions satisfying the assumptions of Theorem 2.1. Let $F = F(t,x,r)$ be a Borel function satisfying the same condition on growth and Lipschitzness as the function f in Theorem 2.1. Then we have the following comparison theorem.

Theorem 5.1 *Assume that almost surely $u_0(x) \leq v_0(x)$ for dx-almost every $x \in [0,1]$. Suppose that for $dt \times dx$-almost every $(t,x) \in [0,T] \times [0,1]$ we have $f(t,x,r) \leq F(t,x,r)$ for all $r \in \mathbf{R}$. Then $Eq(u_0; f, g, \sigma)$ and $Eq(v_0; F, g, \sigma)$ has a unique solution u and v, respectively in the interval $[0,T]$, and almost surely $u(t,x) \leq v(t,x)$ for all $t \in [0,T]$ for dx-almost every $x \in [0,1]$.*

Proof. By Theorem 2.1 the problems $Eq(u_0; f, g, \sigma)$ and $Eq(v_0; F, g, \sigma)$ admit a unique solution u and v, respectively. Let f_n, F_n, g_n be bounded Borel functions of $(t,x,r) \in [0,T] \times [0,1] \times \mathbf{R}$ for every integer $n \geq 1$ such that they are globally Lipschitz in r, $f_n(t,x,r) \leq F_n(t,x,r)$ for $dt \times dx$-every (t,x) and for all r, $f_n = f$, $F_n = F$, $g_n = g$ for $|r| \leq n$, $f_n = F_n = g_n = 0$ for $|r| \geq n+1$. Assume moreover that f_n, F_n and g_n satisfy the same Lipschitz condition and growth condition as f, F and g, respectively with constants independent of n. Let $\{\phi_k\}$ be an orthonormal basis in $\mathbf{H} := L^2([0,1])$ such that ϕ_k is bounded uniformly in $k \geq 1$, and define the Wiener process $W^k(t)$ by (4.1). Fix n and consider the evolution equation

$$du_n(t) = \mathbf{A}_n(t, u_n(t))\, dt + \sum_{k=1}^{n} \mathbf{B}^k(t, u_n(t)) dW^k(t) \qquad u_n(0) = u_0 \qquad (5.1)$$

in the triplet $H^1 \hookrightarrow H \equiv H^* \hookrightarrow H^{-1}$, where $A_n(t)$ and $B^k(t)$ are nonlinear operators mapping H^1 into H^{-1} and into $\mathbf{H}$, respectively, defined by

$$< \mathbf{A}_n(t, \psi), \phi > \quad := \quad -\int_0^1 \psi'(x)\phi'(x)\, dx + \int_0^1 f_n(\psi)(t,x)\phi(x)\, dx$$
$$-\int_0^1 g_n(\psi)(t,x)\phi\prime(x)\, dx$$
$$(\mathbf{B}^k(t, \psi), h) \quad := \quad \int_0^1 \sigma(t,x,\psi(x))\phi_k(x)h(x)\, dx.$$

for $\psi, \phi \in H^1$, $h \in \mathbf{H}$. (We use the notation $< \psi, \phi >$ for the pairing between H^1, H^{-1} and $(,)$ for the scalar product in $\mathbf{H} = L^2([0,1])$.) As in the proof of Theorem 2.1 we can easily see that the monotonicity and coercivity conditions on $(\mathbf{A}_n, \mathbf{B})$ and the semicontinuity and linear growth conditions on $\mathbf{A}_n$ are satisfied, so for every n equation (5.1) has a unique solution $u_n \in C([0, T]; \mathbf{H})$ such that

$$\int_0^T \int_0^1 |u_n(s, x)|^2_{H^1} \, dx ds < \infty \quad (a.s.).$$

(See [Pa], [Ro].) Let v_n denote the solution of (5.1) with v_0 and F_n in place of u_0 and f_n, respectively. Set $w_n := u_n - v_n$. Our aim is to show that almost surely for all $t \in [0, T]$

$$|w_n(t, x)|_+ = 0 \tag{5.2}$$

for dx-almost every $x \in [0, 1]$, where $|p|_+ := max(p, 0)$. To this end we use the following device from [DP]. For every integer $k \geq 1$ we define the functional $\Psi_k : \mathbf{H} \to \mathbf{R}$ by

$$\Psi_k(h) := \int_0^1 \psi_k(h(x)) \, dx,$$

where

$$\psi_k(x) := 1_{x \geq 0} \int_0^x \int_0^y \rho_k(z) \, dz dy,$$

and $\rho(z) := 2kz$ for $z \in [0, \frac{1}{k}]$, $\rho(z) = 0$ for $z \notin [0, \frac{1}{k}]$. Then

$$0 \leq \psi'_k(x) \leq 2|x|_+^2, \qquad 0 \leq \psi''(x) \leq 2 \, 1_{x \geq 0}$$

and $\psi_k \in C^2(\mathbf{R})$ such that $\psi_k(x) \nearrow |x|_+^2$ for $k \to \infty$. One can easily show that Ψ_k is twice Frechet differentiable at every $h \in \mathbf{H}$, the first Frechet derivative $\Psi'_k(h)$ is a continuous linear functional on $\mathbf{H}$, and the second derivative $\Psi''_k(h)$ is a continuous symmetric bilinear form on $\mathbf{H} \times \mathbf{H}$ given by

$$(\Psi'_k(h), h_1) = \int_0^1 \psi'_k(h(x)) h_1(x) \, dx,$$

and

$$\Psi''_k(h)(h_1, h_2) = \int_0^1 \psi''_k(h(x)) h_1(x) h_2(x) \, dx,$$

respectively for h_1, h_2 from $\mathbf{H}$. By the Itô formula from [Pa]

$$\Psi_k(w_n(t)) = \Psi_k(w_n(0)) + \int_0^t A(s) \, ds + \frac{1}{2} \int_0^t B(s) \, ds + M_{nk}(t), \tag{5.3}$$

where

$$A(s) \quad := \quad < \mathbf{A}_n(s, u_n(s)) - \bar{\mathbf{A}}_n(s, v_n(s)), \psi_k'(w_n(s)) >$$

$$B(s) \quad := \quad \sum_{i=1}^{n} (\psi_k''(w_n(s))(\sigma(u_n(s)) - \sigma(v_n(s)))\phi_i, (\sigma(u_n(s)) - \sigma(v_n(s)))\phi_i),$$

$\bar{\mathbf{A}}_n$ is defined like $\mathbf{A}_n$ with F_n in place of f_n, and $M_{nk}(t)$ is a continuous local martingale starting from zero. By the definition of $\mathbf{A}$

$$A(s) = -A^{(1)} + A^{(2)} - A^{(3)}, \tag{5.4}$$

where

$$A^{(1)} \quad := \quad (\tfrac{\partial}{\partial x} w_n(s, x), \tfrac{\partial}{\partial x}(\psi_k'(w_n(s, x)))) = \int_0^1 (\tfrac{\partial}{\partial x} w_n(s, x))^2 \psi_k''(w_n(s, x)) \, dx$$

$$A^{(2)} \quad := \quad (f_n(s, u_n(s)) - F_n(s, v_n(s)), \psi_k'(w_n(s))),$$

$$A^{(3)} \quad := \quad (g_n(s, u_n(s)) - g_n(s, v_n(s)), \tfrac{\partial}{\partial x} \psi_k'(w_n(s)))$$

$$= \quad \int_0^1 (g_n(u_n)(s, x)) - g_n(v_n)(s, x))\psi_k''(w_n(s)) \tfrac{\partial}{\partial x} w_n(s, x) \, dx. \tag{5.5}$$

Since $f_n \leq F_n$, $0 \leq \psi_k'(x) \leq |x|_+$ and F_n is Lipschitz in r, there is a constant K such that

$$A^{(2)} \leq \int_0^1 |(F_n(u_n) - F_n(v_n))(s, x)\psi_k'(w_n)(s, x)| \, dx \leq K \int_0^1 |w_n(s, x)|_+^2 \, dx. \tag{5.6}$$

Since $0 \leq \psi''(x) \leq 2\,1_{x \geq 0}$ by the Lipschitz condition on g there is constant L such that

$$|A^{(3)}| \leq L \int_0^1 |w_n(s, x)|_+^2 \, dx + \frac{1}{2} \int_0^1 (\tfrac{\partial}{\partial x} w_n(s, x))^2 \psi_k''(w_n(s, x)) \, dx. \tag{5.7}$$

By the Lipschitz condition on σ we get a constant L such that

$$B(s) \leq L \int_0^1 |w_n(s, x)|_+^2 \, dx, \tag{5.8}$$

using the boundedness of ϕ_i and the the estimates on ψ_k''. From (5.3) through (5.8) for every n we have a constant K such that for all k

$$E(\Psi_k(w_n(t \wedge \tau))) \leq K \int_0^t E(\||w_n(s \wedge \tau)|_+|_2^2) \, ds$$

for all $t \in [0, T]$ and every stopping time τ. Letting here $k \to \infty$ we get

$$E(\Psi(w_n(t \wedge \tau))) \leq K \int_0^t E(\Psi(w_n(s \wedge \tau))) \, ds,$$

where

$$\Psi(w_n(t)) := \int_0^1 |w_n(t,x)|_+^2 \, dx.$$

Hence we get (5.2) by the Gronwall-Bellman lemma, replacing τ by a sequence of stopping times, which localizes the process $\Psi(w_n(t))$. Consequently, almost surely $u_n(t,x) \leq v_n(t,x)$ for all $t \in [0,T]$ and for dx-almost every $x \in [0,1]$. It remains to prove that u_n and v_n converge to the solutions u and v of $\mathrm{Eq}(u_0; f, g, \sigma)$ and of $\mathrm{Eq}(v_0; F, g, \sigma)$, respectively. To this end note that we can easily get the tightness of u_n in $\mathbf{E} := C([0,T]; L^2([0,1])$ using Corollary 3.3 and Lemma 4.2. Hence we can prove the convergences $u_n(t) \to u(t)$, $v_n(t) \to v(t)$ in probability, uniformly in $t \in [0,T]$ repeating the same argument based on Lemma 4.1 and on Skorokhod representation, which we used in the proof of Theorem 2.1.

6. Equations with non degenerate noise

In order to prove Theorem 2.2 we present a lemma which describes the technique we are going to use in passing to the limit through measurable functionals. This lemma is a useful modification of Lemma 3.3 from [Gy1], which generalizes a method from [Kr2] (see also [Kr3]). To formulate it we need to introduce some notion first. We use the notation λ_d for the Lebesgue measure in $\mathbf{R}^d$.

Definition 6.1 *Let D be a domain in $\mathbf{R}^m$ and let $\zeta = \{\zeta(z) : z \in D\}$ be a random field taking values in $\mathbf{R}^d$. The measure defined on $\mathcal{B}(\mathbf{R}^d)$ by*

$$\mu(A) := E \int_D \chi_A(\zeta(z)) dz$$

is called the occupation measure of ζ.

Definition 6.2 *Let $\{\mu_n\}$ be a sequence of measures on $\mathcal{B}(\mathbf{R}^m)$. We say that μ_n is asymptotically equi-absolutely continuous with respect to a given measure ν on $\mathcal{B}(\mathbf{R}^m)$ if*

$$\lim_{\delta \to 0} \limsup_{n \to \infty} \sup\{\mu_n(A) : A \in \mathcal{B}(\mathbf{R}^d), \nu(A) < \delta\} = 0.$$

Let $\zeta_n := \{\zeta_n(z) : z \in D\}$ be a sequence of $\mathbf{R}^d$-valued random fields on a bounded domain $D \subset \mathbf{R}^m$, let $\{h_n\}_{n=1}^\infty$ be a sequence of Borel functions $h_n : \mathbf{R}^d \to \mathbf{R}$, satisfying the linear growth condition with the same constant for every n. Let ν be a σ-finite measure on the Borel sets of $\mathbf{R}^d$. Assume that the following conditions are satisfied:

(i) ζ_n converges to a random field ζ in probability for λ_m-almost every $z \in D$ such that $|\zeta|_p := \int_D |\zeta(z)|^p \, dz < \infty$ a.s. and the sequence $|\zeta_n|_p$ is bounded in probability for some $p > 1$.

(ii) the occupation measure μ_n of ζ_n is asymptotically absolutely continuous with respect to ν on $\mathbf{R}^d$.

(iii) h_n converges to h in the measure ν on the compacts, i.e.,

$$\lim_{n \to \infty} \nu(\{x \in \mathbf{R}^d : |x| \le R, |h_n(x) - h(x)| \ge \varepsilon\}) = 0$$

for every $\varepsilon > 0$ and $R > 0$.

Lemma 6.3 *Assume (i) through (iii). Then for $n \to \infty$*

$$\int_D |h_n(\zeta_n(z)) - h(\zeta(z))| \, dz \to 0$$

in probability.

Proof. One can prove this result in a similar manner as Lemma 3.1 in [Gy1] is obtained. For the detailed proof we refer to [Gy].

In order to construct via approximations the solution to nonlinear stochastic partial differential equations with non continuous nonlinear drift one uses an estimate for the density of the law of the approximation $u_n(t, x)$ with respect to the Lebesgue-measure. Such an estimate is obtained by rather long computations based on Malliavin calculus. (See, e.g. [BGy]). Making use of the above lemma one does not even need these densities to exist. To explain this let u_0 be an $\mathcal{F}_0$- measurable random element in $L^2([0, 1])$, and let g, σ be Borel functions satisfying the conditions in Theorem 2.2. Let ν be the occupation measure of the $\mathbf{R}^3$-valued random field $\zeta(t, x) := (t, x, v(t, x), (t, x) \in D := [0, T] \times [0, 1]$, where $v = v(t, x)$ is the solution of Eq($u_0; 0, g, \sigma$). Let u_n be a solution of Eq($u_0; \tilde{f}_n, g, \sigma$), where $\tilde{f}_n = \tilde{f}_n(t, x)$ is a $\mathcal{P} \otimes \mathcal{B}([0, 1])$-measurable random field for every integer $n \ge 1$ such that the sequence of random variables

$$\int_0^T \int_0^1 |\tilde{f}_n(t, x)|^2 \, dx dt$$

is bounded in probability. Consider a sequence of Borel functions $h_n = h_n(t, x, r)$ and assume that the following conditions are met:

(A) $u_n(t, x)$ *converges to a random field $u(t, x)$ in probability for every $t \in [0, T]$, $x \in [0, 1]$ such that $\int_0^T \int_0^1 |u(t, x)| \, dx dt < \infty$ a.s. and the sequence of random variables $\int_0^T \int_0^1 |u_n(t, x)| \, dx dt$ is bounded in probability.*

(B) h_n *satisfies the linear growth condition with a constant K independent of n, and it converges to a Borel function h in the measure ν on the compacts of $[0, T] \times [0, 1] \times \mathbf{R}$.*

Proposition 6.4 *Assume (A) and (B). Then for $n \to \infty$*

$$\int_0^T \int_0^1 |h_n(t, x, u_n(t, x)) - h(t, x, u(t, x))| \, dx dt \to 0$$

in probability.

Proof. Define

$$\gamma_n : \ = \ \exp\left(-\int_0^{T \wedge \tau_n^L} \int_0^1 (\tilde{f}_n(s, x)/\sigma(s, x, u_n(s, x)) \, dW(s, x)\right.$$
$$\left. -\frac{1}{2} \int_0^{T \wedge \tau_n^L} \int_0^1 (\tilde{f}_n(s, x)/\sigma(s, x, u_n(s, x)))^2 \, dx ds\right),$$

where

$$\tau_n^L := \inf\{t \geq 0 : \int_0^t \int_0^1 |\tilde{f}_n(s, x)|^2 \, dx ds \geq L\}.$$

Note that

$$\int_0^{T \wedge \tau_n^L} \int_0^1 (\tilde{f}(t, x)/\sigma(t, x, u_n(t, x)))^2 dx dt \leq L/\lambda^2.$$

Therefore $E\gamma_n = 1$ and moreover $E\gamma_n^\alpha \leq C(\alpha)$ for every $\alpha \in \mathbf{R}$, with a constant $C(\alpha)$, which is independent of n. Thus $d\tilde{P} := \gamma_n dP$ defines is a probability measure $d\tilde{P}$ and $d\tilde{W}(t, x) := 1_{t \leq \tau_n^L} \tilde{f}_n(t, x)/\sigma(t, x, u_n(t, x)) \, dx dt + dW(t, x)$ defines an $\mathcal{F}_t$-Brownian sheet under the measure $\tilde{P}$ by Girsanov's theorem. Moreover,

$$E\eta = \tilde{E}(\gamma_n^{-1} \eta) \leq (\tilde{E}\gamma_n^{-2})^{1/2} (\tilde{E}\eta^2)^{1/2} \leq \sqrt{C(-2)} (\tilde{E}\eta^2)^{1/2}$$

for every non-negative random variable η, where $\tilde{E}$ denotes the expectation with respect to $\tilde{P}$. Consequently, u_n solves the problem

$$\frac{\partial}{\partial t} u_n(t, x) = \frac{\partial^2}{\partial x^2} u_n(t, x) + \frac{\partial}{\partial x} g(t, x, u_n(t, x)) + \sigma(t, x, u_n(t, x)) \frac{\partial^2}{\partial t \partial x} \tilde{W}(t, x)$$

$$u_n(t, 0) = u_n(t, 1) = 0 \quad t \in [0, T], \qquad u_n(0, x) = u_0(x) \quad x \in [0, 1]$$

in the stochastic interval $[0, T \wedge \tau_n^L]$. Moreover, for every $A \in \mathcal{B}(\mathbf{R}^3)$

$$E \int_0^T \int_0^1 \chi_A(t, x, u_n(t, x)) dx dt$$

$$\leq \sqrt{C(-2)} \left(\tilde{E} \int_0^{T \wedge \tau_n^L} \int_0^1 \chi_A(t, x, v(t, x)) dx dt \right)^{1/2} + T P(\tau_n^L \leq T)$$

$$\leq \sqrt{C(-2)} (\nu(A))^{1/2} + T P(\tau_n^L \leq T).$$

Hence for the occupation measure μ_n of $\zeta_n(t, x) := (t, x, u_n(t, x))$, $(t, x) \in D := [0, T] \times [0, 1]$ we get

$$\limsup_{n \to \infty} \sup \{ \mu_n(A) : \nu(a) < \delta \} \leq \sqrt{C(-2)} \delta^{1/2}$$

for every $\delta > 0$. Consequently, μ_n is asymptotically equi-absolutely continuous with respect to the measure ν, and we can finish the proof of the proposition by applying Lemma 6.3.

Proposition 6.5 *Let u_0 be an $\mathcal{F}_0$-measurable random element in $L^2([0, 1])$ and let g and σ be a Borel functions satisfying the conditions of Theorem 2.2. Let $f_n = f_n(t, x, r)$ be a sequence of Borel functions mapping $\mathbf{R}_+ \times [0, 1] \times \mathbf{R}$ into $\mathbf{R}$ and let $\xi_n = \xi_n(t, x)$ be a sequence of $\mathcal{P} \otimes \mathcal{B}([0, 1])$-measurable random fields. Assume that*

(a) *$Eq(u_0; f_n + \xi_n, g, \sigma)$ admits a solution u_n converging to a random field u in the measure $dP \otimes dt \otimes dx$ on $\Omega \times [0, T] \times [0, 1]$, as $n \to \infty$, such that $\sup_{t \leq T} \int_0^1 |u(t, x)|^2 dx dt < \infty$ (a.s.) and the sequence of random variables $\int_0^T \int_0^1 |u_n(t, x)|^2 dx dt$ is bounded in probability.*

(b) *$f_n(t, x, r) \to f(t, x, r)$ in the measure ν and $\xi_n(t, x) \to 0$ in the measure $dP \otimes dt \otimes dx$, where f is a Borel function.*

(c) *f_n has linear growth with the same constant for all n, and $|\xi_n(t, x)| \leq \xi(t, x)$ (a.s.) for every $t \geq 0$, $x \in [0, 1]$ where ξ is a $\mathcal{P} \otimes \mathcal{B}([0, 1])$-measurable random field which is almost surely square-integrable on $[0, T] \times [0, 1]$ with respect to $dt \otimes dx$.*

Then u solves $Eq(u_0; f, g, \sigma)$, i.e., there is a modification $\bar{u}$ of u on $[0, T] \times [0, 1]$, such that $\bar{u}$ is a continuous $L^2([0, 1])$-valued stochastic process which is the solution of $Eq(u_0; f, g, \sigma)$.

Proof. Passing to the limit in the equation for u_n one can easily see, using Proposition 6.4 and Lemma 4.6 that u solves $Eq(u_0; f, g, \sigma)$. The existence of the continuous modification of u is obvious by Girsanov's transformation, since $Eq(u_0; 0, g, \sigma)$ admits a unique $L^2([0, 1])$-valued continuous solution. Making use of the comparison theorem and the above proposition one can get the following statement similarly as Theorem in [Gy1] is proved. ■

Theorem 6.6 *Let g and σ be a Borel functions satisfying the conditions of Theorem 2.2. Let $f_n = f_n(t, x, r)$ and $\xi_n = \xi_n(t, x)$ be a sequences of Borel functions and of $\mathcal{P} \otimes \mathcal{B}([0, 1])$-measurable random fields, respectively, satisfying conditions (b), (c) from Proposition 6.5. Assume moreover that f_n is Lipschitz continuous in $r \in \mathbf{R}$ for every $n \geq 1$. Then $Eq(u_0; f_n + \xi_n, g, \sigma)$ admits a unique solution u_n, which converges $dP \otimes dt \otimes dx$-almost everywhere to a random field $u(t, x)$. Moreover u is a solution of $Eq(u_0; f, g, \sigma)$.*

Proof of Theorem 2.2. We call a solution u of $Eq(u_0; f, g, \sigma)$ constructible if there exists a sequence of Lipschitz functions $f_n = f_n(t, x, r)$ and a sequence of $\mathcal{P} \otimes \mathcal{B}([0, 1])$-measurable random fields ξ_n satisfying conditions (b) and (c) from Proposition 6.5 such that the solution u_n of $Eq(u_0; f_n + \xi_n, g, \sigma)$ converges $dP \otimes dt \otimes dx$-almost everywhere to u. It is easy to show that for any Borel function $f = f(t, x, r)$ satisfying the linear growth condition there is a sequence of Lipschitz functions $f_n = f_n(t, x, r)$ satisfying the linear growth condition with the same constant for all n such that it converges to f ν-almost everywhere. (We recall that ν is the occupation measure of $\xi(t, x) := (t, x, v(t, x))$, $(t, x) \in D := [0, T] \times [0, 1]$, where v is the solution of $Eq(u_0; 0, g, \sigma)$). Hence we get the existence of a constructible solution to $Eq(u_0; f, g, \sigma)$ from Theorem 6.6 by taking the sequence f_n and $\xi_n := 0$. Theorem 6.6 implies also the uniqueness of the solution in the class of constructible solutions. To see this let $u^{(1)}$ and $u^{(2)}$ be constructible solutions obtained as the limit of the solutions $u_n^{(1)}$ of $Eq(u_0; f_n^{(1)} + \xi_n^{(1)}, g, \sigma)$ and of the solutions $u_n^{(2)}$ of $Eq(u_0; f_n^{(2)} + \xi_n^{(2)}, g, \sigma)$, respectively, where $f_n^{(1)}$, $f_n^{(2)}$ and $\xi_n^{(1)}$, $\xi_n^{(2)}$ are sequences of of Lipschitz functions and random fields satisfying the same conditions as f_n and ξ_n in Proposition 6.5. Define $f_n := f_n^{(1)}$, $\xi_n := \xi_n^{(1)}$ for odd n and $f_n := f_n^{(2)}$, $\xi_n := \xi_n^{(2)}$ for even n. Then by Theorem 6.6 the solution u_n of $Eq(u_0; f_n + \xi_n, g, \sigma)$ converges to the solution u of $Eq(u_0; f, g, \sigma)$ and since $u^{(1)}$ and $u^{(2)}$ are subsequences of u_n we have $u^{(1)} = u = u^{(2)}$. Hence for the uniqueness in the whole class of solutions we need only to show that every solution u of $Eq(u_0; f, g, \sigma)$ is constructible. To this end let f_n be a sequence of Lipschitz functions converging for ν-almost every (t, x, r) to f such that f_n satisfies the linear growth condition with a constant independent of n. Set $\xi_n := f(t, x, u(t, x)) - f_n(t, x, u(t, x))$. Then by Proposition 6.4 we have $\xi_{n_k} \to 0$ $dP \otimes dt \otimes dx$-almost everywhere for a subsequence $n_k \to \infty$. Note that $u_n := u$ solves $\mathrm{Eq}(u_0; f_n + \xi_n, g, \sigma)$ for every n. Hence u is a constructible solution.

7. Results on approximations

We consider the problem $\mathrm{Eq}(u_0; f, g, \sigma)$ on the time interval $[0, T]$ with some given $T > 0$ in the case $g = 0$. The general case will be studied elsewhere. Replacing the derivatives of u and of W in the space variable x with finite differences we get the following system of stochastic differential equations

$$du^n(t, x_j) = n^2(u^n(t, x_{j+1}) - 2u^n(t, x_j) + u^n(t, x_{j-1}))dt \qquad (7.1)$$

$$+f(t, x_j, u^n(t, x_j)) + n\sigma(t, x_j, u^n(t, x_j))d(W(t, x_{j+1}) - W(t, x_j))$$

$$u^n(t, 0) = u^n(t, 1) = 0 \quad t \in [0, T]$$

$$u^n(0, x_j) = u_0(x_j), \quad x_j := \frac{j}{n}; \ j := 1, 2, \dots, n-1;$$

for positive integers n. It is a natural idea to approximate the solution u^n of the above system by Euler's method. Thus, discretizing the time interval $[0, T]$ by mesh points $t_i = \frac{i}{m}T$ and replacing the stochastic differentials in the equation (1.1) for u^n by the corresponding forward differences one gets

$$u_m^n(t_{i+1}, x_j) = ((I + Tm^{-1}\Delta_n)u_m^n(t_i, \cdot)(x_j) + Tm^{-1}f(t_i, x_j, u_m^n(t_i, x_j))$$

$$+Tm^{-1}\sigma(t_i, x_j, u_m^n(t_i, x_j))d_{nm}W(t_i, x_j),$$

$$u_m^n(t_i, 0) = u_m^n(t_i, 1) = 0 \qquad u_m^n(0, x_j) = u_0(x_j)$$

where $t_i := t_i^m := T\frac{i}{m}$, $x_j := x_j^n := \frac{j}{n}$, I is the identity operator,

$$(\Delta_n \varphi)(x_j) := n^2(\varphi(x_{j+1}) - 2\varphi(x_j) + \varphi(x_{j-1})) =: \Delta_n \varphi(x_j)$$

$$d_{nm}\psi(t_i, x_j) := nm(\psi(t_{i+1}, x_{j+1}) - \psi(t_i, x_{j+1}) - \psi(t_i, x_{j+1}) + \psi(t_i, x_j))$$

for functions φ and ψ defined on $\{x_j : j = 0, 1, \dots, n\}$ and on the lattice

$$\mathcal{L} := \{(t_i, x_j) : i = 0, 1, 2, \dots, m; \ j = 0, 1, \dots, n\},$$

respectively. Note that the above scheme defines a random field u_m^n on $\mathcal{L}$. One can extend the definition of u_m^n from $\mathcal{L}$ in many reasonable ways to get a piecewise constant or a continuous random field on $\mathcal{D} := [0, T] \times [0, 1]$ and to investigate its convergence on $\mathcal{D}$ as $n \to \infty$, $m \to \infty$. We extend u_m^n from

$\mathcal{L}$ by polygonal interpolation. The extension by polygonal interpolation of a function h given on $\mathcal{L}$ is denoted also by h and it is defined as follows:

$$h(t,x) := h(t,x_j) + n(x - x_j)(h(t,x_j) - h(t,x_{j+1}))$$

$$h(t,x_j) := h(t_i,x_j) + \frac{m}{T}(t - t_i)(h(t_{i+1},x_j) - h(t_i,x_j))$$

for $(t,x) \in (t_i,t_{i+1}) \times (x_j,x_{j+1})$, $i := 0,1,\ldots,m-1$; $j = 0,1,2,\ldots,n-1$. It is clear that this extension is continuous on $\mathcal{D}$. This scheme has the good property that at each time point t_{i+1} the approximations $u_m^n(t_{i+1},x_j)$ are defined explicitly by the approximations at the previous time point t_i. Such kind of scheme is often called an explicit scheme in numerical analysis. Note that one cannot apply Euler's method of time discretization directly to equation (1.1). Thus one cannot expect that u_m^n converges for all sequences of positive integers n, m converging to infinity. In the case of the heat equation it is well-known that the sequence of mesh sizes $\delta x := \frac{1}{n}$, $\delta t := \frac{1}{m}T$ should satisfy the condition $\delta t/(\delta x)^2 \leq \frac{1}{2}$. In other words, the condition $\frac{n^2}{m}T \leq \frac{1}{2}$ is needed for the convergence of the approximations u_m^n when n and m tend to infinity. This condition has the bad consequence that if one reduces δx in order to improve accuracy of the approximation one should reduce also δt considerably, and so the amount of computation work increases very rapidly.

We modify the above method of approximations by taking a backward difference instead of a forward one when we approximate the first differential in the right side of equation (7.1). Thus we get the following implicit scheme for the new approximations u^{nm}

$$u^{nm}(t_{i+1},x_j) = u^{nm}(t_i,x_j) + Tm^{-1}(\Delta_n u^{nm}(t_{i+1},\cdot))(x_j)$$

$$+Tm^{-1}f(t_i,x_j,u^{nm}(t_i,x_j)) + Tm^{-1}\sigma(t_i,x_j,u^{nm}(t_i,x_j))d_{nm}W(t_i,x_j)$$

$$u^{nm}(t_i,0) = u(t_i,1) = 0 \qquad u^{nm}(0,x_j) = u_0(x_j)$$

for $i := 0,1,2,\ldots,m-1$, $j := 1,2,\ldots,n-1$ and for integers $n \geq 1$, $m \geq 1$. Note that the operator $I - Tm^{-1}\Delta_n$ is invertible for every $n \geq 1$, $m \geq 1$. Therefore there exists a unique random field $u^{nm} = \{u^{nm}(t,x) : (t,x) \in \mathcal{L}\}$ which solves the scheme. We extend u^{nm} by polygonal interpolation from $\mathcal{L}$ onto $[0,T] \times [0,1]$. We denote this extension also by u^{nm} and we call it implicit approximation. In order to formulate our results on the convergence of the approximations u_m^n and u^{nm} we make some assumptions.

(H) There is a constant C such that for all $t \in [0,T]$, $x,y \in [0,1]$, $p,r \in \mathbf{R}$

$$|f(t,x,r)-f(s,y,p)|+|\sigma(t,x,r)-\sigma(s,y,p)| \leq C(|t-s|^{\frac{1}{4}}+|x-y|^{\frac{1}{2}}+|r-p|).$$

It is well known that (H) implies the existence of a unique solution u of $\text{Eq}(u_0; f, 0, \sigma)$ which is continuous in (t, x). In order to get the convergence of the explicit approximations u_m^n we need the following condition on the sequence (n, m), or in other words, on the sequence of mesh sizes $(\delta t, \delta x) := (\frac{1}{m}T, \frac{1}{n})$

(R)*Let n and m be subsequences of positive integers tending to infinity such that*

$$\frac{n^2}{m}T \le q$$

for some constant $q < \frac{1}{2}$.

Theorem 7.1 *Assume (H). Then the following statements hold:*

(i)*For every $0 < \alpha < \frac{1}{4}$, $0 < \beta < \frac{1}{2}$, $p \ge 1$ and for every $t \in (0, T]$ there is a constant $K := K(p, \alpha, t)$ such that*

$$\sup_{x \in [0,1]} E(|u^{nm}(t, x) - u(t, x)|^{2p}) \le K(m^{-\alpha p} + n^{-\beta p}) \qquad (7.2)$$

for all integers $n \ge 1$, $m \ge 1$.

(ii)*Let $n = n_k$, $m = m_k$ be sequences of integers such that $n_k \ge k^\gamma$, $m_k \ge k^\gamma$ for all $k \ge 1$ for some $\gamma > 0$. Then for $k \to \infty$*

$$\sup_{t \in [0,T]} \sup_{x \in [0,1]} |u^{n_k m_k}(t, x) - u(t, x)| \to 0$$

almost surely.

(iii) *Assume moreover that $u_0 \in C^3([0, 1])$. Then (7.2) holds with $\alpha := \frac{1}{2}$, $\beta := 1$ for all integers $n \ge 1$, $m \ge 1$ for all $t \in [0, T]$ with a constant $K = K(p, \alpha)$ independent of t.*

(iv)*Assume moreover (R). Then the above statements (i) through (iii) hold also for explicit approximations u_m^n in place of implicit approximations u^{nm}.*

For the proof we refer to [Gy2].

References

[BGy]   V. Bally, I. Gyöngy, E. Pardoux, White noise driven parabolic SPDEs with measurable drift. *Journal of Functional Analysis,* **120** (1994), 484-510.

[Ba] H. Bateman, Some recent researches on the motion of fluids. *Monthly Weather Rev.*, **43** (1915), 163-170.

[BP] E. R. Benton, G. W. Platzmann, A table of Solutions of the One-dimensional Burgers Equation. *Quart. Appl. Math.*, **30** (1972), 195-212.

[BCJ] L. Bertini, N. Cancrini, G. Jona-Lasinio, The stochastic Burgers equation. Preprint.

[Bu1] J. M. Burgers, A Mathematical Model Illustrating the Theory of Turbulence. In: von Mises, R., von Karman Th. (eds.) *Advances in Applied Mathematics*, **1** (19489, 171-199.

[Bu2] J. M. Burgers, *The Nonlinear Diffusion Equation*. Dordrecht: D. Reidel Publishing Company 1974.

[PDT] G. Da Prato, A. Debussche, and R. Temam, Stochastic Burgers equation. Preprint n.27 Scuola Normale Superiore di Pisa.

[PG] G. Da Prato, D. Gatarek, Stochastic Burgers equation with correlated noise, *Stochastics and Stochastics Reports*, **52** (1995), 29-41.

[Da] A. M. Davie, Convergence of Numerical Schemes for the Solution of Parabolic Stochastic Partial Differential Equations. Preprint 1996.

[DP] C. Donati-Martin, E. Pardoux, White noise driven SPDEs with reflection, *Probab. Theory Relat. Fields*, **95** (1993), 1-24.

[FR] J. Fritz, B. Rüdinger, Approximation of a one-dimensional stochastic PDE by local mean field type lattice systems, Preprint.

[Fu] T. Funaki, Random motion of strings and related evolution equations, *Nagoya Math. Journal*, **89** (1983), 129-193.

[GRR] A. M. Garsia, E. Rodemich, and H. Rumsey, A real variable lemma and the continuity of some Gaussian processes. *Indiana Univ. Math. J.*, **6** (1970), 565-578.

[Gy1] I. Gyöngy, On non-degenerate quasi-linear stochastic partial differential equations. *Potential Analysis*, **4** (1995), 157-171.

[Gy2] I. Gyöngy, Lattice approximations for stochastic quasi-linear parabolic partial differential equations driven by space-time white noise II. Preprint **32/1995**, Mathematical Institute of the Hungarian Academy of Sciences.

[GyK] I. Gyöngy, N. V. Krylov, Existence of strong solutions for Ito's stochastic equations via approximations, *Probability Theory and Related Fields*, **2** (1996), 143-158.

[GyN] I. Gyöngy, D. Nualart, Implicit scheme for quasi-linear stochastic partial differential equations driven by space-time white noise. To appear in *Potential Analysis*.

[GyP] I. Gyöngy, E. Pardoux, Weak and strong solutions of white noise driven SPDEs. Submitted for publication.

[Gy] I. Gyöngy, *On the stochastic Burgers equations*. Preprint No. 15/1996 Mathematical Institute of the Hungarian Academy of Sciences.

[Ho] E. Hopf, The partial differential equation $u_t + uu_x = \mu u_{xx}$. *Communication on Pure and Applied Math.*, **3** (1950), 201-230.

[Kr1] N. V. Krylov, On L_p-theory of stochastic partial differential equations in the whole space. Preprint 1994.

[Kr2] N. V. Krylov, On stochastic integral of Itô. *Theor. Probability Appl.*, **14** (1969), 330-336.

[Kr3] N. V. Krylov, *Controlled Diffusion Processes*. New York–Heidelberg–Berlin: Springer-Verlag, 1980.

[Li] J. L. Lions, *Quelques Méthods de Resolution des Problémes aux Limites non Linéaries*. Paris: Dunod Gauthier-Villars, 1969.

[Ma] R. Manthey, Existence, uniqueness and continuity of a solution of a reaction-diffusion equation with polynomial nonlinearity and white noise disturbance. *Mathematische Nachrichten*, **125** (1986), 121-133.

[Pa] E. Pardoux, Stochastic partial differential equations and filtering of diffusion processes, *Stochastics*, **3** (1979), 127-167.

[Ro] B. L. Rozovskii, *Stochastic evolution Systems*. Dordrecht: Kluwer, 1990.

[Sk] A. V. Skorokhod, *Studies in the theory of random processes*, New York: Dover Publications, Inc. 1982.

[Wa]   J. B. Walsh, An introduction to stochastic partial differential equations. In: Hennequin, P. L. (ed.), *École d'été de Probabilités de St. Flour XIV*, Lect. Notes Math. **1180**, 265–437, Berlin–Heidelberg–New York: Springer 1986

Department of Probability Theory and Statistics
Eötvös Loránd University,
Budapest, Hungary

Received November 28, 1996

Parallel Numerical Solution of a Class of Volterra Integro–Differential Equations

Gerd Heber and Christoph Lindemann

1. Introduction

A discrete–event stochastic system makes state transitions when events associated with the occupied state occur; events occur only at an increasing sequence of random times. The underlying stochastic process of a discrete–event system has to record the state of the system as it evolves over continuous time. The *generalized semi–Markov process* provides a framework for specifying and analyzing stochastic processes underlying most kinds of discrete–event systems; see e.g., Glynn [Gl]. A generalized semi–Markov process is a continuous–time stochastic process that makes a state transition when one or more "events" associated with the occupied state occur. Events associated with a state compete to trigger the next state transition, and each set of trigger events has its own distribution for determining the next state. At each state transition of the generalized semi–Markov process, new events may be scheduled. For each of these new events, a clock indicating the time until the event is scheduled to occur is set according to an independent (stochastic) mechanism. If a scheduled event is not in the set of events that triggers a state transition but is associated with the next state, its clock *continues* to run; if such an event is not associated with the next state, it is *cancelled*, and the corresponding clock reading is discarded. The generalized semi–Markov process is a continuous–time stochastic process $\{S(t) : t \geq 0\}$ that records the state of the system as it evolves over time.

In previous work, we introduced a numerical method for computing limiting distributions for finite–state generalized semi–Markov processes with exponential and deterministic events [LS]. The method rests on observation, at equidistant time points, of the continuous time Markov process and analysis of the resulting general state space Markov chain (GSSMC). This embedded GSSMC and the continuous time Markov process have the same limiting distribution, provided that limits for the Markov process exist. Such numerical methods are applicable for the steady–state analysis of networks of queues, deterministic and stochastic Petri nets, stochastic process algebras, and other discrete–event stochastic systems with an underlying stochastic process which can be represented as a generalized semi–Markov process with exponential and deterministic events.

In general, the method for steady state analysis of generalized semi–Markov processes with exponential and deterministic events requires the numerical solution of a system of Volterra integral equations of the second type which comprises multi–dimensional integrals. While a considerable amount of results for the efficient numerical solution of such Volterra equations with one–dimensional integrals is available (see e.g., [BH]), to the best of our knowledge, the numerical solution of such Volterra integral equations with multi–dimensional integrals and their parallelization has not been considered yet.

In this paper, we describe a parallel implementation for the numerical solution of the stationary equations of the GSSMC which allows their efficient numerical solution on a cluster of modern workstations or a multi-processor system. The key observation is that we can employ a wavefront approach for solving the set of systems of linear equations that result from discretizing the integral expressions in the Volterra equations by an appropriate quadrature rule.

This remainder of this paper is organized as follows. Section 2 describes the transition kernel of the GSSMC. The stationary equations for the GSSMC that constitute a system of Volterra integral equations are introduced in Section 3. Section 4 presents the method for parallel numerical solution of this system of Volterra integral equations. The final section contains some concluding remarks and outlines current work.

2. The Transition Kernel of the GSSMC

We consider GSMPs comprising of a finite number of states denoted by N and exponential and deterministic events. We assume that all deterministic events have the same delay $D > 0$. Enumerate the deterministic events of the GSMP by $t_1, t_2, \ldots, t_M$. Let $C_m(t)$ be the clock reading associated with deterministic event t_m at time t $(1 \leq m \leq M)$. In any state in which deterministic event t_m is not active, we define $C_m(t) = 0$. Then, using the approach in [LS], we consider a discrete–time stochastic process $\{X(nD) : n \geq 0\}$ by observing of the corresponding continuous-time process $\{S(t), C_1(t), C_2(t), \ldots, C_M(t) : t \geq 0\}$ at a sequence $\{nD : n \geq 0\}$ of fixed times. According to [LS], we define the step size as the delay of deterministic events. That is:

$$X(nD) = (S(nD), C_1(nD), C_2(nD), \ldots, C_M(nD)) \qquad (2.1)$$

Heuristically, $S(nD)$ represents the state and $C_m(nD)$ represents the m–th component of the clock-reading vector (remaining activation time of deterministic events t_m) at time nD. The memoryless property of the exponential distribution implies that $\{X(nD) : n \geq 0\}$ is a GSSMC, i.e., it satisfies the Markov property. If $\{X(nD) : n \geq 0\}$ is an aperiodic, positive recurrent chain with a regeneration set; that is, $\{X(nD) : n \geq 0\}$ is a Harris ergodic chain [Fe], the discrete–time process $\{X(nD) : n \geq 0\}$ has a unique stationary distribution. Otherwise, the presented approach allows the computation of time–averaged distributions. Using such a "stochastic skeleton approach", for the case that a stationary distribution exists, we can conclude that the stationary or time–averaged distribution of the discrete–time process $\{S(nD) : n \geq 0\}$ is equal to the stationary or time–averaged distribution of the continuous-time process $\{S(t) : t \geq 0\}$ [LS]. If a stationary distribution exists, that is

$$\lim_{t \to \infty} P\{S(t) = i\} = \lim_{n \to \infty} P\{S(nD) = i\} \qquad (2.2)$$

The stationary analysis of the GSSMC $\{X(nD) : n \geq 0\}$ is based on its transition kernel which specifies probabilities for state transitions in the GSSMC. Recall that we assume that the GSMP comprises of N states. For ease of exposition, we restrict the discussion to GSMPs in which at most two deterministic events may be concurrently active. Then, the subset of states in which only exponential events are active is denoted by S_{exp}. Similarly, the subsets of states in which one deterministic event and two deterministic events are (concurrently) active are denoted by S_{det1} and S_{det2}, respectively. We enumerate these states as follows:

$$\begin{aligned}
S_{\text{exp}} &= \{1, 2, \ldots, N_1\} \\
S_{\text{det1}} &= \{N_1 + 1, N_1 + 2, \ldots, N_1 + N_2\} \\
S_{\text{det2}} &= \{N_1 + N_2 + 1, N_1 + N_2 + 2, \ldots, N\}
\end{aligned} \qquad (2.3)$$

The transition kernel is a square matrix of dimension N and, in general, its ij-entries are functions of clock readings associated with the current state i and intervals for clock readings associated with the new state j. Due to the construction of the GSSMC $\{X(nD) : n \geq 0\}$ the transition kernel can be written as $\mathbf{P}(c_1, c_2, a_1, a_2)$. Subsequently, its ij-entries are defined as conditional probabilities that the next state is j with clock readings

$C_1 \in [0, a_1]$ and $C_2 \in [0, a2]$ given that the current state is i with clock readings $C_1 = c_1$ and $C_2 = c_2$. That is:

$$
\begin{aligned}
p_{ij}(c_1, c_2, a_1, a_2) \quad = \quad & P\{S((n+1)D) = j, \, C_1((n+1)D) \leq a_1, \\
& C_2((n+1)D) \leq a_2 \mid S(nD) = i, \, C_1(nD) = c_1, \\
& C_2(nD) = c_2\}
\end{aligned}
\tag{2.4}
$$

As shown in [LS], all ij-entries of the transition kernel can be numerically determined by transient analysis of appropriately defined continuous–time Markov chains with discrete state space. Such a Markov chain is associated with each state of the GSMP. The numerical transient analysis can be efficiently performed with the randomization technique [GM].

Using equation (2.4), the general form of the transition kernel $\mathbf{P}(c_1, c_2, a_1, a_2)$ for the GSSMC underlying a GSMP of the considered class can be written as a composition of 9 quadratic submatrices $\mathbf{P}_{ij}(\,.\,)$ of appropriate dimension. For the case $c_1 \leq c_2$ and $a_1 \leq a_2$, the transition kernel $\mathbf{P}(c_1, c_2, a_1, a_2)$ has the form:

$$
\left[
\begin{array}{c|c|c}
\mathbf{P}_{11}(a_2) & \mathbf{P}_{12}(a_1, a_2) & \mathbf{P}_{13}(a_1, a_2) \\
\hline
\mathbf{P}_{21}(c_1, a_2) & \mathbf{P}_{22}(c_1, a_1, a_2) & \mathbf{P}_{23}(c_1, a_1, a_2) \\
\hline
\mathbf{P}_{31}(c_1, a_2) & \mathbf{P}_{32}(c_1, c_2, a_1, a_2) & \mathbf{P}_{33}(c_1, c_2, a_1, a_2)
\end{array}
\right]
\begin{array}{l}
1 \\
\vdots \\
N_1 \\
N_1 + 1 \\
\vdots \\
N_1 + N_2 \\
N_1 + N_2 + 1 \\
\vdots \\
N
\end{array}
$$

For $c_1 \geq c_2$ and $a_1 \leq a_2$ the transition kernel $\mathbf{P}(c_1, c_2, a_1, a_2)$ is of similar form. The only difference lies in that submatrix $\mathbf{P}_{31}(\,.\,)$ may depend on c_2 instead of c_1, i.e., for $c_1 \geq c_2$ and $a_1 \leq a_2$ this submatrix is of the form $\mathbf{P}_{31}(c_2, a_2)$. It is important to observe that the transition kernel is symmetric with respect to $c_1 \leq c_2$ and $c_1 \geq c_2$. Thus, it is sufficient to compute the kernel matrix just for the former case. We employ this symmetry property in Section 4 for the solution of the system of two-dimensional Volterra integral equations.

3. The Stationary Equations of the GSSMC

The stationary equations of the embedded GSSMC underlying the considered class of GSMPs comprise a system of Volterra integro–differential equations with two–dimensional integral expressions. Using (2.3), we define stationary probabilities π_i for $i \in S_{\exp}$, $\pi_i(a_1)$ for $i \in S_{\det 1}$, and $\pi_i(a_1, a_2)$ for $i \in S_{\det 2}$ of the GSSMC underlying a GSMP in which at most two deterministic events may be concurrently active.

$$
\begin{aligned}
\pi_i &= \lim_{n \to \infty} P\{S(nD) = i\} \\
\pi_i(a_1) &= \lim_{n \to \infty} P\{S(nD) = i,\ C_1(nD) \le a_1\} \\
\pi_i(a_1, a_2) &= \lim_{n \to \infty} P\{S(nD) = i,\ C_1(nD) \le a_1,\ C_2(nD) \le a_2\}
\end{aligned}
\tag{3.1}
$$

As shown in [LS] for the considered class of GSMPs, the stationary equations of the underlying embedded chain $\{X(nD) : n \ge 0\}$ are of the form

$$
\begin{aligned}
\pi_i = {} & \sum_{j=1}^{N_1} \pi_j \cdot p_{ji}(a_2) + \sum_{j=N_1+1}^{N_1+N_2} \int_0^{a_1} \frac{\mathrm{d}\pi_j(c_1)}{\mathrm{d}c_1} \cdot p_{ij}(c_1, a_2)\,\mathrm{d}c_1 \\
& + 2 \cdot \sum_{j=N_1+N_2+1}^{N} \int_0^{a_1} \int_0^{c_2} \frac{\partial^2 \pi_j(c_1, c_2)}{\partial c_1 \partial c_2} \cdot p_{ji}(c_1, a_2)\,\mathrm{d}c_1 \mathrm{d}c_2
\end{aligned}
\tag{3.2}
$$

for $1 \le i \le N_1$ and $0 < a_1, a_2 \le D$ with $c_1 \le c_2$ and $a_1 = a_2$.

$$
\begin{aligned}
\pi_i(a_1) = {} & \sum_{j=1}^{N_1} \pi_j \cdot p_{ji}(a_1, a_2) + \sum_{j=N_1+1}^{N_1+N_2} \int_0^{a_1} \frac{\mathrm{d}\pi_j(c_1)}{\mathrm{d}c_1} \cdot p_{ji}(c_1, a_1, a_2)\,\mathrm{d}c_1 \\
& + 2 \cdot \sum_{j=N_1+N_2+1}^{N} \int_0^{a_1} \int_0^{c_2} \frac{\partial^2 \pi_j(c_1, c_2)}{\partial c_1 \partial c_2} \cdot p_{ji}(c_1, c_2, a_1, a_2)\,\mathrm{d}c_1 \mathrm{d}c_2
\end{aligned}
\tag{3.3}
$$

for $N_1 + 1 \le i \le N_1 + N_2$ and $0 < a_1, a_2 \le D$ with $c_1 \le c_2$ and $a_1 = a_2$.

$$
\begin{aligned}
\pi_i(a_1, a_2) = {} & \sum_{j=1}^{N_1} \pi_j \cdot p_{ji}(a_1, a_2) + \sum_{j=N_1+1}^{N_1+N_2} \int_0^{a_1} \frac{\mathrm{d}\pi_j(c_1)}{\mathrm{d}c_1} \cdot p_{ji}(c_1, a_1, a_2)\,\mathrm{d}c_1 \\
& + 2 \cdot \sum_{j=N_1+N_2+1}^{N} \int_0^{a_1} \int_0^{c_2} \frac{\partial^2 \pi_j(c_1, c_2)}{\partial c_1 \partial c_2} \cdot p_{ji}(c_1, c_2, a_1, a_2)\,\mathrm{d}c_1 \mathrm{d}c_2 \\
& + \sum_{j=N_1+N_2+1}^{N} \int_{a_1}^{a_2} \int_0^{a_1} \frac{\partial^2 \pi_j(c_1, c_2)}{\partial c_1 \partial c_2} \cdot \tilde{p}_{ji}(c_2, a_1, a_2)\,\mathrm{d}c_1 \mathrm{d}c_2
\end{aligned}
\tag{3.4}
$$

for $N_1 + N_2 + 1 \leq i \leq N$ and $0 < a_1, a_2 \leq D$ with $c_1 \leq c_2$ and $a_1 \leq a_2$.
In Equation (3.4), expressions of the form $\tilde{p}_{ji}(c_2, a_1, a_2)$ are derived from the kernel elements $p_{ji}(c_1, c_2, a_1, a_2)$ by projection from the two–dimensional plane $[0, a_1] \times [0, a_2]$ on the one–dimensional line $[a_1, a_2]$. Equations (3.2) to (3.4), in general, comprise a system of integro-differential equations with constant boundary conditions:

$$
\begin{aligned}
\pi_i(0) &= 0 \quad \text{for } N_1 + 1 \leq i \leq N_1 + N_2 \\
\pi_i(c_1, 0) &= 0 \quad \text{for } N_1 + N_2 + 1 \leq i \leq N \text{ and } 0 \leq c_1 \leq D \quad (3.5) \\
\pi_i(0, c_2) &= 0 \quad \text{for } N_1 + N_2 + 1 \leq i \leq N \text{ and } 0 \leq c_2 \leq D
\end{aligned}
$$

The system of integro–differential equations (3.2) to (3.4) together with the boundary conditions (3.5) and the normalization equation (3.6) uniquely determine the stationary probabilities of a GSMP of the considered class.

$$
\sum_{j=1}^{N_1} \pi_j + \sum_{j=N_1+1}^{N_1+N_2} \pi_j(D) + \sum_{j=N_1+N_2+1}^{N} \pi_j(D, D) = 1 \quad (3.6)
$$

As shown in [LS], in the system of integro-differential equations (3.2) to (3.4) we can replace all (partial) derivatives of stationary probabilities by (partial) derivatives of kernel elements. Since the former are unknown functions, whereas the latter can be numerically determined from transient state probabilities of continuous-time Markov chains with discrete state space using the Chapman-Kolmogorov equation (see [Fe]), this transformation simplifies the numerical solution of the stationary equations, considerably. Thus, the system of integro-differential equations (3.2) to (3.4) can be transformed, using integration by parts and the boundary conditions into a system of two-dimensional Volterra integral equations of the second type.

4. Parallel Numerical Solution of the Stationary Equations

Numerical solution methods for Volterra equations are based an appropriate discretization scheme for integral expression using some quadrature formula [BH]. Let M the number of discretization steps in each direction which is determined depending on the given error tolerance. For ease of exposition, we assume that M is even. Δ denotes the length of the discretization step. Then, we have to solve a linear system of equations of the form $\mathbf{A}_{m,n} \pi(m\Delta, n\Delta) = b_{m,n}$ for each mesh point $(m\Delta, n\Delta)$. Note that in general vectors $b_{m,n}$ in this set of systems of linear equations are determined from solutions of linear systems previously solved. However, we can employ a kind of wavefront computation scheme for a parallel numerical solution of the system of Volterra equations derived from (3.2) – (3.4). The following provides a high-level algorithmic description of the proposed computation scheme.

Initial Step

(a) solve the system of linear equations corresponding to a discretization of (3.2) – (3.4) for $a_1 = a_2 = \Delta$.

(b) solve the system of linear equations corresponding to a discretization of (3.4) for $a_1 = \Delta$ and $a_2 = 2\Delta$.

Expanding the Wavefront ($k = 2, 3 \ldots, \frac{m}{2}$ and $l = 0, 1 \ldots, k - 1$)

(a) solve the system of linear equations corresponding to a discretization of (3.3) and (3.4) for $a_1 = a_2 = k\Delta$.

(b) solve the systems of linear equations corresponding to discretizations of (3.4) for $a_1 = (k - l)\Delta$ and $a_2 = (k + l)\Delta$.

(c) solve the systems of linear equations corresponding to discretizations of (3.4) for $a_1 = (k - l)\Delta$ and $a_2 = (k + l + 1)\Delta$.

Turnover Point of the Wavefront ($k = \frac{M}{2}$ and $l = 0, 1 \ldots, \frac{M}{2} - 1$)

(c) solve the system of linear equations corresponding to discretizations of (3.4) for $a_1 = (\frac{M}{2} - l)\Delta$ and $a_2 = (\frac{m}{2} + l + 1)\Delta$.

Shrinking the Wavefront ($k = \frac{M}{2} + 1, \frac{M}{2} + 2, \ldots, M$ and $l = 0, 1 \ldots, M - k$)

(a) solve the system of linear equations corresponding to a discretization of (3.3) and (3.4) for $a_1 = a_2 = k\Delta$.

(b) solve the systems of linear equations corresponding to discretizations of (3.4) for $a_1 = (k - l)\Delta$ and $a_2 = (k + l)\Delta$.

(c) solve the systems of linear equations corresponding to discretizations of (3.4) for $a_1 = (k - l)\Delta$ and $a_2 = (k + l + 1)\Delta$.

Steps of the form (a) contain two-dimensional triangular integrals and require the computation of two-dimensional Volterra equations. In the initial step, the matrix $\mathbf{A}_{1,1}$ of the linear system has dimension N and computes stationary probabilities π_i for $i \in S_{exp}$, $\pi_i(\Delta)$ for $i \in S_{det1}$, and $\pi_i(\Delta, \Delta)$ for $i \in S_{det2}$. In subsequent steps of the form (a), the correspond-

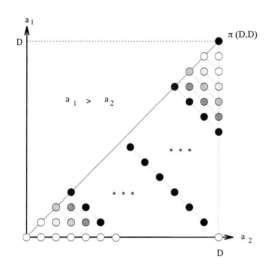

ing matrices $\mathbf{A}_{k,k}$ have dimension $N - N_1$ and compute stationary probabilities $\pi_i(k\Delta)$ for $i \in S_{det1}$ and $\pi_i(k\Delta, k\Delta)$ for $i \in S_{det2}$. Steps of the form
(b) and (c) contain two-dimensional rectangular integrals and, thus, reduce
to the computation of one-dimensional Volterra equations. The matrices
$\mathbf{A}_{m,n}$ of these steps have dimension $N - N_1 - N_2$ and compute stationary
probabilities $\pi_i((k-l)\Delta, (k+l)\Delta)$ for $i \in S_{det2}$.

For the solution of each linear equations, we propose to employ a *generalized minimum residual* method (GMRES [St]) preconditioned using solutions of linear systems of neighborhood mesh points. The employment
of GMRES has the advantage that this method additionally allows the ex-
ploitation of fine–grain parallelism in the solution of the linear systems.
Recall that even if there is some dependence in the linear systems of equations derived from the discretization of the system of Volterra integral equations, we can solve some linear systems in parallel. The figure shown above
illustrates the parallel computation scheme. Each circle in the plane represents a mesh point $(m\Delta, n\Delta)$ for which a system of linear equations has
to be solved. Observe that all linear systems at the front of the wave (cir-
cles connected by a dotted line) can be solved in parallel. Thus, all linear
systems in steps of the form (b) and (c) in the high-level description of
the method are independent from each other and, thus, can be solved in
parallel.

5. Conclusions

This paper presented a parallel numerical algorithm for solving a class of Volterra integral equations that arise from the stationary analysis of generalized semi-Markov processes (GSMPs) with exponential and deterministic events. Practical applications of such generalized semi-Markov processes stem from performance modeling of computer and communication systems. In this domain, the size of the state space of the GSMP, N, is typically very large for practical applications (i.e., $N \geq 100,000$). Therefore, a parallelization of the numerical solution method is essential for the practical applicability in performance modeling.

In current work, we are implementing the presented numerical method using the special-purpose language PROMOTER [Sch]. PROMOTER is a high-level programming language designed for simplifying the task of programming parallel applications for distributed memory architectures. PROMOTER is implemented in top of the object–oriented language C++.

References

[LS]   C. Lindemann and G.S. Shedler, *Numerical Analysis of Deterministic and Stochastic Petri Nets with Concurrent Deterministic Transitions*, *Performance Evaluation* **27** 1996, in press.

[BH]   H. Brunner and P.J. van der Houwen, *Numerical Solution of Volterra Equations*, North Holland, 1986.

[Fe]   W. Feller, *An Introduction to Probability Theory and its Application*, Vol. 2, 2nd Edition, John Wiley & Sons, 1971.

[Gl]   P.W. Glynn, A GSMP Formalism for Discrete-event Systems, *Proc. IEEE* **77** 1989, 14-23.

[GM]   D. Gross and D.R. Miller, The Randomization Technique as a Modeling Tool and Solution Procedure for Transient Markov Processes, *Operations Research* **32** 1984, 345-361.

[Sch]   A. Schramm, Irregular Applications in PROMOTER, *Proc. 2nd Int. Conf. on Programming Models for Massively Parallel Systems*, Berlin, Germany, 1995.

[St]   W. J. Stewart, *Numerical Solution of Markov Chains*, Princeton
        University Press 1994.

Authors address:

GMD Research Institute for Computer Architecture and

Software Technology (GMD FIRST),

Rudower Chaussee 5,

12489 Berlin, Germany.

e-mail: {heber, lind}@first.gmd.de

Received October 22, 1996

On the Laws of the Oseledets Spaces
of Linear Stochastic Differential Equations

Peter Imkeller

1. Introduction

In many cases, the asymptotic properties of random dynamical systems generated by (non-linear) stochastic differential equations are determined by their invariant manifolds, and thus, to first order, by the linear invariant manifolds related to their linearizations. These were first described in the famous multiplicative ergodic theorem by Oseledets [Os] and are called "Oseledets spaces". Although they are the direct analogue of eigenspaces for real matrices in the deterministic setting, they can explicitly be given only in rare cases (see Arnold [Ar]), and are hardly accessible. In this note we shall for simplicity from the very beginning consider linear stochastic differential equations instead of linearizations for non-linear ones, and describe some smoothness results for the laws of their Oseledets spaces, using the approach of Malliavin's calculus.

To be more precise, let $A_0, ..., A_m$ be $d \times d$-matrices with real coefficients, let $W = (W^1, ..., W^m)$ be a canonical two-sided m−dimensional Wiener process on $(\Omega, \mathbf{F}, \mathbf{P})$, i.e. both $(W_t)_{t \geq 0}$ and $(W_{-t})_{t \geq 0}$ are canonical m-dimensional Brownian motions. Consider the Stratonovitch equation

$$dx_t = \sum_{i=0}^{m} A_i x_t \circ dW_t^i, \tag{1.1}$$

where we formally put $dW_t^0 = dt$ in order to simplify notation. Now assume for a moment that the random term in (1.1) is turned off, and for simplicity that A_0 is symmetric and has eigenvalues $\lambda_1 > ... > \lambda_r$ and corresponding eigenspaces $E_1, ..., E_r$. Then the eigenspaces can be characterized dynamically as those linear subspaces of $\mathbf{R}^d$ such that for trajectories of the solution of (1.1) started therein, the corresponding eigenvalues are observed as the asymptotic exponential growth rate both at ∞ and at $-\infty$:

$$\lambda_i = \lim_{t \to \pm\infty} \frac{1}{t} \ln \|\Phi_t x\| \quad \text{iff} \quad 0 \neq x \in E_i, \tag{1.2}$$

$1 \leq i \leq r$, where $\Phi_t = \exp(A_0 t)$, $t \in \mathbf{R}$, is the fundamental solution of
(1.1). Now if we switch the noise term on again, there is still a fundamental
solution $(\Phi_t)_{t \in \mathbf{R}}$ of (1.1), i.e. a random family of linear isomorphisms of
$\mathbf{R}^d$, describing the dynamics of the system, and the analogue of the above
characterization of eigenspaces is just given by the theorem of Oseledets:
there exist *Lyapunov numbers* $\lambda_1 > ... > \lambda_r$ and random linear subspaces
$E_1, ..., E_r$ of $\mathbf{R}^d$ (the *Oseledets spaces*) such that (1.2) holds true $\mathbf{P}$−a.s.. In
the proof of this theorem (see Arnold [Ar]) the spaces E_i are constructed
as intersections of the spaces appearing in two "flags" of nested random
linear subspaces of $\mathbf{R}^d$

$$V_1^+ \supset ... \supset V_r^+, \quad V_1^- \subset ... \subset V_r^-$$

determined by the inequalities

$$x \in V_i^\pm \quad \text{iff} \quad \lambda_i \overset{\leq}{\underset{\geq}{}} \lim_{t \to \pm\infty} \frac{1}{t} \ln \|\Phi_t x\|.$$

These inequalities indicate in particular that the "forward (+)" flag is measurable with respect to the history of W on $\mathbf{R}_+$, the "backward" one with
respect to its history on $\mathbf{R}_-$. Hence

$$E_i = V_i^+ \cap V_i^-, \quad 1 \leq i \leq r,$$

is in general measurable just with respect to the whole history of W (see
Arnold [Ar]). The independence of $(W_t)_{t \geq 0}$ and $(W_{-t})_{t \geq 0}$ implies, moreover, that forward and backward flags are independent.

   For this reason, it turns out that one essentially has to study the law of
forward (or equivalently backward) flag, and combine the obtained smoothness results for the two by a purely analytical result. We thus consider, say,
the forward flag. In order to be able to profit from the algebraic structure
of the set of linear operators on $\mathbf{R}^d$ we identify linear subspaces of $\mathbf{R}^d$ with
the orthogonal projectors on them, this way embedding the Grassmannian
manifolds into the space of linear operators in $\mathbf{R}^{d \times d}$. First of all (and indeed for almost the rest of this note) we fix one of the subspaces appearing
in the forward flag, say Q, of rank, say, k with $1 \leq k \leq d$, and deal with
the smoothness of its law. More precisely, we shall work with the following
version of the Grassmannian of k−dimensional linear subspaces of $\mathbf{R}^d$ (for
details see [Im])

$$G_k(d) = \{OE_k O^* : O \in \mathcal{O}(d)\},$$

where, with respect to the canonical basis $e_1, ..., e_d$ of $\mathbf{R}^d$ we define

$$E_k e_i = \begin{cases} e_i, & i \leq k, \\ 0, & i > k. \end{cases}$$

As we shall see in the following section, the main idea to deal with the law of Q consists in identifying the space with the "stationary state" of a flow of diffeomorphisms on the manifold $G_k(d)$. To obtain this flow, we shall decompose the linear flow $(\Phi_t)_{t \in \mathbf{R}}$ into a "radial" and "angular" part living on $G_k(d)$. For this reason we make use of the precise description of $TG_k(d)$. By I we usually denote the identity operator (matrix) in $\mathbf{R}^{d \times d}$.

Lemma 1.1 *For $P \in G_k(d)$ we have*

$$T_P \, G_k(d) = \{(I - P) \, A \, P + P \, A^* \, (I - P) : \quad A \in \mathbf{R}^{d \times d}\}.$$

Proof. Let L denote the space on the right hand side of the asserted equation,

$$L' = \{(I - P) \, B \, P + P \, B^* \, (I - P) : \quad B \in \mathbf{R}^{d \times d} \text{skew symmetric}\}.$$

Since for any $A \in \mathbf{R}^{d \times d}$ we may define the skew symmetric $B = (I - P) \, A \, P + P \, A^* \, (I - P)$ and have

$$(I - P) \, A \, P + P \, A^* \, (I - P) = (I - P) \, B \, P + P \, B^* \, (I - P),$$

we obtain $L = L'$. Therefore, it essentially remains to show $T_P G_k(d) \subset L'$. For this purpose, fix a smooth curve $(P_t)_{t \in I}$ through $P_0 = P = O \, E_k \, O^*$, where I is an open interval containing 0.

To $(P_t)_{t \in \mathbf{R}}$ there corresponds at least one smooth curve $(O_t)_{t \in \mathbf{R}}$ through $O_0 = O$. Now, $O'_0 = D \, O$ with a skew symmetric D. Hence

$$
\begin{aligned}
P'_0 &= O'_0 \, E_k \, O^*_0 + O_0 \, E_k \, (O^*_0)' = D \, O \, E_k \, O^* + O \, E_k \, O^* \, D^* \\
&= D \, P + P \, D^* = (I - P) \, D \, P + P \, D^* \, (I - P) \in L',
\end{aligned}
$$

since D is skew symmetric. $\blacksquare$

2. A decomposition of the linear flow

Remember we fixed Q as an element of the forward flag, taking its values in $G_k(d)$. We shall now give a multiplicative decomposition of the linear flow $(\Phi_t)_{t \in \mathbf{R}}$ given by the fundamental solution of (1.1). This decomposition can be considered the generalization of the radial-angular decomposition of the one-point motion (see Khasminskii [Kh], or Arnold, Oeljeklaus, Pardoux [AOP]). To obtain the angular component we just have to project the vector fields of (1.1) onto $T G_k(d)$. So for $A \in \mathbf{R}^{d \times d}$ let

$$
\begin{aligned}
h_A(P) &= (I - P) A P + P A^* (I - P), \\
g_A(P) &= (I - P) A P - P A^* (I - P) + P A P + (I - P) A (I - P),
\end{aligned}
$$

$P \in G_k(d)$. Then the following decomposition is obtained.

Theorem 2.1 *Let $P \in G_k(d)$, and $(P_t)_{t \in \mathbf{R}}$ be the solution of*

$$
dP_t = \sum_{i=0}^{m} h_{A_i}(P_t) \circ dW_t^i, \quad P_0 = P, \tag{2.1}
$$

$(R_t)_{t \in \mathbf{R}}$ *the solution of*

$$
dR_t = \sum_{i=0}^{m} g_{A_i}(P_t) R_t \circ dW_t^i, \quad R_0 = I. \tag{2.2}
$$

Then we have $\Phi_t P = P_t R_t$, $(I - P_t) \Phi_t = R_t (I - P)$, $t \in \mathbf{R}$.

Proof. By definition for $A \in \mathbf{R}^{d \times d}, P \in G_k(d)$

$$
h_A(P) + P g_A(P) = (I - P) A P + P A P = A P. \tag{2.3}
$$

Well known results about stochastic differential equations (on manifolds) imply that the processes $(P_t)_{t \in \mathbf{R}}$ and $(R_t)_{t \in \mathbf{R}}$ are well defined. For $t \in \mathbf{R}$ let $X_t = P_t R_t$. Then according to the rules of Stratonovitch calculus and (2.3)

$$
\begin{aligned}
dX_t &= dP_t R_t + P_t dR_t \\
&= \sum_{i=0}^{m} [h_{A_i}(P_t) + P_t g_{A_i}(P_t)] R_t \circ dW_t^i = \sum_{i=0}^{m} A_i X_t \circ dW_t^i.
\end{aligned}
$$

Moreover, $X_0 = P$. Hence by strong uniqueness $X_t = \Phi_t P$ for all $t \in \mathbf{R}$. The "dual" equation is proved by a similar argument. ∎

Now denote by $P_t(\omega, P) = P_t(\omega), \omega \in \Omega, t \in \mathbf{R}$, where $(P_t)_{t \in \mathbf{R}}$ is the solution of (2.1) starting in $P \in G_k(d)$. A version of $P_t(.,.)$ can be chosen which is a diffeomorphism of $G_k(d)$ for any $t \in \mathbf{R}$ (see Arnold [Ar]). For the canonical shift on Ω by $t \in \mathbf{R}$ we write $\theta_t : \Omega \to \Omega, \omega \mapsto (s \mapsto \omega(s+t) - \omega(t))$.

Corollary 2.2 *Let $(\hat{P}_t)_{t \in \mathbf{R}}$ be the solution of*

$$
d\hat{P}_t = \sum_{i=0}^{m} h_{A_i}(\hat{P}_t) \circ dW_t^i, \quad \hat{P}_0 = Q, \tag{2.4}
$$

$(\hat{R}_t)_{t\in\mathbf{R}}$ *the solution of*

$$d\hat{R}_t = \sum_{i=0}^{m} g_{A_i}(\hat{P}_t)\,\hat{R}_t \circ dW_t^i, \quad \hat{R}_0 = I. \tag{2.5}$$

Then we have $\Phi_t\, P = \hat{P}_t\,\hat{R}_t$, $(I - \hat{P}_t)\,\Phi_t = \hat{R}_t\,(I - Q)$, $t \in \mathbf{R}$. *Moreover*

$$\hat{P}_t = Q \circ \theta_t,$$

$t \in \mathbf{R}$.

Proof. The first part follows immediately from the theorem and the fact that Stratonovitch integrals obey the substitution rule, i.e. roughly speaking in the solutions appearing in Theorem 2.1 Q can be substituted for P. To obtain the second assertion, note that according to Arnold [Ar], p. 137, or by the very characterization of the spaces of the forward flag above we must have for $t \in \mathbf{R}$

$$\Phi_t\, Q(\mathbf{R}^d) = Q \circ \theta_t(\mathbf{R}^d).$$

Hence the first part and surjectivity of $\hat{R}_t$ yield

$$\hat{P}_t(\mathbf{R}^d) = Q \circ \theta_t(\mathbf{R}^d).$$

Since both $\hat{P}_t$ and $Q \circ \theta_t$ are orthogonal projectors, we obtain the desired equation. ∎

The last result of Corollary 2.2 states that Q is the "stationary state" of the flow $(P_t(.,.))_{t\in\mathbf{R}}$. This observation can be put in a slightly more stringent form by using the invariance of $\mathbf{P}$ with respect to $\theta_t, t \in \mathbf{R}$. For the details see [Im].

Theorem 2.3 *Let* ρ *be the law of* Q, *and* $T : \Omega \times G_k(d) \to G_k(d), (\omega, P) \mapsto P_t(\omega, P)$. *Then we have*

$$\rho = (\mathbf{P} \otimes \rho) \circ T^{-1}.$$

As a consequence of Corollary 2.2 or Theorem 2.3 the law of Q can be studied via the law of $P_t(., P)$, *uniformly in* P. This is done by applying Malliavin's calculus for the random element Q.

3. Malliavin's calculus for orthogonal projectors

We first introduce some basic notation of Malliavin's calculus. See Nualart [Nu] for a more detailed account. For $T \geq 0, 1 \leq j \leq m$, we shall denote by D^j the derivative operator (the H-*derivative*) which, for a smooth random variable of the form

$$F = f(W_{t_1}, ..., W_{t_n}), \quad f \in C_b^\infty((\mathbf{R}^m)^n), \quad t_1, ..., t_n \in \mathbf{R}_+,$$

takes the form

$$D_s^j F = \sum_{i=1}^n \frac{\partial}{\partial x_{i,j}} f(W_{t_1}, ..., W_{t_n}) 1_{[0,t_i]}(s).$$

For each $p \geq 1, \mathbf{D}_{p,1}([0,T])$ will denote the Banach space of random variables on the Wiener space, defined as the closure of the set of smooth random variables with respect to the norm

$$\|F\|_{p,1} = \|F\|_p + \sum_{1 \leq j \leq m} E([\int_0^T |D_s^j F|^2 \, ds]^{\frac{p}{2}})^{\frac{1}{p}},$$

to which D^j extends in a natural way. For $k \in \mathbf{N}$, we define the H−derivative of order k by k−fold iteration of the above derivative on the space of smooth random variables. For such a random variable F, the expression $D_{s_1...s_k}^{j_1...j_k} F$ for $1 \leq j_i \leq m, 1 \leq i \leq k, s_1, ..., s_k \in \mathbf{R}_+$ is self explanatory. For $T \geq 0, p \geq 1, k \in \mathbf{N}$, we denote by $\mathbf{D}_{p,k}([0,T])$ the Banach space given by the completion of the set of smooth random variables with respect to the norm

$$\| F \|_{p,k} = \| F \|_p + \sum_{1 \leq j \leq k} \sum_{1 \leq l_i \leq m, 1 \leq i \leq j} [E([\int_0^T (D_{s_1...s_j}^{l_1...l_j} F)^2 \, ds_1...ds_j]^{\frac{p}{2}})]^{\frac{1}{p}}.$$

As a Fréchet derivative, the H−derivative is a derivation in the algebraic sense and thus obeys the usual rules, in particular Leibniz' rule.

Theorem 3.1 *For any $T \geq 0, 1 \leq j \leq m, p \geq 1, k \in \mathbf{N}$ we have $P_t, Q \in \mathbf{D}_{p,k}([0,T])$. For $0 \leq r \leq t$ we have*

$$D_r^j P_t = h_{A_j^{r,t}}(P_t), \quad D_r^j Q = -h_{A_j^r}(Q),$$

where $A_j^r = \Phi_r^{-1} A_j \Phi_r, A_j^{r,t} = \Phi_t A_j^r \Phi_t^{-1}$.

Proof. We concentrate on the harder second part. First of all, the H-differentiability of P_t is classical (see for example Nualart [Nu]). Using the relationship between P_t and Q discussed above and passing to the limit $t \to \infty$ the H-differentiability of Q and thus of $\hat{P}_t$ is proved. Alternatively, one may use the results of [AI], where it is explicitly shown. Using the differentiability properties of $\hat{P}_t, \hat{R}_t$ we shall now derive the formula for the derivative of Q.

Observe first that by independence of increments of the Wiener process and the equation $\hat{P}_t = Q \circ \theta_t$ we have $D_r^j \hat{P}_t = 0$. Moreover, the rule for the derivation of stochastic integrals and (1.1) yield the simple equation $D_r^j \Phi_t = \Phi_t A_j^r$. Hence Corollary 2.2 gives

$$
\begin{aligned}
0 &= (I - \hat{P}_t) \Phi_t A_j^r Q + (I - \hat{P}_t) \Phi_t D_r^j Q \\
&= \hat{R}_t (I - Q) A_j^r Q + \hat{R}_t (I - Q) D_r^j Q.
\end{aligned}
$$

Now $\hat{R}_t$ is invertible. Hence

$$
(I - Q) D_r^j Q \, Q = -(I - Q) A_j^r Q.
$$

It finally remains to recall that $D_r^j Q$ takes its values in $T_Q G_k(d)$. ∎

Theorem 2.3 indicates that the criteria for smoothness of the law of Q can be deduced from well known criteria for the smoothness of the law of $P_t(., P)$. Indeed, as is made rigorous in [Im], the invariance of **P** with respect to θ_t once again yields that the criteria for positivity and integrability of the Malliavin matrix of Q can be read off corresponding criteria for the Malliavin matrix of $P_t(., P)$, uniformly in P. Recalling the classical relationship between Hörmander's hypoellipticity condition and regularity of laws of stochastic differential equations it is therefore not surprising to find that the law of Q is smooth under Hörmander type conditions on the vector fields $h_{A_i}, 0 \le i \le m$. To state them, let

$$
\begin{aligned}
\mathcal{L}^k = \; &\mathrm{span}\{h_{A_1}, .., h_{A_m}, [h_{A_{i_p}}, [h_{A_{i_{p-1}}}, [..[h_{A_{i_1}}, h_{A_{i_0}}]..] : \\
&0 \le i_j \le m, 1 \le j \le p, \, p \in \mathbf{N}\}.
\end{aligned}
$$

$\mathcal{L}^k$ is the ideal generated by $h_{A_1}, .., h_{A_m}$ in the Lie algebra generated by $h_{A_0}, .., h_{A_m}$.

Theorem 3.2 *Suppose Q takes its values in $G_k(d)$, and that*

$$
\mathcal{L}_P^k = T_P G_k(d) \quad \textit{for} \quad P \in G_k(d).
$$

Then the law of Q possesses a C^∞−density with respect to Riemannian volume on $G_k(d)$.

Now recall that Q was taken from the forward flag of random linear subspaces which, together with an independent backward one, determines the Oseledets spaces. It is clear that if we require the Hörmander condition of Theorem 3.2 for any dimension $k, 1 \leq k \leq d$, we obtain that the whole flag possesses a smooth law on the respective flag manifold. One more step is necessary to pass from the independent flags via intersections to the Oseledets spaces. This step is purely analytical and makes use of a quite general argument based upon the co-area formula of Federer [Fe], was pointed out to me by M. Zähle, and relates smoothness of the law of the intersection of spaces of flags to smoothness of the flags. For details see [Im]. Here is the main result of this note.

Theorem 3.3 *Suppose that for $1 \leq k \leq d$ we have*

$$\mathcal{L}_P^k = T_P G_k(d) \quad for \quad P \in G_k(d).$$

Then the law of $(E_1, ..., E_r)$ possesses a C^∞−density with respect to Riemannian volume on the respective product of Grassmannian manifolds.

Remark. 1. The conditions of Theorem 3.3 are fulfilled for example if the Lie group generated by $A_0, ..., A_m$ is the general or special linear group over $\mathbf{R}^d$.

2. The classification theory of semisimple Lie groups indicates that the Hörmander condition of Theorem 3.2 for $1 \leq k \leq d$ is a consequence of the same condition, but only for $k = 1, 2$. But we cannot substantiate this conjecture at this place.

References

[Ar]   L. Arnold, *Random dynamical systems*. Preliminary version of book. Institut für Dynamische Systeme, Universität Bremen (1995).

[AI]   L. Arnold, and P. Imkeller, Furstenberg-Khasminskii formulas for Lyapunov exponents via anticipative calculus, *Stochastics and Stochastics Reports*, **54** (1995), 127-168.

[AOP] L. Arnold, E. Oeljeklaus, E. Pardoux, *Almost sure and moment stability for linear Ito equations.* in: Lyapunov Exponents. L. Arnold, V. Wihstutz (eds.). **LNM 1186** (1986), 129-159. Springer: Berlin.

[Ba] P. Baxendale, *The Lyapunov spectrum of a stochastic flow of diffeomorphisms.* in: Arnold, L., Wihstutz, V. (eds.): Lyapunov exponents. **LNM 1186** (1986), 322-337.

[BH] N. Bouleau, F. Hirsch, *Dirichlet forms and analysis on Wiener space.* W. de Gruyter, Berlin 1991.

[Co] A. Coquio, Calcul de Malliavin, existence et régularité de la densité d'une probabilité invariante d'une diffusion sur une variété Riemannienne compacte. *Preprint,* Inst. Fourier, Grenoble (1994).

[Fe] H. Federer, *Geometric measure theory.* Springer: Berlin 1969.

[He] S. Helgason, *Differential geometry, Lie groups, and symmetric spaces.* Academic Press: New York 1978.

[IW] N. Ikeda, S. Watanabe, *Stochastic differential equations and diffusion processes.* North Holland (2 nd edition): 1989.

[Im] P. Imkeller, *On the smoothness of the laws of random flags and Oseledets spaces of random dynamical systems generated by linear stochastic differential equations.* Preprint, MU Lübeck (1996).

[Kh] R. Z. Khasminskii, *Stochastic stability of differential equations.* Sijthoff and Noordhoff, Alphen 1980.

[Nu] D. Nualart, *The Malliavin calculus and related topics.* Springer: Berlin 1995.

[Os] V. I. Oseledets, A multiplicative ergodic theorem. Lyapunov characteristic numbers for dynamical systems. *Trans. Moscow Math. Soc.,* **19** (1968), 197-231.

[Ru] D. Ruelle, Ergodic theory of differentiable dynamical systems. *Publ. Math.,* **IHES 50** (1979), 275-306.

[St1] D. W. Stroock, The Malliavin Calculus. Functional Analytic Approach. *J. Funct. Anal.,* **44** (1981), 212-257.

[Ta] S. Taniguchi, Malliavin's stochastic calculus of variations for manifold valued Wiener functionals. *Z. Wahrscheinlichkeitstheorie verw. Geb.,* **65** (1983), 269-290.

[Wa]   S. Watanabe, *Stochastic differential equations and Malliavin calculus.* Tata Institut of Fundamental Research. Springer: Berlin 1984.

Institut für Mathematik
Humboldt-Universität zu Berlin
Unter den Linden 6
10099 Berlin
Germany

Received September 11, 1996

On Stationarity of Additive Bilinear State-space Representation of Time Series

Márton Ispány[1]

Abstract

In this paper necessary and sufficient conditions are proved for the existence of strictly and weakly stationary solution of additive bilinear state–space stochastic systems.

1. Introduction

In recent times there has been a growing interest in various nonlinear time series models. There are many environments, in which the resulting data sets cannot be modeled by the linear ARMA models. Wegman et al [WST] provide several examples on the area of meteorology, hydrology, biology e.t.c., which are clearly nonlinear and the reader can find an excellent theoretical setup of nonlinear time series in Tong's book [To]. The purpose of this paper to derive necessary and sufficient conditions for the existence of strict and second order stationary solution of so–called additive bilinear state–space stochastic system. This kind of state–space representation is a special case of bilinear time series models investigated by several author (see [GA] and [SG]). However, as we shall see, the additive bilinear state–space representation enable us to describe more general nonlinear time series than bilinear one.

This paper is organized as follows. Having defined the additive bilinear state–space system we introduce the concept of controllability and observability, which play the same fundamental role from stationarity point of view as in the linear systems theory. Instead of the linear Gaussian case, where the conditions for strict and second order stationarity agree with each other, in nonlinear case there is a huge class of time series models, which have strictly stationary but have not second order stationary solution. In section 3 we prove that for a controllable and observable additive bilinear system the Lyapunov exponents, which describe the evolution of our system

---

[1]This research was supported by Hungarian Foundation for Scientific Researches under Grant No. OTKA–1951/1996.

in time as they work in the dynamical systems theory, completely characterize the existence of strictly stationary, asymptotically stable solution. In section 4 it is shown that under appropriate moment condition for the innovation process the existence of second order stationary solution depends on the spectral properties of a particular matrix defined by the coefficients of our model. Finally, we make some remarks and some open problems are raised.

2. The additive bilinear state–space representation

It is said that the real stochastic process $\{y_t, t \in \mathbb{Z}\}$ has an additive bilinear state space representation if the following state space stochastic model is valid

$$
\begin{aligned}
X_t &= AX_{t-1} + \sum_{j=1}^{m} B_j X_{t-1} w_{t-j} + cw_t, \\
y_t &= d^\top X_t, \qquad t \in \mathbb{Z},
\end{aligned}
\tag{2.1}
$$

where $A, B_j \in \mathbb{R}^{n \times n}$, $j = 1, \ldots m$, $c, d \in \mathbb{R}^n$, the state vector X_t is a n–dimensional random variable and $\{w_t, t \in \mathbb{Z}\}$ is a stochastic process on the probability space $(\Omega, \mathcal{A}, \mathbb{P})$. In general, we assume that $\{w_t, t \in \mathbb{Z}\}$ is an ergodic strictly stationary process. We have to point to the fact that usually the state–space representation (2.1) is Markovian nor in wide neither in strong sense because the stochastic state space hypothesis (iii) does not hold (see p. 109 in [Ca]).

Remark 2.1 *If $\{y_t, t \in \mathbb{Z}\}$ fulfills the general bilinear model $BL(l, r, m, q)$ defined by*

$$
y_t = \sum_{j=1}^{l} a_j y_{t-j} + \sum_{j=0}^{r} c_j w_{t-j} + \sum_{i=1}^{m} \sum_{j=1}^{q} b_{ij} y_{t-i} w_{t-j},
$$

then it is known that $\{y_t, t \in \mathbb{Z}\}$ has additive bilinear state space representation (see [To] p. 137).

The following example shows that the additive bilinear state–space representation can be used to describe much more complicated non–linear stochastic processes.

Remark 2.2 *Consider the stochastic difference equation*

$$
y_t = ay_{t-1} + b_1 y_{t-1} w_{t-1} + b_2 y_{t-2} w_{t-1} w_{t-2} + cw_t.
$$

Then the stochastic process $\{y_t, t \in \mathbb{Z}\}$ has additive bilinear state–space representation (2.1), where

$$X_t = \begin{pmatrix} y_t \\ y_{t-1} w_{t-1} \end{pmatrix}, \qquad c = d = \begin{pmatrix} 1 \\ 0 \end{pmatrix},$$

$$A = \begin{pmatrix} a & 0 \\ 0 & 0 \end{pmatrix}, \qquad B_1 = \begin{pmatrix} b_1 & b_2 \\ 1 & 0 \end{pmatrix}.$$

We introduce the concepts of controllability and observability by straightforward generalization of their definition in simple bilinear case (see Sec. 4.4 in [Ru] in deterministic case and [Fr] p. 784 in stochastic case). We recall that the vector $v \in \mathbb{R}$ is said to be cyclic for $\mathcal{M}$, where $\mathcal{M} = \{M_i, i \in \mathbb{I}\}$ is a finite set of $n \times n$ matrices, if

$$\bigvee \{M_{i_1}^{n_1} \cdot \ldots \cdot M_{i_r}^{n_r} v \mid i_j \in \mathbb{I}, \ n_j \in \mathbb{N}, \ j = 1, \ldots, r, \ r \in \mathbb{N}\} = \mathbb{R}^n,$$

where $\bigvee \mathcal{S}$ denotes the closed subspace in $\mathbb{R}^n$ spanning by subset $\mathcal{S} \subset \mathbb{R}^n$.

Definition 2.3 *The additive bilinear state–space representation (M) is said to be controllable (observable) if the vector c (d) is cyclic for the set of matrices $\{A, B_j, j = 1 \ldots, m\}$. The closed subspaces in $\mathbb{R}^n$ defined by*

$$\mathcal{C} = \bigvee \{B_{k_1}^{l_1} \cdot \ldots \cdot B_{k_r}^{l_r} c \mid l_1, \ldots, l_r \geq 1, \ 0 \leq k_1, \ldots, k_r \leq m\},$$

$$\mathcal{O} = \bigvee \{B_{k_1}^{l_1} \cdot \ldots \cdot B_{k_r}^{l_r} d \mid l_1, \ldots, l_r \geq 1, \ 0 \leq k_1, \ldots, k_r \leq m\},$$

where $B_0 = A$, are called the subspace of controllable and the subspace of observable states, respectively.

To check the controllability and observability the following proposition is useful, which is a simple generalization of Theorem 4.13 and Theorem 4.14 in [Ru].

Proposition 2.4 *Let $v \in \mathbb{R}^n$, $\{M_i, i \in \mathbb{I}\}$ is a finite set of $n \times n$ matrices and define recursively the matrices*

$$v_1 = v, \qquad v_j = [M_i v_{j-1}, i \in \mathbb{I}],$$

$$V_j = [v_1 \ldots v_j],$$

where the symbol $[M_i, i \in \mathbb{I}]$ denotes a matrix consisting of matrices M_i, $i \in \mathbb{I}$ as blocks. Then the vector v is cyclic for $\{M_i, i \in \mathbb{I}\}$ if and only if $\operatorname{rank} V_n = n$.

Iterating the model it can be seen that the solution of (2.1) has the following formal infinite series representation

$$y_t = d^\top c w_t + \sum_{r=1}^{\infty} \sum_{l_1,\dots,l_r \geq 1} \sum_{0 \leq k_i \neq k_{i+1} \leq m} d^\top B_{k_1}^{l_1} \cdot \dots \cdot B_{k_r}^{l_r} c$$

$$\prod_{\{i:k_i \neq 0\}} \prod_{s=0}^{l_i-1} w_{t-k_i-\sum_{j=1}^{i-1} l_j - s} w_{t-\sum_{j=1}^{r} l_j}, \qquad (2.2)$$

where the superscript $\top$ denotes the transpose. We shall investigate the almost sure and mean square convergence of this infinite series.

3. Strictly stationary solution

First we define the asymptotically stable and geometrically stable solution of the state space representation (2.1).

Definition 3.1 *Let $\{z_t(X_0), t \in \mathbb{N}\}$ denote the resulting output process initializing the state–space stochastic system (2.1) at 0 with the random variable X_0, i.e., $z_t = d^\top X_t$, $t = 0, 1, \dots$, where X_t satisfies the recursive formula*

$$X_t = AX_{t-1} + \sum_{j=1}^{m} B_j X_{t-1} w_{t-j} + c w_t, \qquad t = 1, 2, \dots .$$

The state space stochastic system (2.1) is said to be asymptotically stable if it has asymptotically stable solution, i.e., there exists a strictly stationary process $\{y_t, t \in \mathbb{Z}\}$ satisfying (2.1) and $\varepsilon > 0$ such that $y_n - z_n(x_0) \to 0$ in probability as $n \to \infty$ for all $\|x_0\| < \varepsilon$; geometrically stable if

$$|y_n - z_n(X_0)| \leq C \gamma^n \qquad a.s.$$

for all initial random state vector X_0 and $n \in \mathbb{N}$, where $0 < \gamma < 1$ is an absolute constant and C may depend on ω and X_0.

Then our main theorem on strictly stationary solution of the model (2.1) is the following

Theorem 3.2 *Let $\{w_t, t \in \mathbb{Z}\}$ be an ergodic strictly stationary process and $\mathbb{E} \ln_+ |w_t| < \infty$. If the upper Lyapunov exponent of the stochastic process $\mathbf{A}_t = A + \sum_{j=1}^{m} B_j w_{t-j+1}$, $t \in \mathbb{Z}$, defined by*

$$\gamma = \lim_{n \to \infty} \frac{1}{n} \mathbb{E} \ln \|\mathbf{A}_n \cdot \dots \cdot \mathbf{A}_1\|$$

is negative, then the infinite series (2.2) converges almost surely for all $t \in \mathbb{Z}$. The stochastic process $\{y_t, t \in \mathbb{Z}\}$ defined by (2.2) is the unique strictly stationary solution of the state–space stochastic system (2.1). Moreover, this solution is geometrically stable.

Conversely, if the state space stochastic system (2.1) is controllable, observable and asymptotically stable, then the upper Lyapunov exponent γ is negative.

Proof. Since the model (2.1) has random coefficient autoregressive representation of the form

$$
\begin{aligned}
X_t &= \mathbf{A}_{t-1} X_{t-1} + c w_t, \\
y_t &= d^\top X_t, \qquad t \in \mathbb{Z},
\end{aligned}
\tag{3.1}
$$

if we iterate the first equation of (3.1) we have

$$
X_t = \sum_{i=0}^{k-1} \mathbf{B}_{t,i} c w_{t-i} + \mathbf{B}_{t,k-1} X_{t-k},
$$

where $\mathbf{B}_{t,0} = I$, $\mathbf{B}_{t,k} = \mathbf{A}_{t-1} \cdot \ldots \cdot \mathbf{A}_{t-k}$, $k = 1, 2, \ldots$. Applying the Fürstenberg–Kesten theorem or the subadditive ergodic theorem (see [FK] or [Ki]) we have

$$
\frac{1}{k} \ln \|\mathbf{B}_{t,k}\| = \frac{1}{k} \ln \|\mathbf{A}_{t-1} \cdot \ldots \cdot \mathbf{A}_{t-k}\| \longrightarrow \gamma < 0
$$

almost surely as $k \to \infty$. (We note that $\mathbb{E} \ln_+ |w_t| < \infty$ implies $\mathbb{E} \ln_+ |\mathbf{A}_t| < \infty$, therefore γ is well defined.) Hence

$$
\|\mathbf{B}_{t,k}\|^{\frac{1}{k}} \longrightarrow e^\gamma < 1
$$

almost surely as $k \to \infty$. On the other hand, by the moment lemma and the Borel–Cantelli lemma

$$
\mathbb{P} \left(\limsup_{k \to \infty} \left\{ |w_{t-k}|^{\frac{1}{k}} > a \right\} \right) = 0.
$$

for all $a > 1$. Hence

$$
\limsup_{k \to \infty} \|\mathbf{B}_{t,k} c w_{t-k}\|^{\frac{1}{k}} \le e^\gamma < 1
$$

and by the Cauchy's root test the infinite series

$$
y_t = \sum_{k=1}^{\infty} d^\top \mathbf{A}_{t-1} \cdot \ldots \cdot \mathbf{A}_{t-k} c w_{t-k} + d^\top c w_t
\tag{3.2}
$$

converges absolutely almost surely. If we replace $\mathbf{A}_{t-i}$ with its definition it
can be seen that (2.2) converges almost surely. The uniqueness of strictly
stationary solution follows from inequality

$$\left| y_t - \sum_{i=0}^{k-1} \mathbf{B}_{t,i} cw_{t-i} \right| = |d^\top \mathbf{B}_{t,k-1} X_{t-k}| \leq \left[tr\left(\mathbf{B}_{t,k-1}\mathbf{B}_{t,k-1}^\top\right)\right]^{\frac{1}{2}} \cdot |y_{t-k}|,$$

where the r.h.s. converges to 0 in probability by Slutzky's theorem (see
Problem 3.10.3 in [LR]) since $tr(\mathbf{B}_{t,k-1}\mathbf{B}_{t,k-1}^\top)$ tends to 0 almost surely as
$k \to \infty$ and $|y_{t-k}|$ is fix in distribution by strict stationarity. Finally, the
geometric stability follows from the following estimation. For all ω with
probability 1

$$|y_n - z_n(X_0)| = \left| \sum_{k=n}^\infty d^\top \mathbf{B}_{n,k} cw_{t-k} - d^\top \mathbf{B}_{n,n-1} X_0 \right|$$
$$\leq \|d\| \left(\frac{e^\gamma}{1-e^\gamma} + \|X_0\| \right) \cdot e^{\gamma(n-1)}$$

if n is enough large.

To prove the second part of our theorem we note that

$$\lim_{n\to\infty} |d^\top \mathbf{A}_n \cdot \ldots \cdot \mathbf{A}_1 c| = 0$$

in probability if there exists asymptotically stable solution. By controllability and observability we get

$$\lim_{n\to\infty} \|\mathbf{A}_n \cdot \ldots \cdot \mathbf{A}_1\| = 0$$

in probability. Then a simple modification of Lemma 3.4 in [BP] shows
that the upper Lyapounov exponent γ is negative.   ∎

Since the negativity of γ implies the asymptotic stability of y_t defined
by (2.2) the following corollary is evident.

Corollary 3.3 *Let the state–space stochastic system* (2.1) *is controllable
and observable. Then* (2.1) *has asymptotically stable solution if and only if
γ is negative.*

If we consider the first order additive bilinear state–space representation, where $\{w_t, t \in \mathbb{Z}\}$ is an i.i.d. sequence satisfying an additional moment
condition, then the following theorem can be proven, which is a generalization of Theorem 2.1 in [L1] under weaker moment condition. For any

square matrix $M = (m_{ij})$ $|M|^\delta$ denotes the matrix $(|m_{ij}|^\delta)$. It easy to see that $|\cdot|^\delta$ is submultiplicative, i.e., $|M_1 M_2|^\delta \le |M_1|^\delta |M_2|^\delta$ if $0 < \delta \le 1$. Moreover, denote by $r(M)$ the spectral radius of M.

Theorem 3.4 *Let $m = 1$, $\{w_t, t \in \mathbb{Z}\}$ is an i.i.d. sequence with $\mathbb{E}|w_t|^\delta < \infty$, $0 < \delta \le 1$. Define the matrix $\Gamma = \mathbb{E}|A + B_1 w_1|^\delta$. Then $r(\Gamma) < 1$ implies that the upper Lyapounov exponent γ is negative, i.e., the state–space representation (2.1) has unique geometrically stable, strictly stationary solution.*

Proof. Choose a matrix norm $\| \cdot \|$ such that $\|M\|^\delta \le \||M|^\delta\|$ (e.g. $\|M\| = \sum_{ij} |m_{ij}|$). Since $r(\Gamma) < 1$ there exists $0 < \lambda < 1$ such that $\limsup_n \|\Gamma^n\|^{\frac{1}{n}} < \lambda$. By independence of the innovation and submultiplicativity of $|\cdot|^\delta$ we get

$$\|\Gamma^n\| = \left\| \prod_{i=1}^n \mathbb{E}|A + B_1 w_i|^\delta \right\| \ge \mathbb{E} \left\| \prod_{i=1}^n (A + B_1 w_i) \right\|^\delta .$$

Therefore

$$0 > \ln \lambda \ge \limsup_{n \to \infty} \frac{1}{n} \ln \mathbb{E} \left\| \prod_{i=1}^n (A + B_1 w_i) \right\|^\delta$$

$$\ge \limsup_{n \to \infty} \frac{1}{n} \mathbb{E} \ln \left\| \prod_{i=1}^n (A + B_1 w_i) \right\|^\delta = \gamma \cdot \delta,$$

where we used the Jensen inequality and the fact that every matrix norm is equivalent. ∎

4. Second order stationary solution

Definition 4.1 *The state–space stochastic system (2.1) is said to be asymptotically stable in mean square sense if it has asymptotically stable solution in mean square sense, i.e., there exists a second order stationary process $\{y_t, t \in \mathbb{Z}\}$ satisfying (2.1) and $\varepsilon > 0$ such that $\mathbb{E}(y_n - z_n(x_0))^2 \to 0$ as $n \to \infty$ for all $\|x_0\| < \varepsilon$; geometrically stable in mean square sense if*

$$\mathbb{E}|y_n - z_n(X_0)|^2 \le C\gamma^n$$

for all $n \in \mathbb{N}$ and initial random state vector X_0 in L^2, which is independent of the σ algebra $\sigma\{w_s, s \ge -m\}$, where $0 < \gamma < 1$ is an absolute constant and C may depend on X_0.

Denote by symbol $\otimes$ the usual tensor product for matrices. Then our main theorem on second order stationary solution of (2.1) is the following

Theorem 4.2 *Let $\{w_t, t \in \mathbb{Z}\}$ be a sequence of i.i.d. random variables such that $\mathbb{E}w_t^{2(m+1)} < \infty$ and $\mathbb{E}w_t^{2k+1} = 0$, $k = 0, \dots, m$. If*

$$r\left(A \otimes A + \sigma^2 \sum_{j=1}^m B_j \otimes B_j\right) < \frac{1}{m}, \qquad (4.1)$$

where $\sigma^2 = \mathbb{E}w_t^2$, then the infinite series (2.2) converges almost surely as well as in the mean square. $\{y_t, t \in \mathbb{Z}\}$ is the unique solution of the state space representation (2.1). Moreover, this solution is geometrically stable in mean square sense.

Conversely, if the state–space representation (2.1) is controllable and observable, and has second order stationary, asymptotically stable solution in mean square sense, then the all eigenvalues of the matrix

$$A \otimes A + \sigma^2 \sum_{j=1}^m B_j \otimes B_j \qquad (4.2)$$

are less than 1 in modulus.

Proof. It can be proved that the upper Lyapounov exponent is negative if condition (4.1) holds. Hence the infinite series (2.2) converges almost surely and this series is the unique solution by Theorem 3.2. It remains to prove that the infinite series converges in mean square. Let

$$\xi(\mathbf{k}, \mathbf{l}) = \prod_{\{i : k_i \neq 0\}} \prod_{s=0}^{l_i-1} w_{t-k_i - \sum_{j=1}^{i-1} l_j - s} w_{t - \sum_{j=1}^r l_j}$$

where $\mathbf{k}$ and $\mathbf{l}$ denote the multiindices $(k_1, \dots, k_r)$ and $(l_1, \dots, l_r)$, respectively. Define an equivalence relation on pairs of multiindices by

$$(\mathbf{k}, \mathbf{l}) \sim (\mathbf{k}', \mathbf{l}') \iff \mathbb{E}\xi(\mathbf{k}, \mathbf{l})\xi(\mathbf{k}', \mathbf{l}') \neq 0$$

and let

$$c(\mathbf{k}, \mathbf{l}) = \#\{(\mathbf{k}', \mathbf{l}') \mid (\mathbf{k}', \mathbf{l}') \sim (\mathbf{k}, \mathbf{l})\}.$$

It can be seen that an arbitrary w_t, $t \in \mathbb{Z}$, can occurs at most to the power $2(m+1)$ in all expression of the form $\xi(\mathbf{k}, \mathbf{l})\xi(\mathbf{k}', \mathbf{l}')$ and there exist absolute constants C_1 and C_2 such that

$$C_1 \sigma^{2k} \leq \mathbb{E}w_t^{2k} \leq C_2 \sigma^{2k}, \qquad k = 0, \dots, m.$$

Therefore, by independence of innovation and the Cauchy–Schwartz inequality we get

$$\mathbb{E}y_t^2 \leq C_2 \sum_{\mathbf{k},\mathbf{l}} c(\mathbf{k},\mathbf{l})|d^T B_{k_1}^{l_1} \cdots\cdots B_{k_r}^{l_r} c|^2 \cdot \sigma^2 \sum_{k_i \neq 0} l_i .$$

Thus we have to show that the r.h.s. of this inequality is finite and to prove that fact we have to estimate $c(\mathbf{k},\mathbf{l})$. A simple combinatorial argument shows that $(\mathbf{k},\mathbf{l}) \sim (\mathbf{k}',\mathbf{l}')$ iff $|\mathbf{l}| = |\mathbf{l}'|$, where $|\mathbf{l}| = \sum_i l_i$ and $c(\mathbf{k},\mathbf{l}) \leq m^{|\mathbf{l}|}$. Therefore it is enough to prove that the infinite series

$$\sum_{\mathbf{k},\mathbf{l}} m^{|\mathbf{l}|} |d^T B_{k_1}^{l_1} \cdots\cdots B_{k_r}^{l_r} c|^2 \cdot \sigma^2 \sum_{k_i \neq 0} l_i \tag{4.3}$$

converges. But the convergence of this series follows from the following lemma since the spectral radius of the matrix $m(A \otimes A + \sigma^2 \sum_j B_j \otimes B_j)$ is less than 1. $\blacksquare$

Lemma 4.3 *Let $\{M_i, i \in \mathbb{I}\}$ be a finite set of $n \times n$ matrices, $c,d \in \mathbb{R}^n$ and consider the following statements:*

(i)
$$r\left(\sum_{i \in \mathbb{I}} M_i \otimes M_i\right) < 1 ;$$

(ii)
$$\sum_{r=1}^{\infty} \sum_{l_1,\ldots,l_r \geq 1} \sum_{i_j \neq i_{j+1}} |d^T M_{i_1}^{l_1} \cdots\cdots M_{i_r}^{l_r} c|^2 < \infty;$$

(iii)
$$\sum_{|\mathbf{l}|=l} |d^T M_{i_1}^{l_1} \cdots\cdots M_{i_r}^{l_r} c|^2 \to 0 \qquad as\ l \to \infty.$$

Then (i) implies that (ii) and (iii) hold for all $c,d \in \mathbb{R}^n$. Conversely, if c and d are cyclic for $\{M_i, i \in \mathbb{I}\}$, then (ii) and (iii) implies (i), respectively.

We note that the proof of this lemma is based on the spectral radius formula and a simple linear algebraic manipulation. The geometric stability in mean square sense follows from an estimation of the form

$$\mathbb{E}|y_n - z_n(X_0)|^2 \leq 2C_2 \left(\sum_{|\mathbf{l}| \geq n} c(\mathbf{k},\mathbf{l})|d^T B_{k_1}^{l_1} \cdots\cdots B_{k_r}^{l_r} c|^2 \sigma^2 \sum_{k_i \neq 0} l_i \right.$$

$$\left. + \sum_{|\mathbf{l}|=n-1} c(\mathbf{k},\mathbf{l})|d^T B_{k_1}^{l_1} \cdots\cdots B_{k_r}^{l_r} c|^2 \sigma^2 \sum_{k_i \neq 0} l_i \cdot \mathbb{E}\|X_0\|^2 \right)$$

and the fact that there exists $\lambda > 1$ such that

$$\sum_{|\mathbf{l}|=n} c(\mathbf{k},\mathbf{l})|d^T B_{k_1}^{l_1} \cdot \ldots \cdot B_{k_r}^{l_r} c|^2 \sigma^{2\sum_{k_i \neq 0} l_i} \leq \frac{1}{\lambda^n}$$

for all enough large n by Lemma 4.3. Namely, let $1 < \lambda < r^{-1}(m(A \otimes A + \sigma^2 \sum_j B_j \otimes B_j))$.

Finally, since the asymptotical stability in mean square sense implies that $\mathbb{E}|z_n(\alpha c) - z_n(0)|^2 \to 0$, where $\alpha < \varepsilon/\|c\|$ we get

$$\sum_{|\mathbf{l}|=n} |d^T B_{k_1}^{l_1} \cdot \ldots \cdot B_{k_r}^{l_r} c|^2 \sigma^{2\sum_{k_i \neq 0} l_i} \to 0$$

as $n \to \infty$ and the Lemma 4.3 proves the second part of theorem.

For first order additive bilinear state–space representation (i.e., when $m = 1$) it can be seen that condition (4.1) is necessary and sufficient for second order asymptotically stable stationarity. Thus, we have the following

Theorem 4.4 *Let $\{w_t, t \in \mathbb{Z}\}$ be a sequence of i.i.d. random variables such that $\mathbb{E}w_t^4 < \infty$, $\mathbb{E}w_t = \mathbb{E}w_t^3 = 0$ and suppose that the first order additive bilinear state space representation (M) is controllable and observable. Then (M) has asymptotically stable solution in mean square sense if and only if*

$$r(A \otimes A + \sigma^2 B_1 \otimes B_1) < 1. \tag{4.4}$$

5. Closing remarks and open problems

First we note that the second part of Theorem 3.2 and 4.2 does not remain true if we drop the assumption of controllability or observability if $n > 1$. However, if $\mathcal{C} = \mathcal{O}$, then the converse part of our theorems remain true restricting the coefficient matrices of our model to the subspace $\mathcal{C} = \mathcal{O}$.

The upper Lyapunov exponent used to appear in several papers giving sufficient condition for the existence of strictly stationary solution of a stochastic difference equation, but usually it takes a simpler form. E.g. if $m = n = 1$ then $\gamma = \mathbb{E}\ln|a + b_1 w_t|$ (see Corollary 2.2 in [L1] and Theorem 1 in [Qu]). Another frequent sufficient condition for strict stationarity in the first order case is $\mathbb{E}\ln|A + B_1 w_t| < 0$ because this expectation is greater or equal than γ (see Theorem 1 in [Mo]). This condition is also well known in the theory of generalized autoregressive processes (see [Br], [BP] and [Ve]). Other sufficient conditions can be found in [L2], [LB] and [BSW], which can be derived from upper Lyapunov exponent and requires a knowledge of the maximum eigenvalue in modulus of an $n^2 \times n^2$ matrix. The Theorem 3.4 is useful from computational point of view because it is sufficient to calculate of the spectral radius of a $n \times n$ matrix. In the condition (4.1) for the existence of second order stationary solution the upper bound $1/m$ seems to be strong. Our conjecture is that the necessary condition is sufficient, too. To prove this a sharper estimation would be necessary for $c(\mathbf{k}, \mathbf{l})$. Moreover, the necessary condition in Theorem 3 appears on the area of multiple bilinear models (see [Te]), as well.

The infinite series (2.2) and (3.2) give two kind of representation for y_t. We can think of (2.2) as nonlinear Wold decomposition of y_t and of (3.2) as linear Wold decomposition of y_t with random coefficient. Since the response characteristic $d^T B_{k_1}^{l_1} \cdot \ldots \cdot B_{k_r}^{l_r} c$ of state–space stochastic system (2.1) is enough general we can hope that all nonlinear model depending on finite number of parameters can be described by the additive bilinear state–space representation.

Finally, two open problems are raised. In the Theorem 3.2 it is proved that if $\gamma < 0$, then there is strictly stationary solution and it can be seen that if $\gamma > 0$, then there is no strictly stationary solution, the model is exponentially explosive. The same is true for condition (4.4) from second order stationarity point of view. But what happens when $\gamma = 0$ or the spectral radius equals to 1? We know that there is no asymptotically stable solution, however, it may exist nonstable or chaotic solution. If some additional conditions are valid, then such kind of solution does not exist (e.g. when the innovation is Gaussian and there is an explicit representation of L^2–functionals by Wiener–Itô representation, or the model can be transformed into the generalized autoregressive model), but the answer to this question is unknown in general. The second problem is: if there is asymptotically stable strictly stationary solution, which is ergodic since the innovation is ergodic, is it geometrically ergodic? The main problem is that (2.1) is non–Markovian if $m > 1$, thus the standard Markovian framework (see Appendix in [To]) can not be applied. For Markovian bilinear state–space representation (when $m = 1$) a condition for geometrical ergodicity can be found in [Mo].

References

[BSW] M. Bhaskara Rao, T. Subba Rao and A. M. Walker, On the existence of some bilinear time series models, *J. of Time Series Analysis* **4** (1983), 95–110.

[BP]   Ph. Bougerol and D. Picard, Strict stationarity of generalized autoregressive processes, *The Ann. of Prob.* **20** (1992), 1714–1730.

[Br]   A. Brandt, The stochastic equation $Y_{n+1} = A_n Y_n + B_n$ with stationary coefficients, *Adv. in Appl. Prob.* **18** (1986), 211–220.

[Ca]   P. E. Caines, *Linear Stochastic Systems*, J. Wiley, New York, 1988.

[Fr]   A. E. Frazho, Schur contraction and stochastic bilinear systems, in:Proc. of 23rd Conference on Decision and Control, Las Vegas (1984), 781–786.

[FK]   H. Fürstenberg and H. Kesten, Products of random matrices, *Ann. Math. Stat.* **31** (1960), 457–469.

[GA]   C. W. J. Granger and A. P. Andersen, *An Introduction to Bilinear Time Series Models*, Vandenhoeck and Ruprecht, Gottingen, 1978.

[Ki]   J. F. C. Kingman, The ergodic theory of subadditive stochastic processes, *J. Roy. Statist Soc. Ser. B* **30** (1968), 499–510.

[L1]   J. Liu, A simple condition for the existence of some stationary bilinear time series, *J. of Time Series Analysis* **10** (1989), 33–39.

[L2]   J. Liu, A note on causality and invertibility of a general bilinear time series model, *Adv. in Applied Prob.* **22** (1990), 247–250.

[LB]   J. Liu and P. J. Brockwell, On the general bilinear time series model, *J. of Applied Prob.* **25** (1988), 553–564.

[LR]   R. G. Laha and V. K. Rohatgi, *Probability Theory*, J. Wiley, New York, 1979.

[Mo]   A. Mokkadem, Conditions suffisantes d'existence et d'ergodicité géométrique des modelès bilinéaires, *C.R.A.S. Paris* **301** (1985), 375–377.

[Qu]   B. G. Quinn, Stationarity and invertibility of simple bilinear models, *Stochastic Processes and their Appl.* **12** (1982), 225–230.

[Ru]   W. J. Rugh, *Nonlinear System Theory. The Volterra–Wiener Approach*, The John Hopkins University Press, Baltimore, 1981.

[SG]   T. Subba Rao and M. M. Gabr, *An introduction to bispectral analysis and bilinear time series models*, Lect. Notes in Stat. *24*, Springer–Verlag, New York, 1984.

[Te]   Gy. Terdik, Second order properties for multiple–bilinear models, *J. of Multivariate Analysis* **35** (1990), 295–307.

[To]   H. Tong, *Non–linear Time Series*, Oxford University Press, Oxford, 1990.

[Ve]   W. Vervaat, On a stochastic difference equation and a representative of non–negative infinitely divisible random variables, Adv. in Appl. Prob. **11** (1979), 750–783.

[WST] E. J. Wegman, S. C. Schwartz and J. B. Thomas (eds.), *Topics in Non–Gaussian Signal Processing*, Spinger–Verlag, New York, 1990.

Department of Probability Theory
Lajos Kossuth University of Debrecen
H–4010, Debrecen, Pf. 58.
Hungary

Received September 30,1996

On Convergence of Approximations
of Ito–Volterra Equations

Alexander Kolodii

1. Introduction

Let $(\Omega, \mathfrak{F}, (\mathfrak{F}_s)_{s \geqslant 0}, P)$ be a filtered probability space satisfying the usual assumptions; $\mathbb{C}$ be the space of continuous functions $x : R_+ \mapsto R^d$ with the topology of uniform convergence on compact subsets of R_+; $\mathfrak{C}_t$ be the minimal σ-algebra containing a cylindrical subsets of $\mathbb{C}$ with coordinates in $[0, t]$, $t \geqslant 0$.

Consider the stochastic integral equation

$$\xi(t) = h(t) + \int_0^t a(t, s, \xi) d\alpha(s) + \int_0^t b(t, s, \xi) d\mu(s) , \quad t \geqslant 0 ; \qquad (1.1)$$

where $(h(s))_{s \geqslant 0}$ is an $(\mathfrak{F}_s)_{s \geqslant 0}$-adapted process with sample path in $\mathbb{C}$; $(\alpha(s))_{s \geqslant 0}$ is an R^{m_1}-valued $(\mathfrak{F}_s)_{s \geqslant 0}$-adapted locally bounded variation continuous process; $(\mu(s))_{s \geqslant 0}$ is an R^{m_2}-valued continuous local $(\mathfrak{F}_s)_{s \geqslant 0}$-martingale; a $d \times m_1$-matrix $a(t, s, \omega, x)$ and a $d \times m_2$-matrix $b(t, s, \omega, x)$ are assumed to be defined when $t \in R_+$, $s \in R_+$, $\omega \in \Omega$, $x \in \mathbb{C}$; $a(\cdot, \cdot, \cdot, \cdot)$ and $b(\cdot, \cdot, \cdot, \cdot)$ as functions on $R_+ \times [0, v] \times \Omega \times \mathbb{C}$ are $\mathfrak{B}(R_+) \widehat{\otimes} \mathfrak{B}([0, v]) \widehat{\otimes} \mathfrak{F}_v \widehat{\otimes} \mathfrak{C}_v$-measurable for every $v > 0$.

The problem of existence and uniqueness of solutions of stochastic Ito–Volterra equations has been studied by many authors for various types of equations and under various conditions (see, for instance, [KV], [Kld], [Pr], [Tu]). Our aim in this paper is to prove the convergence in law of discrete approximation to the weak solution of equation (1.1). A similar problem for stochastic differential equations has been studied among others in [GS].

Below we use the following notation. For $x \in \mathbb{C}$,

$$\|x\|_t = \sup_{0 \leqslant s \leqslant t} |x(s)| , \quad \Delta(t, x, u) = \sup_{0 \leqslant s \leqslant v \leqslant s+u \leqslant t} |x(v) - x(s)| , \quad t \geqslant 0, u \geqslant 0.$$

$\mathbb{T}_k$ is the set of stopping times $\tau \leqslant k$, $k \in \mathbb{N}$. Φ is the set of pairs (φ, p) such that p is a number, $p > 2$, $(\varphi(t))_{t \geqslant 0}$ is an increasing continuous function, $\int_0^1 \varphi(t)t^{-1-1/p}dt < \infty$. $\overline{\mu}$ is the trace of characteristic of μ. $\overline{\alpha}(s) = \sum_{j=1}^{m_1} \int_0^s |d\alpha_j|$. $C(z)$ denotes a nonnegative constant depending only on a parameter z.

2. Continuous modifications and moments of stochastic integrals

Consider the stochastic integrals

$$\zeta(t) = \int_0^t \beta(t,s)d\mu(s) , \quad \zeta'(t) = \int_0^t [\beta(t,s) - \beta(s,s)]d\mu(s) , \quad t \geqslant 0 ,$$

where a $d \times m_2$-matrix $\beta(t,s,\omega)$ is assumed to be defined when $t, s \in R_+$, $\omega \in \Omega$; $\beta(\cdot,\cdot,\cdot)$ as a function on $R_+ \times [0,v] \times \Omega$ is $\mathcal{B}(R_+)\widehat{\otimes}\mathcal{B}([0,v])\widehat{\otimes}\mathfrak{F}_v$-measurable for every $v > 0$.

Lemma 2.1 *Assume that there exist $(\varphi, p) \in \Phi$ and a nonnegative $(\mathfrak{F}_s)_{s \geqslant 0}$-adapted measurable processes $(B_k(s))_{s \geqslant 0}$, $k \in \mathbb{N}$, such that under $t \vee t' \leqslant k$,*

$$\overline{\zeta}_k(t) \triangleq \int_0^t B_k^2(s)d\overline{\mu}(s) < \infty \qquad a.s. ;$$

$$|\beta(t,s)| \leqslant B_k(s) ; \quad |\beta(t',s) - \beta(t,s)| \leqslant B_k(s)\varphi(|t' - t|) .$$

Then the process ζ has a continuous modification. Moreover if $\tau \in \mathbb{T}_k$, $E(\overline{\zeta}_k(\tau))^{p/2} < \infty$ then

$$E\|\zeta\|_\tau^p \leqslant C(k,p,\varphi)E(\overline{\zeta}_k(\tau))^{p/2} ; \tag{2.1}$$

$$E\Delta^p(\tau,\zeta',u) \leqslant C(k,p,\varphi,u)E(\overline{\zeta}_k(\tau))^{p/2} ; \tag{2.2}$$

where $C(k,p,\varphi,u) \to 0$ as $u \to 0$.

Remark 2.2 *If $\beta(t,s) = \beta(s)$ does not depend on t then in Lemma 2.1 one may suppose $B_k(s) = |\beta(s)|$, $\varphi(s) = 0$. In this case the inequality (2.1) coincides with Burkholder–Gundy inequality for stochastic integrals:*

$$E\left\| \int_0^\cdot \beta(s)d\mu(s) \right\|_\tau^p \leqslant C(p)E\left(\int_0^\tau |\beta(s)|^2 d\overline{\mu}(s) \right)^{p/2} . \tag{2.3}$$

To prove Lemma 2.1 we shall use the following statement which follows from the results of paper [Ibr].

Lemma 2.3 *Let* $(\eta(t))_{t \geqslant 0}$ *is an* R^d*-valued process,*

$$w_{t,p}(u, \eta) = \sup_{0 \leqslant s \leqslant v \leqslant s+u \leqslant t} (E|\eta(v) - \eta(s)|^p)^{1/p}, \qquad u \geqslant 0, \ t \geqslant 0, \ p > 1.$$

If $\int_0^1 w_{t,p}(u, \eta) u^{-1-1/p} du \ < \ \infty$ *then* η *has a continuous modification. Moreover*

$$(E\|\eta\|_t^p)^{1/p} \ \leqslant \ C(p,t) \left(\int_0^1 w_{t,p}(u, \eta) u^{-1-1/p} du \ + \ \sup_{s \leqslant t}(E|\eta(s)|^p)^{1/p} \right);$$

$$\left(E\Delta^p(t, \eta, u) \right)^{1/p} \ \leqslant \ C(p,t) \left(w_{t,p}(u, \eta) u^{-1/p} \ + \ \int_0^u w_{t,p}(u, \eta) u^{-1-1/p} du \right).$$

Proof of Lemma 2.1. (a) Assume that $E\left(\overline{\zeta}_k(k) \right)^{p/2} < \infty$ for some k. Using the inequality (2.3), under $t \vee t' \leqslant k$, we have

$$E|\zeta'(t)|^p \leqslant 2^p C(p) E\left(\overline{\zeta}_k(k) \right)^{p/2};$$

$$E|\zeta'(t') - \zeta'(t)|^p \leqslant 2^p C(p) \varphi^p(|t' - t|) E\left(\overline{\zeta}_k(k) \right)^{p/2}.$$

From Lemma 2.3 it follows that $(\zeta'(s))_{s \in [0,k]}$ has a continuous modification, inequality (2.2) holds for $\tau = k$, and

$$E\|\zeta'\|_k^p \leqslant C'(k, p, \varphi) E\left(\overline{\zeta}_k(k) \right)^{p/2}.$$

Hence the inequality (2.1) holds for $\tau = k$. Since, for $\tau \in \mathbb{T}_k$,

$$\|\zeta\|_\tau \leqslant \left\| \int_0^{\cdot \wedge \tau} \beta(\cdot, s) d\mu(s) \right\|_k,$$

$$\Delta(\tau, \zeta', u) \leqslant \Delta\left(k, \int_0^{\cdot \wedge \tau} [\beta(\cdot, s) - \beta(s, s)] d\mu(s), u \right),$$

it follows that the inequalities (2.1) and (2.2) hold.

(b) In the general case we consider the stopping times

$$\tau_{k,n} = \inf\{t : \overline{\zeta}_k(t) > n\} \wedge k, \qquad k, n \in \mathbb{N}.$$

Let $J_{k,n}(\cdot)$ denotes the indicator of interval $[0, \tau_{k,n}]$. It then follows from the first part of proof that

$$\zeta_{k,n}(t) \triangleq \int_0^t J_{k,n}(s)\beta(t,s)d\mu(s) , \qquad t \in [0,k] ,$$

has a continuous modification and $\|J_{k,n}(\zeta_{k,n} - \zeta_{k,n+1})\|_k = 0$. Hence, there exists a continuous processes $(\zeta_k(t))_{t \in [0,k]}$ such that $\|J_{k,n}(\zeta_{k,n} - \zeta_k)\|_k = 0$. Since

$$J_{k,n}(t)\zeta_k(t) = J_{k,n}(t) \int_0^t \beta(t,s)d\mu(s) , \qquad t \in [0,k] ,$$

it follows that $\zeta_k(t) = \zeta(t)$, $t \in [0,k]$. Thus there exists a continuous modification of ζ.

3. Discrete approximations of Ito–Volterra equations

Let $\mathcal{D}$ denote the set of partitions $\delta = \{s_0 = 0 < s_1 < s_2 < \dots , \lim_j s_j = \infty\}$; $|\delta| = \sup_j(s_j - s_{j-1})$. For any $\delta \in \mathcal{D}$ we define the approximation of equation (1.1) as follows:

$$\xi_\delta(t) = h(t) + \int_0^t a_\delta(t,s,\xi_\delta)d\alpha(s) + \int_0^t b_\delta(t,s,\xi_\delta)d\mu(s) , \qquad t \geqslant 0 ;$$

where $a_\delta(t,s,x) = a(t, \lambda_\delta(s), \nu_\delta(x))$; $b_\delta(t,s,x) = b(t, \lambda_\delta(s), \nu_\delta(x))$;
$\lambda_\delta(s) = s_{j-1}$ for $s \in [s_{j-1}, s_j[, j \in \mathbb{N}$; $\nu_\delta(x, u) = x(0)$ for $u \in [0, s_1[$;

$$\nu_\delta(x,u) = \frac{s_{j+1} - u}{s_{j+1} - s_j}x(s_{j-1}) + \frac{u - s_j}{s_{j+1} - s_j}x(s_j) \quad \text{for} \quad u \in [s_j, s_{j+1}[, \ j \in \mathbb{N} .$$

Note that $\xi_\delta(s_j)$ depends only on

$$\{\eta(0), \eta(s_i), \xi_\delta(s_{i-1}), [\alpha(s_i) - \alpha(s_{i-1})], [\mu(s_i) - \mu(s_{i-1})]; \ i = \overline{1,j}\} .$$

We fix any $(\varphi, p) \in \Phi$ and consider the following conditions.
(H1) For every fixed (s,x), $(a(t,s,x))_{t\geqslant 0}$ is the continuous function. There exists a constant $G \geqslant 0$ such that

$$|a(t,s,x)| + |b(t,s,x)| \leqslant G(1 + \|x\|_s) ;$$

$$|b(t',s,x) - b(t,s,x)| \leqslant G\varphi(|t' - t|)(1 + \|x\|_s) .$$

(H2) There exist an increasing nonnegative functions $(\psi(s))_{s \geqslant 0}$ and $(L(t))_{t \geqslant 0}$ such that $\lim\limits_{s \to 0} \psi(s) = \psi(0) = 0$,

$$|a(t, s, x) - a(t, u, x)| + |b(t, s, x) - b(t, u, x)| \leqslant \psi(|s - u|)L(\|x\|_s) \; ;$$

$$|b(t', s, x) - b(t', u, x) - b(t, s, x) + b(t, u, x)| \leqslant \varphi(|t' - t|)\psi(|s - u|)L(\|x\|_s) \; .$$

(H3) For $n \in \mathbb{N}$, there exists an increasing nonnegative function $(\varrho_n(v))_{v \geqslant 0}$ such that $\lim\limits_{v \to 0} \varrho_n(v) = \varrho_n(0) = 0$ and under $\|x\|_s \vee \|x'\|_s \leqslant n$,

$$|a(t, s, x) - a(t, s, x')| + |b(t, s, x) - b(t, s, x')| \leqslant \varrho_n(\|x - x'\|_s) \; ;$$

$$|b(t', s, x) - b(t', s, x') - b(t, s, x) + b(t, s, x')| \leqslant \varphi(|t' - t|)\varrho_n(\|x - x'\|_s) \; .$$

Remark 3.1 *Under condition* **(H1)**, $(\xi_\delta(s))_{s \geqslant 0}$ *is the continuous* $(\mathfrak{F}_s)_{s \geqslant 0}$-*adapted process.*

Remark 3.2 *Under conditions* **(H1)** *and* **(H3)**, *there exists a weak solution* ξ *of the equation (1.1) (see [Kld]).*

Theorem 3.3 *Under condition* **(H1)**, *the family of processes*

$$\left(\xi_\delta(\cdot), \; \eta_\delta(\cdot) \triangleq \int_0^{\cdot} a_\delta(\cdot, s, \xi_\delta)d\alpha(s), \; \zeta_\delta(\cdot) \triangleq \int_0^{\cdot} b_\delta(\cdot, s, \xi_\delta)d\mu(s) \right)_{\delta \in \mathcal{D}}$$

is tight in $\mathbb{C} \times \mathbb{C} \times \mathbb{C}$.

Proof. Define the stopping times $\tau_\ell = \inf\{t : \|h\|_t \vee \overline{\alpha}(t) \vee \overline{\mu}(t) > \ell\}$, $\ell \in \mathbb{N}$. Let $J_\ell(\cdot)$ is the indicator of interval $[0, \tau_\ell]$. Note that, for $\tau \in \mathbb{T}_k$,

$$E\|J_\ell\xi_\delta\|_\tau^p \leqslant E \sup_{t \leqslant \tau} \left\{ \ell + \left| \int_0^t J_\ell(s)a_\delta(t, s, \xi_\delta)d\alpha(s) \right| + \right.$$

$$\left. + \left| \int_0^t J_\ell(s)b_\delta(t, s, \xi_\delta)d\mu(s) \right| \right\}^p \; .$$

Using the inequality (2.1) and Hölder's inequality, we obtain

$$E\|J_\ell\xi_\delta\|_\tau^p \leqslant C' + C''E \int_0^\tau J_\ell(s)\|J_\ell\xi_\delta\|_s^p d\big(\overline{\alpha}(s) + \overline{\mu}(s)\big)$$

for every $\tau \in \mathbb{T}_k$, where constants C' and C'' do not depend on δ. It then follows by the stochastic Gronwall's lemma [Mln] that

$$\sup_\delta E\|J_\ell \xi_\delta\|_k^p < \infty .$$

Hence, $\sup_\delta E\|J_\ell \eta_\delta\|_k^p < \infty$, $\sup_\delta E\|J_\ell \zeta_\delta\|_k^p < \infty$. Therefore

$$\lim_{L \to \infty} \sup_\delta P\{\|\xi_\delta\|_k \vee \|\eta_\delta\|_k \vee \|\zeta_\delta\|_k > L\} = 0 \quad \forall k \in \mathbb{N} .$$

Define the stopping times $\sigma_{n,\delta} = \inf\{t : \|\xi_\delta\|_t > n\}$. Let $J_{\ell,n,\delta}$ the indicator of $[0, \sigma_{n,\delta} \wedge \tau_\ell]$. Consider the process

$$\zeta_\delta'(t) = \int_0^t b_\delta(s, s, \xi_\delta) d\mu(s) , \quad t \geqslant 0 .$$

Using the inequality (2.2), we obtain

$$E\Delta^p(\sigma_{n,\delta} \wedge \tau_\ell \wedge k , \zeta_\delta - \zeta_\delta' , u) \leqslant$$

$$\leqslant E\Delta^p\left(k , \int_0^{\cdot} J_{\ell,n,\delta}(s)[b_\delta(\cdot, s, \xi_\delta) - b_\delta(s, s, \xi_\delta)] d\mu(s) , u \right) \leqslant$$

$$\leqslant C(k, u) G^p (1 + n)^p \ell^{p/2} .$$

Hence,

$$\lim_{u \to 0} \sup_\delta E\Delta^p(\sigma_{n,\delta} \wedge \tau_\ell \wedge k , \zeta_\delta - \zeta_\delta' , u) = 0 .$$

Using the standard methods (see [GS]), we have

$$\lim_{u \to 0} \sup_\delta E\Delta^p(\sigma_{n,\delta} \wedge \tau_\ell \wedge k , \zeta_\delta' , u) = 0 .$$

Since $\lim_{n \to \infty} \sup_\delta P\{\sigma_{n,\delta} < k\} = 0$ and

$$P\{\Delta(k, \zeta_\delta, u) > \varepsilon\} \leqslant P\{\sigma_{n,\delta} \wedge \tau_\ell < k\} + P\{\Delta(\sigma_{n,\delta} \wedge \tau_\ell \wedge k , \zeta , u) > \varepsilon\} ,$$

it follows that

$$\lim_{u \to 0} \sup_\delta P\{\Delta(k, \zeta, u) > \varepsilon\} = 0 \quad \forall \varepsilon > 0, k \in \mathbb{N} .$$

Hence, $(\zeta_\delta)_{\delta \in \mathcal{D}}$ is tight. The tightness of $(\eta_\delta)_{\delta \in \mathcal{D}}$ may be proved similarly.

Theorem 3.4 *Let the conditions* **(H1)**, **(H2)**, **(H3)** *hold, and the equation (1.1) have a unique weak solution* ξ. *Then* ξ_δ *converges in law to* ξ *as* $|\delta| \to 0$.

Proof. Consider any sequence of partitions $(\delta_m)_{m \in \mathbb{N}}$ such that $\lim_{m \to \infty} |\delta_m| = 0$. Since $(\xi_{\delta_m}, h, \alpha, \mu, \eta_{\delta_m}, \zeta_{\delta_m})_{m \in \mathbb{N}}$ is tight, it follows that there exists a subsequence $(\xi_{\delta_{m(j)}}, h, \alpha, \mu, \eta_{\delta_{m(j)}}, \zeta_{\delta_{m(j)}})_{j \in \mathbb{N}}$ converging in law. We will use the convention that $m(j) = m$. Using the Skorohod's theorem (about the weak convergence of laws in the Polish space) and standard arguments (see, for instance, [GS]), we may assert that there exist a probability space $(\Omega^*, \mathfrak{F}^*, (\mathfrak{F}_s^*)_{s \geqslant 0}, P^*)$ and $(\mathfrak{F}_s^*)_{s \geqslant 0}$-adapted continuous processes $\xi_m^*, h^*, \alpha^*, \mu^*, \eta_m^*, \zeta_m^*, \xi^*, \eta^*, \zeta^*$ such that

a) the laws of $\xi_m^*, h^*, \alpha^*, \mu^*, \eta_m^*, \zeta_m^*$ coincide with the laws of $\xi_{\delta_m}, h, \alpha, \mu, \eta_{\delta_m}, \zeta_{\delta_m}$;

b) $(\alpha^*(s))_{s \geqslant 0}$ is the locally bounded variation continuous process; $(\mu^*(s))_{s \geqslant 0}$ is the continuous local $(\mathfrak{F}_s^*)_{s \geqslant 0}$-martingale;

c) $\xi_m^*, \eta_m^*, \zeta_m^*$ converge P^*–a.s. to ξ^*, η^*, ζ^* in the topology of space $\mathbb{C}$.

Now the proof of Theorem 3.4 follows from the equalities

$$\eta^*(t) = \int_0^t a(t, s, \xi^*) d\alpha^*(s), \quad \zeta^*(t) = \int_0^t b(t, s, \xi^*) d\mu^*(s), \quad P^*\text{-a.s.}, \ t \geqslant 0.$$

We will prove the second equality. The first equality may be proved similarly. It suffices to show that for any $\varepsilon > 0$, $k \in \mathbb{N}$,

$$\lim_{m \to \infty} P\left\{ \left\| \int_0^\cdot b_{\delta_m}(\cdot, s, \xi_{\delta_m}) d\mu(s) - \int_0^\cdot b(\cdot, s, \xi_{\delta_m}) d\mu(s) \right\|_k > \varepsilon \right\} = 0; \quad (3.1)$$

$$\lim_{m \to \infty} P^*\left\{ \left\| \int_0^\cdot b(\cdot, s, \xi_m^*) d\mu^*(s) - \int_0^\cdot b(\cdot, s, \xi^*) d\mu^*(s) \right\|_k > \varepsilon \right\} = 0. \quad (3.2)$$

Define $\beta(t, s) = J_{\ell,n,\delta_m}(s)[b_{\delta_m}(t, s, \xi_{\delta_m}) - b(t, s, \xi_{\delta_m})]$ and note that

$$|\beta(t, s)| \leqslant B_{\ell,n,m}(s), \quad |\beta(t', s) - \beta(t, s)| \leqslant B_{\ell,n,m}(s)\varphi(|t' - t|);$$

where $B_{l,n,m}(s) = J_{\ell,n,\delta_m}(s)\big[\varrho_n\big(\Delta(s, \xi_{\delta_m}, 2|\delta_m|)\big) + \psi(|\delta_m|)L(n)\big]$. By the inequality (2.1) and Lebesgue dominated convergence theorem

$$E\left\| \int_0^\cdot J_{\ell,n,\delta_m}(s)[b_{\delta_m}(\cdot, s, \xi_{\delta_m}) - b(\cdot, s, \xi_{\delta_m})] d\mu(s) \right\|_k^p \leqslant$$

$$\leqslant C(k,p,\varphi)E\left(\int\limits_0^k B_{\ell,n,\delta_m}^2(s)d\overline{\mu}(s)\right)^{p/2} \to 0 \qquad \text{as} \quad m \to \infty \,.$$

Hence, (3.1) holds.

Define $\tau_\ell^* = \inf\{t : \overline{\mu}^*(t) > \ell\}$, $\sigma_{n,m}^* = \inf\{t : \|\xi_m^*\|_t \vee \|\xi^*\|_t > n\}$. Let $J_{\ell,n,m}^*(\cdot)$ be the indicator of interval $[0,\tau_\ell^* \wedge \sigma_{n,m}^*]$. Note that

$$\lim_{\ell,n\to\infty} \sup_m P^*\{\tau_\ell^* \wedge \sigma_{n,m}^* < k\} = 0 \qquad \forall k \,;$$

and

$$E^*\left\|\int\limits_0^{\cdot} J_{\ell,n,m}^*(s)[b(\cdot,s,\xi_m^*) - b(\cdot,s,\xi^*)]d\mu^*(s)\right\|_k^p \leqslant$$

$$\leqslant C(k,p,\varphi)E^*\left(\int\limits_0^k J_{\ell,n,m}^*(s)\varrho_n^2(\|\xi_m^*-\xi^*\|_s)d\overline{\mu}(s)\right)^{p/2} \to 0 \qquad \text{as} \quad m \to \infty \,.$$

Therefore (3.2) holds.

Remark 3.5 *If the conditions* **(H1)**, **(H2)**, **(H3)** *hold with* $\varrho_n(v) = vC(n)$ *then the equation (1.1) has the unique strong solution* ξ *(see [Kld]), and*

$$\lim_{|\delta|\to 0} P\{\|\xi_\delta - \xi\|_k > \varepsilon\} = 0 \quad \forall \varepsilon > 0, k \in \mathbb{N} \,.$$

References

[GS]   I.I. Gihman, A.V. Skorohod, *Stochastic Differential Equations and Their Applications*, Naukova Dumka, Kiev, 1982.

[Ibr]   I.A. Ibragimov, On conditions for smoothing the trajectories of random functions, *Theory Probab. Appl.* **28** (1983), 229–250.

[KV]   M.L. Kleptsyna, A.Ju. Veretennikov, On the strong solutions of Ito–Volterra stochastic equations, *Theory Probab. Appl.* **29** (1984), 154–158.

[Kld]   A.M. Kolodii, On conditions for existence of solutions of integral equations with stochastic line integrals, *Probability Theory and Mathematical Statistics, Proc. 6th Vilnius Conference, 1993*, ed. B. Grigelionis et al. (1994), 405–422.

[Mln]   A.V. Mel'nikov, Gronwall's lemma and stochastic equations driven by components of semimartingales, *Matemat. Zametki* **32** (1982), 411–423.

[Pr]   P.P. Protter, Volterra equations driven by semimartingales, *Annals of Probability* **13** (1985), 519 – 530.

[Tu]   C. Tudor, On weak solutions of Volterra stochastic equations, *Bollettino Unione Mathematica Italiana* **B1** (1987), 1033–1054.

Volgograd State University
400062, Volgograd, Russia

Received September 27, 1996

Non-isotropic Ornstein-Uhlenbeck Process
and White Noise Analysis

Izumi Kubo

1. Introduction

White noise analysis is usually set up on a big Gel'fand triple based on a basic Gel'fand triple. The author introduced various constructions of the triple [KK2][Ku2]. A recent one is the construction by using weighted number operators in [Ku3][Ku4]. The first purpose of this note is to create the big space by help of *a non-isotropic Ornstein-Uhlenbeck process* whose generator is a weighted number operator. The process is called by I. Shigekawa [Sh] simply an Ornstein-Uhlenbeck process.

After this, we will see the relationships between the space $(\mathcal{E})$ of test functionals in Hida calculus and the Sobolev space $W^{\infty,\infty-}$ in Malliavin calculus. Moreover we will clarify finite dimensional test functionals in both calculi.

In Section 2, we will prepare basic spaces and a weighted number operator with weight A. In Section 3, we will introduce a non-isotropic Ornstein-Uhlenbeck Process with parameter A. In Section 4, we will construct a Gel'fand triple by using the semigroup of the process. In Section 5, we will compare two spaces $(\mathcal{E})$ and $W^{\infty,\infty-}(\mathbb{R})$ of test functionals. In Section 6, finite dimensional test functionals in both calculi will be characterized.

2. Basic spaces and weighted number operators

Suppose that E_0 is a separable real Hilbert space with an inner product $(\cdot,\cdot)_0$ and A is a positive self-adjoint operator such that $A^{-\kappa}$ is a Hilbert-Schmidt operator for some $\kappa > 0$ and that $\|A^{-1}\|_0 < 1$. For $p \geq 0$, let E_p be the domain of A^p and let associate an inner product $(\xi,\eta)_p \equiv (A^p\xi, A^p\eta)_0$ for $\xi, \eta \in E_p$. For $p < 0$, an inner product $(\xi,\eta)_p \equiv (A^p\xi, A^p\eta)_0$ can be introduced for $\xi, \eta \in E_0$. Let E_p be the completion of E_0 with respect to the norm $\|\cdot\|_p \equiv (\cdot,\cdot)_p^{1/2}$. Then obviously, E_{-p} is the dual of E_p and $E_q \subset E_p$ for $q > p$. For any $q \geq p + \kappa$, the injection $\iota_{p,q} : E_q \mapsto E_p$ is of Hilbert-Schmidt type with $\|\iota_{p,q}\|_{H.S.} = \|A^{p-q}\|_{H.S.}$.

Let $\mathcal{E}$ be the projective limit of E_p and $\mathcal{E}^*$ be the inductive limit. Then we have a Gel'fand triple

$$\mathcal{E} \subset E_0 \subset \mathcal{E}^* \quad \text{with} \quad \mathcal{E} \equiv \lim_{p \to \infty} E_p, \ \mathcal{E}^* \equiv \lim_{-\infty \leftarrow p} E_p.$$

Let μ on $\mathcal{E}^*$ be the measure of Gaussian white noise characterized by its characteristic functional

$$\int_{\mathcal{E}^*} \exp[i\langle x, \xi \rangle] d\mu(x) = \exp[-\tfrac{1}{2}\|\xi\|_0^2] \quad \text{for any} \ \xi \in \mathcal{E}. \tag{2.1}$$

Denote by (E_0) the complex L^2-space $L^2(\mathcal{E}^*, \mu)$.

Let $\mathcal{P}(\mathcal{E})$ be the set of all polynomials in $\{\langle x, \eta \rangle; \eta \in \mathcal{E}\}$ over $\mathbb{C}$. The Gâteaux derivative is defined by

$$D_z \varphi(x) \equiv \lim_{h \to 0} \frac{1}{h} \big(\varphi(x + hz) - \varphi(x) \big) \quad \text{for} \ z \in \mathcal{E}, \ \varphi \in \mathcal{P}(\mathcal{E}).$$

For $\varphi(x) = f(\langle x, \eta_1 \rangle, \cdots, \langle x, \eta_m \rangle) \in \mathcal{P}(\mathcal{E})$, the *weighted number operator with weight A* is defined by

$$\mathcal{N}_A \varphi(x) \ \equiv \ -\sum_{i,j=1}^{m} (\eta_i, \log A\eta_j)_0 \frac{\partial^2}{\partial u_i \partial u_j} f(\langle x, \eta_1 \rangle, \cdots, \langle x, \eta_m \rangle)$$

$$\tag{2.2}$$

$$+ \sum_{i=1}^{m} \langle x, \log A\eta_i \rangle \frac{\partial}{\partial u_i} f(\langle x, \eta_1 \rangle, \cdots, \langle x, \eta_m \rangle).$$

It is easily seen that for a c.o.n.s. $\{\eta_k\}_{k=1}^\infty (\subset \mathcal{E})$ of E_0,

$$\mathcal{N}_A \varphi(x) = -\sum_{k=1}^{\infty} \Big(D_{\sqrt{\log A}\eta_k} D_{\sqrt{\log A}\eta_k} - \langle x, \sqrt{\log A}\eta_k \rangle D_{\sqrt{\log A}\eta_k} \Big) \varphi(x).$$

The equality guarantees that $\mathcal{N}_A$ does not depend on the choice of $\{\eta_k\}$ and on the representation of φ.

Since $A^{-\kappa}$ is of Hilbert-Schmidt type, A has a c.o.n.s. of eigenvectors $\{\zeta_k\}_{k=1}^\infty$; $A\zeta_k = \lambda_k \zeta_k$. Then $\zeta_k \in \mathcal{E}$ for any $k \geq 1$, and

$$\|A^{-\kappa}\|_{H.S.}^2 = \sum_{k=1}^{\infty} \lambda_k^{-2\kappa} < \infty$$

For the c.o.n.s., it holds that

$$\mathcal{N}_A \varphi(x) = -\sum_{k=1}^{\infty} \log \lambda_k \Big(D_{\zeta_k} D_{\zeta_k} - \langle x, \zeta_k \rangle D_{\zeta_k} \Big) \varphi(x).$$

We see that $\mathcal{N}_A$ is positive and symmetric:

$$(\mathcal{N}_A \varphi, \varphi)_{(E_0)} = \int_{\mathcal{E}^*} \Big\| \sum_{i=1}^{m} \frac{\partial}{\partial u_i} f(\langle x, \eta_1 \rangle, \cdots, \langle x, \eta_1 \rangle) \sqrt{\log A}\eta_i \Big\|_0^2 d\mu(x).$$

If the domain of a positive self adjoint operator A' includes $\mathcal{E}$, we can define $\mathcal{N}_{A'}$ similarly. Then we have a commutation relation;

$$[\mathcal{N}_A, \mathcal{N}_{A'}]\varphi = \sum_{i=1}^{m} \langle x, [\log A, \log A']\eta_i\rangle \frac{\partial}{\partial u_i} f(\langle x, \eta_1\rangle, \cdots, \langle x, \eta_1\rangle).$$

If $A' \geq A$, then

$$e^{\mathcal{N}_{A'}} \geq e^{\mathcal{N}_A}.$$

The usual number operator is $\mathcal{N} = \mathcal{N}_{eI}$.

3. Non-isotropic Ornstein-Uhlenbeck process

On a suitable probability space $(\Omega, \mathcal{B}, P)$ there exists a continuous Gaussian process $\{X(t, x, \omega); t \geq 0, x \in \mathcal{E}^*\}$ in $\mathcal{E}^*$ such that its mean and covariance functionals are

$$\mathbf{E}\Big[\langle X(t, x, \omega), \xi\rangle\Big] = \langle x, A^{-t}\xi\rangle,$$

$$\text{Cov}\Big[\langle X(t, x, \omega), \xi\rangle, \langle X(s, y, \omega), \eta\rangle\Big] = \langle (A^{-|t-s|} - A^{-t-s})\xi, \eta\rangle$$

and that

$$X(t, x, \omega) = X(t, 0, \omega) + A^{-t}x.$$

We call the process X a *non-isotropic Ornstein-Uhlenbeck process with parameter A*.

For a standard Brownian motion $B(t, \omega)$ on $\mathcal{E}^*$, the process is a solution of a linear stochastic differential equation

$$\begin{aligned} dX(t, x, \omega) &= -(\log A)X(t, x, \omega)dt + \sqrt{2\log A}dB(t, \omega), \\ X(0, x, \omega) &= x. \end{aligned} \tag{3.1}$$

and it can be given by the stochastic integral

$$X(t, x, \omega) = \sqrt{2\log A} \int_0^t A^{-t+u}dB(u, \omega) + A^{-t}x.$$

Notice that if $p \leq -\kappa$, then $\mu(E_p) = 1$ and hence $L^2(\mathcal{E}^*, \mu)$ is naturally equivalent to $L^2(E_p, \mu)$. The following proposition is proved in various ways.

Proposition 3.1 *The process $\{X(t, x), x \in \mathcal{E}^*\}$ is a diffusion on $\mathcal{E}^*$ with the invariant measure μ and its generator is $-\mathcal{N}_A$ on $L^2(E_p, \mu)$. More precisely, for $p \leq -\kappa$, the process $\{X(t, x), x \in E_p\}$ is a diffusion on E_p.*

Let $\{V_t; t \geq 0\}$ denote the semigroup of $\{X(t, x)\}$.

Proposition 3.2 (1) The semigroup satisfies

$$V_t = \exp[-t\mathcal{N}_A], \quad \|V_t\varphi\|_{(E_0)} \leq \|\varphi\|_{(E_0)},$$

$$\int_{\mathcal{E}''} V_t\varphi(x)d\mu(x) = 1 \quad for \quad t \geq 0.$$

(3.2)

(2) For $t \geq \kappa$, V_t is of Hilbert-Schmidt type on (E_0) with

$$\|V_t\|_{H.S.} = \prod_{k=1}^{\infty}(1 - \lambda_k^{-2t})^{-1/2}.$$

(3) For $t > s > 0$, $V_s \geq V_t \geq 0$.

Put $(E_p) \equiv V_p(E_0)$ for $p \geq 0$ and put $(\mathcal{E}) \equiv \cap_p(E_p)$. For suitable φ (e.g. $\varphi \in \mathcal{P}(\mathcal{E})$ or $\in (E_\kappa)$),

$$V_t\varphi(x) = \mathbf{E}\left[\varphi(X(t, x, \omega))\right]$$

$$= \int_{\mathcal{E}_*} \varphi(A^{-t}x + \sqrt{1 - A^{-2t}}y)d\mu(y).$$

(3.3)

Replacing A in (3.3) to e^A, Shigekawa [Sh] defined a semigroup T_t by the expression (3.3) on an abstract Wiener space and *an Ornstein-Uhlenbeck operator* $\mathcal{L}_A$ as its generator. In the other word,

$$\mathcal{N}_A = -\mathcal{L}_{\log A}.$$

For the proof, we introduce notation. Let $H_n(u)$ be the Hermite polynomial defined by the generating function

$$\sum_{n=0}^{\infty} \frac{t^n}{n!} H_n(u) = \exp[ut - \frac{1}{2}t^2].$$

(3.4)

Put

$$\mathbb{N}_0 \equiv \mathbb{N} \cup \{0\}, \quad \vec{\mathbb{N}}_0 \equiv \{\vec{n} = (n_1, n_2, \cdots, n_m, 0, 0, \cdots), n_j \in \mathbb{N}_0, \quad m \in \mathbb{N}\},$$

$$\vec{\lambda}^{t\vec{n}} \equiv \prod_{k=1}^{\infty} \lambda_k^{tn_k}, \quad \vec{n}! \equiv \prod_{k=1}^{\infty} n_k!, \quad H_{\vec{n}}(x) \equiv \prod_{k=1}^{\infty} H_{n_k}(\langle x, \zeta_k \rangle),$$

where $\{\zeta_k\}$ is the eigenvectors of A in Section 2. Then we see the following equalities

$$\left(H_{\vec{n}}(x), H_{\vec{m}}(x)\right)_{(E_0)} = \delta_{\vec{n},\vec{m}}\vec{n}!,$$

(3.5)

$$\mathcal{N}_A H_{\vec{n}}(x) = \sum_{k=1}^{\infty} n_k \log \lambda_k n_k H_{\vec{n}}(x) = \log \vec{\lambda}^{\vec{n}} H_{\vec{n}}(x), \qquad (3.6)$$

and

$$e^{t\mathcal{N}_A} H_{\vec{n}}(x) = \vec{\lambda}^{t\vec{n}} H_{\vec{n}}(x), \ t \in \mathbb{R}. \qquad (3.7)$$

Moreover, $\{H_{\vec{n}}(x); \ \vec{n} \in \vec{\mathbb{N}}_0\}$ is a complete orthogonal basis in (E_0).

For $\xi \in \mathcal{E}$, put

$$f(\xi; x) \equiv \exp[\langle x, \xi \rangle - \frac{1}{2}\|\xi\|_0^2].$$

Then we see

$$(f(\xi; x), f(\eta; x))_{(E_0)} = e^{(\xi, \eta)_0} \qquad (3.8)$$

by (2.1) and hence (3.5) holds by (3.4). Since $f(\xi; x) \notin \mathcal{P}(\mathcal{E})$, the following discussion must be carefully justified. But it is valid. By (2.2),

$$\mathcal{N}_A f(\eta; x) = (\eta, \log A\eta)_0 f(\eta; x) + \langle x, \log A\eta \rangle f(\eta; x) = \frac{d}{dt} f(A^t \eta; x)\big|_{t=0}.$$

Hence we have

$$e^{t\mathcal{N}_A} f(\eta; x) = f(A^t \eta; x).$$

This and (3.4) imply (3.6) and (3.7).

Proof of Proposition 3.2 By (3.1), Itô's formula shows that $-\mathcal{N}_A$ is the generator. Since the eigenvalues of $\mathcal{N}_A$ are positive by (3.6), positivities in (1) and (3) are clear. Since

$$(V_t H_{\vec{n}}(x), 1)_{(E_0)} = \vec{\lambda}^{-t\vec{n}}(V_t H_{\vec{n}}(x), 1)_{(E_0)} = 0$$

for $\vec{n} \neq 0$, the last equality in (3.2) is obvious. By (3.7), we have

$$\|V_t\|_{H.S.}^2 = \sum_{\vec{n} \in \vec{\mathbb{N}}_0} \vec{\lambda}^{-2t\vec{n}} = \prod_{k=1}^{\infty}(1 - \lambda_k^{-2t}) < \infty,$$

which shows (2). $\blacksquare$

4. Gel'fand triple

Let $\{V_t, t \geq 0\}$ be the semigroup of the non-isotropic Ornstein-Uhlenbeck process with parameter A, which controls the basic Gel'fand triple $\mathcal{E} \subset E_0 \subset \mathcal{E}^*$ as in the previous section.

For $p \geq 0$, put

$$\left(E_p\right) \equiv V_p\left(E_0\right). \tag{4.1}$$

For $\varphi = V_p\varphi', \psi = V_p\psi' \in \left(E_p\right)$ define their inner product by

$$\left(\varphi, \psi\right)_p \equiv \left(\varphi', \psi'\right)_{\left(E_0\right)}. \tag{4.2}$$

For $p < 0$, $\varphi, \psi \in \left(E_0\right)$ (or $\in \mathcal{P}(\mathcal{E})$), define their inner product by

$$\left(\varphi, \psi\right)_p \equiv \left(V_p\varphi, V_p\psi\right)_{\left(E_0\right)}. \tag{4.3}$$

Let $\left(E_p\right)$ be the completion of $\left(E_0\right)$ (or $\mathcal{P}(\mathcal{E})$, resp.) with respect to the inner product $\|\cdot\|_p$. Define norms $\|\varphi\|_p \equiv \left(\varphi, \varphi\right)_p^{1/2}$ for $\varphi \in \left(E_p\right)$ and for any $p \in \mathbb{R}$.

Theorem 4.1 *(1)* $\left(E_q\right) \subset \left(E_p\right)$, $q \geq p$, $\left(E_p\right)^{*} = \left(E_{-p}\right)$ *for $p \in \mathbb{R}$.*

(2) $\left(\mathcal{E}\right) \subset \left(E_0\right) \subset \left(\mathcal{E}^*\right)$ *is a Gel'fand triple, where*

$$\left(\mathcal{E}\right) \equiv \lim_{p \to \infty} \left(E_p\right) \quad and \quad \left(\mathcal{E}^*\right) \equiv \lim_{-\infty \leftarrow p} \left(E_p\right).$$

Proof. By Proposition 3.2 (3), The first assertion of (1) is obvious. The second equality is true by

$$\left(V_p\varphi, V_{-p}\psi\right)_{\left(E_0\right)} = \left(\varphi, \psi\right)_{\left(E_0\right)}$$

for $\varphi, \psi \in \mathcal{P}(\mathcal{E})$. Proposition 3.2 (2) implies (2). ∎

We will see that this construction is the usual one. Several equivalent constructions are known:

(1) Wiener-Itô decomposition.

(2) Fock spaces.

(3) Reproducing kernel Hilbert spaces.

(4) Second quantization.

(5) Lee's analytic functional space.

(6) Yokoi's inverse Gaussian transform.

(7) H.K.P.S.–theory (increasing order).

(8) K.K.–theory (compositions with generalized functions).

(9) Weighted number operators.

Here we refer one of the oldest methods using reproducing kernels [KT][IK]. The S-*transform* is defined by

$$(S\Phi)(\xi) \equiv \int_{\mathcal{E}^*} \Phi(x + \xi) d\mu(x) = \big(\Phi(x), f(\xi; x)\big)_0 \qquad (4.4)$$

for $\xi \in \mathcal{E}, \Phi \in (E_0)$. Then $\mathcal{K}_0 \equiv S(E_0)$ is the reproducing kernel complex Hilbert space with the reproducing kernel $K_0(\xi, \eta) \equiv \exp[(\xi, \eta)_0], \ \xi, \eta \in \mathcal{E}$. Let $\mathcal{K}_p$ be the reproducing kernel Hilbert space with the reproducing kernel

$$K_p(\xi, \eta) \equiv \exp\big[(\xi, \eta)_{-p}\big], \quad \xi, \eta \in \mathcal{E}, \ p \in \mathbb{R}.$$

The following is a Gel'fand triple

$$\mathcal{K} \subset \mathcal{K}_0 \subset \mathcal{K}^* \quad \text{with} \quad \mathcal{K} \equiv \lim_{p \to \infty} \mathcal{K}_p, \quad \mathcal{K}^* \equiv \lim_{-\infty \leftarrow p} \mathcal{K}_p.$$

Define a Gel'fand triple $(\mathcal{E}) \subset (E_0) \subset (\mathcal{E}^*)$ as the inverse image of $\mathcal{K} \subset \mathcal{K}_0 \subset \mathcal{K}^*$ by S.

Since the S-transform of $f(\eta; x)$ is

$$Sf(\eta; \cdot)(\xi) = \big(f(\eta; \cdot), f(\xi; \cdot)\big)_{(E_0)} = \exp[(\eta, \xi)_0] \qquad (4.5)$$

by (3.7), we have

$$
\begin{aligned}
(Sf(\eta, \cdot), Sf(\zeta, \cdot))_{\mathcal{K}_p} &= (\exp[(\eta, \cdot)_0], \exp[(\zeta, \cdot)_0])_{\mathcal{K}_p} \\
&= \exp[(\eta, \zeta)_{-p}] \\
&= \exp[(A^{-p}\eta, A^{-p}\zeta)_0] \\
&= \big(f(A^{-p}\eta, \cdot), f(A^{-p}\zeta, \cdot)\big)_{(E_0)} \\
&= \big(V_p f(\eta, \cdot), V_p f(\zeta, \cdot)\big)_{(E_0)} \\
&= \big(f(\eta, \cdot), f(\zeta, \cdot)\big)_{(E_p)}.
\end{aligned}
$$

by (4.2), (4.3) and (4.5). This means that the new (E_p) has the same inner product as the old one.

5. Sobolev space

 Inspired by a paper of J. Potthoff [Po] on Meyer's equivalence, I. Shigekawa discussed a Sobolev space controlled by A on an abstract Wiener space. Here we reformulate his space on the Gel'fand triple $\mathcal{E} \subset E_0 \subset \mathcal{E}^*$ controlled by A. Define an operator $d\Gamma(A)_n$ on $E_0^{\otimes n}$ by

$$d\Gamma(A)_n \eta_1 \otimes \cdots \otimes \eta_n \equiv \sum_{k=1}^{n} \eta_1 \otimes \cdots \otimes A\eta_k \otimes \cdots \otimes \eta_n.$$

The norm of $F \in L^p(\mu, \mathcal{E}^* \mapsto E_0^{\otimes n})$ is

$$\|F\|_{L^p} \equiv \left(\int_{\mathcal{E}^*} \|F(x)\|_{E_0^{\otimes n}}^p \, d\mu(x) \right)^{1/p}.$$

 Since $D_z\varphi(x)$ is linear in $z \in \mathcal{E}^*$, its restriction on E_0 defines a linear map $D : \mathcal{P}(\mathcal{E}) \longmapsto \mathcal{P}(\mathcal{E}) \otimes E_0$; that is,

$$D_\zeta \varphi(x) = (D\varphi(x), \zeta)_0 \quad \text{for} \ \zeta \in E_0.$$

D is called E_0-*derivative* D. Shigekawa introduced D_A by

$$D_A \equiv \sqrt{A}D, \quad i.e. \ (D_A\varphi(x), \zeta)_0 = D_{\sqrt{A}\zeta}\varphi(x)$$

For $\varphi(x) = f(\langle x, \eta_1 \rangle, \cdots, \langle x, \eta_m \rangle)$, we have

$$
\begin{aligned}
\|D_A\varphi\|_{E_0}^2 &= \sum_{k=1}^{\infty} |D_{\sqrt{A}\zeta_k}\varphi(x)|^2 \\
&= \sum_{k=1}^{\infty} \sum_{i,j=1}^{m} \langle \eta_i, \sqrt{A}\zeta_k \rangle \langle \eta_j, \sqrt{A}\zeta_k \rangle \frac{\partial}{\partial u_i} f \frac{\partial}{\partial u_j} f \\
&= \sum_{i,j=1}^{m} (\sqrt{A}\eta_i, \sqrt{A}\eta_j)_0 \frac{\partial}{\partial u_i} f \frac{\partial}{\partial u_j} f.
\end{aligned}
$$

Hence

$$
\begin{aligned}
\|D_A\varphi\|_{L^p}^p &= \int_{\mathcal{E}^*} \|D_A\varphi\|_{E_0}^p \, d\mu(x) \\
&= \int_{\mathcal{E}^*} \left(\sum_{i,j=1}^{m} \langle \sqrt{A}\eta_i, \sqrt{A}\eta_j \rangle \frac{\partial}{\partial u_i} f \frac{\partial}{\partial u_j} f \right)^{p/2} d\mu(x).
\end{aligned}
$$

 Define *an Ornstein-Uhlenbeck semigroup* T_t by

$$T_t\varphi(x) \equiv \int_{\mathcal{E}^*} f(e^{-tA}x + \sqrt{1 - e^{2tA}}y) d\mu(y),$$

and let $\mathcal{L}_A$ be the generator of T_t, which is called *an Ornstein-Uhlenbeck operator*. It is represented as

$$\mathcal{L}_A\varphi(x) = \sum_{i,j=1}^{m} (\eta_i, A\eta_j)_0 \frac{\partial^2}{\partial u_i \partial u_j} f - \sum_{i=1}^{m} \langle x, A\eta_i \rangle \frac{\partial}{\partial u_i} f$$

for $\varphi = f(\langle x, \eta_1 \rangle, \cdots, \langle x, \eta_m \rangle)$. He showed:

Theorem 5.1 *(Shigekawa [Sh], Theorem 3.1)*

$$\|D_A\varphi\|_{L^p} \lesssim \|\sqrt{1 - \mathcal{L}_A}\varphi\|_{L^p} \lesssim \|D_A\varphi\|_{L^p} + \|\varphi\|_{L^p}, \ \varphi \in \mathcal{P}(\mathcal{E}).$$

Theorem 5.2 *(Shigekawa [Sh], Theorem 3.6)*

$$\{\|(I - \mathcal{L}_A + d\Gamma(A)_n)^{k/2}\varphi\|_{L^p}; k = 0, 1, 2, \cdots\}$$

and

$$\{\|(1 + d\Gamma(A)_{n+m})^\ell D_A^m \varphi\|_{L^p}; \ell, m = 0, 1, 2, \cdots\}$$

are equivalent systems of seminorms.

Define norms by

$$\|\varphi\|_{s,p} \equiv \|(I - \mathcal{L}_A + d\Gamma(A^*)_n)^{s/2}\varphi\|_{L^p}.$$

Then Sobolev space $W^{s,p}(E_0^{\otimes n})$ is the completion of $\mathcal{P}(\mathcal{E}) \otimes E_0^{\otimes n}$ with respect to $\|\varphi\|_{s,p}$.

6. Spaces $(\mathcal{E})$ and $W^{\infty,\infty-}(\mathbb{R})$

For the scalar case, we compare those spaces. Actually, we are interested in $e^{t\mathcal{N}_A}$ and $(1 - \mathcal{L}_A)^s$. They have common eigenfunctions;

$$e^{t\mathcal{N}_A} H_{\vec{n}}(x) = \vec{\lambda}^{t\vec{n}} H_{\vec{n}}(x),$$

$$(1 - \mathcal{L}_A)^s H_{\vec{n}}(x) = \left(1 + \sum_{k=1}^{\infty} n_k \lambda_k\right)^s H_{\vec{n}}(x).$$

From these, the relationship between $(\mathcal{E})$ and $W^{2,\infty-}(\mathbb{R})$ is clear. For example, obviously

$$(\mathcal{E}) \subset W^{2,\infty-}(\mathbb{R}).$$

In [KK1], Kubo and Kuo discussed finite dimensional test functionals and generalized functionals. For fixed $\{\eta_i, 1 \le i \le m\}$, a generalized functional $\Phi \in (\mathcal{E}^*)$ is said to be *finite dimensional based on* $\{\eta_i, 1 \le i \le m\}$, if

Φ is a limit point of polynomials in $\{\langle x, \eta_i \rangle, 1 \le i \le m\}$ in the topology of $(\mathcal{E}^*)$. A finite dimensional test functional $\varphi \in (\mathcal{E})$ based on $\{\eta_i, 1 \le i \le m\}$ is similarly defined. By Theorem 3.5 of [KK1], we can see a characterization theorem:

Theorem 6.1 φ *is a test functional based on linearly independent* $\{\eta_i, 1 \le i \le m\}$, *if and only if there exits an entire function* $f(\vec{z})$ *in* $\vec{z} = (z_1, \cdots, z_m) \in \mathbb{C}^m$ *such that*

$$\varphi(x) = f(\langle x, \eta_1 \rangle, \cdots, \langle x, \eta_m \rangle)$$

and

$$\sup_{\vec{z} \in \mathbb{C}^m} e^{-a|\vec{z}|^2} |f(\vec{z})| < \infty$$

for any $a > 0$.

We wish to give a similar characterization theorem for $W^{\infty,\infty-}(\mathbb{R})$. For simplicity, we observe one dimensional $\varphi = f(\langle x, \eta \rangle) \in W^{\infty,\infty-}(\mathbb{R})$ based on $\eta \in \mathcal{E}$ with $\|\eta\|_0 = 1$. For $\varphi(x) = f(\langle x, \eta \rangle)$, we have

$$D_A^n \varphi(x) = f^{(n)}(\langle x, \eta \rangle) \otimes (\sqrt{A}\eta)^{\otimes n}.$$

Hence

$$(1+d\Gamma(A)_n)^\ell D_A^n \varphi(x) = f^{(n)}(x) \otimes \sum_{\ell_1 + \cdots + \ell_n \le \ell} \frac{\ell!}{\ell_1! \cdots \ell_n!} \sqrt{A} A^{\ell_1} \eta \otimes \cdots \otimes \sqrt{A} A^{\ell_n} \eta.$$

Put

$$\rho(u) \equiv (2\pi)^{-1/4} \exp[-\frac{1}{4}u^2].$$

Then we can evaluate norms as

$$\|\varphi\|_{L^p} = \left(\int_\mathbb{R} |f(u)|^p \rho(u)^2 du \right)^{1/p},$$

$$\|D_A^n \varphi\|_{L^p} = \|\sqrt{A}\eta\|_0^n \left(\int_\mathbb{R} |f^{(n)}(u)|^p \rho(u)^2 du \right)^{1/p},$$

and

$$\|(1 + d\Gamma(A)_n)^\ell D_A^n \varphi(x)\|_{L^p}$$

$$= \| \sum_{\ell_1 + \cdots + \ell_n \le \ell} \frac{\ell!}{\ell_1! \cdots \ell_n!} \sqrt{A} A^{\ell_1} \eta \otimes \cdots \otimes \sqrt{A} A^{\ell_n} \eta \|_{E_0^{\otimes n}}$$

$$\times \left(\int_\mathbb{R} |f^{(n)}(u)|^p \rho(u)^2 du \right)^{1/p}.$$

Therefore, $\varphi(x) = f(\langle x, \eta \rangle)$ belongs to $W^{\infty,\infty-}(\mathbb{R})$, if and only if

$$\int_{\mathbb{R}} |f^{(n)}(u)|^p \rho(u)^2 du < \infty \quad \text{for any } n \geq 0, \ p \geq 1. \tag{6.1}$$

Now we give another equivalent condition to (6.1). Define new norms by

$$|||f|||_{n,\epsilon} \equiv \sup\{|f^{(k)}(u)|\rho(u)^{\epsilon}; u \in \mathbb{R}, 0 \leq k \leq n\}.$$

Theorem 6.2 *For $\varphi(x) = f(\langle x, \eta \rangle)$, the following conditions are equivalent.*

(i) $\varphi \in W^{\infty,\infty-}(\mathbb{R})$,

(ii) $\int_{\mathbb{R}} |f^{(n)}(u)|^p \rho(u)^2 du < \infty$ *for any $n \geq 0$, $p \geq 1$,*

(iii) $|||f|||_{n,\epsilon} < \infty$ *for any $n \geq 0$, $\epsilon > 0$.*

For the proof, we need some lemmata. The first one is well known.

Lemma 6.3 *If*

$$\int_{\mathbb{R}} (1 + u^2)^s |g^{(k)}(u)|^2 du < \infty \quad \text{for } k = 0, 1$$

for a fixed $s \geq 1$, then $g(u)$ has a continuous version $\tilde{g}$ which satisfies

$$\lim_{|u| \to \infty} (1 + u^2)^{s/2} \tilde{g}(u) = 0.$$

We simply denote by the same symbol g its continuous version. Since $(1 + u^2)^s$ belongs to $L^p(\mathbb{R}, \omega^2(u)du)$, and $\int_{\mathbb{R}} \omega^2(u) du = 1$, we can easily see the following lemma.

Lemma 6.4 *If*

$$\int_{\mathbb{R}} |f^{(k)}(u)|^4 \rho(u)^2 du < \infty \quad \text{for any } k \leq n,$$

then $f(u)$ is $(n-1)$-times continuously differentiable and $f(u)\rho(u)$ belongs to $C_b^{(n-1)}$, where

$$C_b^{(n-1)} \equiv \{g; \ \sup\{|g^{(k)}(u)|; u \in \mathbb{R}, \ 0 \leq k \leq n-1\} < \infty\}.$$

Proof. By the Schwarz inequality,

$$\int_{\mathbb{R}} (1+u^2)^s |f^{(k)}(u)|^2 \rho(u)^2 du$$

$$\leq \left(\int_{\mathbb{R}} (1+u^2)^{2s} \rho(u)^2 du \int_{\mathbb{R}} |f^{(k)}(u)|^4 \rho(u)^2 du \right)^{1/2} < \infty$$

for any $k \leq n$ and any $s \geq 0$. Since $(f(u)\rho(u))^{(n)}$ can be written as

$$\frac{d^n}{du^n}(f(u)\rho(u)) = \sum_{k=0}^{n} p_k(u) f^{(k)}(u)\rho(u)$$

with suitable polynomials $p_k(u)$, we have

$$\int_{\mathbb{R}} (1+u^2)^s \left| \frac{d^n}{du^n}(f(u)\rho(u)) \right|^2 du < \infty.$$

By Lemma 6.3, we have the assertion. ∎

Lemma 6.5 *For a fixed $p \in \mathbb{N}$, if*

$$\int_{\mathbb{R}} |f^{(k)}(u)|^{4p} \rho(u)^2 du < \infty \quad \text{for } 0 \leq k \leq n,$$

then $f^{(k)}(u)\rho(u)^{1/p}$ belongs to $C_b^{(n-k-1)}$ for $0 \leq k \leq n-1$.

Proof. Notice the following expression

$$\frac{d^n}{du^n}(f(u)^p \rho(u)) = \sum_{k_1+\cdots+k_p \leq n} p_{k_1,\cdots,k_p}(u) f^{(k_1)}(u) \cdots f^{(k_p)}(u)\rho(u)$$

with suitable polynomials $p_{k_1,\cdots,k_p}(u)$ and the inequality

$$\int_{\mathbb{R}} |f_1 \cdots f_p| \omega^2(u) du \leq \prod_{j=1}^{p} \left(\int_{\mathbb{R}} |f_j|^p \omega^2(u) du \right)^{1/p}$$

for $f_j \in L^p(\mathbb{R}, \omega^2(u)du), 1 \leq j \leq p$. Hence, similarly to Lemma 6.4, we have

$$\int_{\mathbb{R}} (1+u^2)^s \left| \frac{d^\ell}{du^\ell}(f^{(k)}(u)^p \rho(u)) \right|^2 du < \infty \quad \text{for } 0 \leq k+\ell \leq n.$$

Therefore, we can show that $f^{(k)}(u)^p \rho(u)$ belongs to $C_b^{(n-k-1)}$ for $0 \leq k \leq n-1$. Thus we have

$$\sup\left\{ \left| \frac{d^\ell}{du^\ell}(f^{(k)}(u)\rho(u)^{1/p}) \right|; u \in \mathbb{R} \right\} < \infty$$

for $0 \leq k + \ell \leq n - 1$.   ■

Proof of Theorem 6.2. The equivalence of (i) and (ii) are shown already. Now suppose (ii). For $\epsilon > 0$ take p as $p\epsilon > 1$. Since $f^{(k)}(u)^p \rho(u)$, $0 \leq k \leq n - 1$, are bounded by Lemma 6.5 and since $\rho(u)^\epsilon \leq (2\pi)^{(p-\epsilon)/4} \rho(u)^{1/p}$, it holds that

$$\sup\{|f^{(k)}(u)\rho(u)^\epsilon|; u \in \mathbb{R}\} < \infty \quad \text{for } 0 \leq k \leq n - 1.$$

Thus (iii) holds. Conversely, suppose that (iii) holds. For given $p \geq 1$, take $\epsilon > 0$ so that $p\epsilon < 2$. Then

$$\int_{\mathbb{R}} |f^{(n)}|^p \rho(u)^2 du \leq \int_{\mathbb{R}} \|f\|_{n,\epsilon}^p \rho(u)^{2-p\epsilon} du < \infty.$$

■

Theorem 6.6 *For* $\varphi(x) = f(\langle x, \eta_1 \rangle, \cdots, \langle x, \eta_m \rangle)$ *with linearly independent* $\eta_1, \cdots, \eta_m$*, the following conditions are equivalent.*

(i) $\varphi \in W^{\infty, \infty-}(\mathbb{R})$,

(ii) $\int_{\mathbb{R}} |f^{(n)}(\vec{u})|^p \rho(\vec{u})^2 d\vec{u} < \infty$ *for any* $n \geq 0$, $p \geq 1$,

(iii) $\|f\|_{n,\epsilon} < \infty$ *for any* $n \geq 0$, $\epsilon > 0$, *where*

$$\|f\|_{n,\epsilon} \equiv \sup\left\{\left|\frac{\partial^{k_1 + \cdots + k_m}}{\partial u_1^{k_1} \cdots \partial u_m^{k_m}} f(\vec{u})\right| \rho(\vec{u})^\epsilon; \vec{u} \in \mathbb{R}^m, 0 \leq k_1 + \cdots + k_m \leq n\right\},$$

with $\vec{u} = (u_1, \cdots, u_m)$, $\rho(\vec{u}) \equiv \rho(u_1) \cdots \rho(u_m)$, $d\vec{u} = du_1 \cdots du_m$.

An infinite dimensional version of Theorem 6.1 is stated in Theorem 4.6 of [Ku2]. A more explicit statement is given as follows by characterization theorems [KPS], [KK2], [Ku2] of the space $S(\mathcal{E})$, and the relation of the spaces $S(\mathcal{E})$ and $(\mathcal{E})$ in [KT], [IKT], [Ku2].

Theorem 6.7 *A functional* $\varphi(x)$ *in* $x \in \mathcal{E}^*$ *belongs to* $(\mathcal{E})$*, if and only if* φ *can be extended to an entire functional on* $\mathcal{E}_c^*$ *and for any* p *and any* $a > 0$*, the following holds;*

$$\sup_{z \in \mathcal{E}_c^*} |\varphi(z)| \exp[-a\|z\|_p^2] < \infty.$$

Let $\mathcal{S}(\mathbb{R}^m)$ and $\mathcal{S}'(\mathbb{R}^m)$ be the space of Schwartz test functions and Schwartz distributions, respectively. Then, Theorem 6.6 can be rewritten as follows :

Theorem 6.8 *For* $\varphi(x) = f(\langle x, \eta_1 \rangle, \cdots, \langle x, \eta_m \rangle)$ *with linearly independent* $\eta_1, \cdots, \eta_m$, φ *belongs to* $W^{\infty,\infty-}(\mathbb{R})$, *if and only if* $f \in \bigcap_{\epsilon>0} \mathcal{S}(\mathbb{R}^m) \rho(\vec{u})^{-\epsilon} = \bigcap_{\epsilon>0} \left(C^{\infty}(\mathbb{R}^m) \cap \mathcal{S}'(\mathbb{R}^m) \right) \rho(\vec{u})^{-\epsilon}$.

The above theorems suggest the following speculations:

(1) *A finite dimensional* $\Phi(x) \in W^{\infty,1+}(\mathbb{R})$ *based on* η *with* $\|\eta\|_0 = 1$ *can be represented as*

$$\Phi(x) = F(\langle x, \eta \rangle) \tag{6.2}$$

with $F \in \bigcap_{p>0} \mathcal{S}'(\mathbb{R}) \rho(u)^{p-2}$.

Such kinds of compositions with distributions were discussed in [Ku1], [Wa], [KK1], [Ku2]. Since the equivalence of the systems of norms $\{ \||f\||_{n,\epsilon}; n \geq 0, \epsilon > 0 \}$ and

$$\left\{ \left(\sum_{k=0}^{n} \int_{\mathbb{R}} |f^{(k)}(u)|^p \rho(u)^2 du \right)^{1/p}; \; n \geq 0, \; p \geq 1 \right\}$$

is seen by the above discussions, we can show that Φ given by (6.2) belongs to $W^{\infty,1+}(\mathbb{R})$. However, a deeper result of distribution theory is necessary to see whether the assertion is valid or not.

(2) *A functional* $\varphi(x)$ *in* $x \in \mathcal{E}^*$ *belongs to* $W^{\infty,\infty-}(\mathbb{R})$, *if and only if for any* p *and any* $a > 0$,

$$\sup_{x \in \mathcal{E}^*} |(1 - \mathcal{L}_A)^s \varphi(x)| \exp[-a\|x\|_p^2] < \infty \quad \text{for any } s.$$

Of course, a functional $\psi(x)$, which satisfies

$$\|(1 - \mathcal{L})^s \psi\|_{L^p} < \infty \quad \text{for any } p \leq 1, s,$$

is not necessarily continuous on $\mathcal{E}^*$. However, $\varphi(x) \in W^{\infty,\infty-}(\mathbb{R})$ is more restricted and may have the property stated in the assertion.

References

[Hi1]   T. Hida, *Analysis of Brownian Functionals.* Carleton Mathematical Lecture Notes, **13** (1975).

[Hi2]   T. Hida, *Brownian Motion,* Springer–Verlag (1980).

[IK]   Y. Ito and I. Kubo, Calculus on Gaussian white noise and Poisson white noise, *Nagoya Math. J.*, **111** (1988), 41–84.

[IKT]   Y. Ito, I. Kubo and S. Takenaka, *Calculus on Gaussian white noise and Kuo's Fourier Transform,* in "White noise analysis", World Scientific 1990, 180-207.

[Ko]   Yu. G. Kondratiev, Nuclear spaces of entire functions in problems of infinite-dimensional analysis, *Soviet Math. Dokl.*, **22** (1980), 588–592.

[KSW]   Yu. G. Kondratiev, L. Streit, W. Westerkamp, *Generalized functions in infinite dimensional analysis,* (preprint, 1995)

[Ku1]   I. Kubo, *Ito formula for generalized Brownian functional,* Lecture Notes in Control and Information Sciences, **49** (1983), 156–166

[Ku2]   I. Kubo, *The structure of Hida distributions,* in "Mathematical Approach to Fluctuations **I**", World Scientific (1994), 49–114.

[Ku3]   I. Kubo, *A direct setting of white noise calculus,* in "Stochastic Analysis on Infinite Dimensional Spaces", Longman, (1994), 152–166.

[Ku4]   I. Kubo, *Generalized functionals in white noise analysis*, in "Probability Theory and Mathematical Statistics", World Scientific, (1996), 237–243.

[KK1]   I. Kubo and H.-H. Kuo, Finite dimensional Hida distributions, *J. Func. Anal.*, **128** (1995), 1–47.

[KK2]   I. Kubo and H.-H. Kuo, *A simple proof of Hida distribution characterization theorem,* in "Exploring Stochastic low", International Science Publisher (1996), 243–250.

[KT]   I. Kubo and S. Takenaka, Calculus on Gaussian white noise I, II, III, IV, *Proc. Japan Acad.*, **56A** (1980), 376–380, *ibid.* **56A** (1980), 411–416, *ibid.* **57A** (1981), 433–437, *ibid.* **58A** (1982), 186–189.

[Kuo]   H.-H. Kuo, *White noise distribution theory,* CRC Press (1996)

[KPS]   H.-H. Kuo, J. Potthoff and L. Streit, A characterization of white noise test functionals, *Nagoya Math. J.*, **121** (1991), 185–194.

[Le]   Y.-J. Lee, Analytic version of test functionals, Fourier transform and a characterization of measures in white noise calculus, *J. Func. Anal.*, **100** (1991), 459–380.

[Ma]   P. Malliavin, *Stochastic calculus of variation and Hypoelliptic operators,* in "Proc. Intern. Symp. SDE", Kyoto (1976), 195–263.

[Me]   P. Meyer, *Note sur les processus d'Ornstein-Uhlenbeck,* in "Seminaire de probabilitiés XVI", Lecture Notes in Math. **920** (1982), 95–133.

[Po]   J. Potthoff, On Meyer equivalence, *Nagoya Math. J.*, **111** (1988), 99–109.

[PS]   J. Potthoff and L. Streit, A characterization of Hida distributions, *J. Func. Anal.*, **101** (1991), 212–229.

[Sh]   I. Shigekawa, Sobolev spaces over the Wiener space based on an Ornstein-Uhlenbeck operator, *J. Math. Kyoto Univ.*, **32-4** (1992), 731–748.

[Yo]   Y. Yokoi, Simple setting for white noise calculus using Bargain space and Gauss transform, *Hiroshima Math. J.*, (1993), 97–121.

[Wa]   S. Watanabe, *Malliavin's calculus in terms of generalized Wiener functionals,* Lecture Notes in Control and Information Sciences **49**, Springer–Verlag (1983), 284–290 .

Department of Mathematics,
Faculty of Science,
Hiroshima University,
Higashi-Hiroshima, 739 JAPAN
E-mail: kubo@math.sci.hiroshima-u.ac.jp

Received September 30, 1996

Stochastic Processes with Independent Increments on a Lie Group and their Selfsimilar Properties

Hiroshi Kunita

1. Introduction

Convolution semigroups on a Lie group have been studied details. One of the fundamental facts is Hunt's representation formula of its infinitesimal generator (Hunt [Hu]). In Applebaum-Kunita [AK] we applied it to the study of Lévy processes on a Lie group and obtained stochastic differential equations (SDE) which govern them.

In this paper we study convolution hemigroups (temporally inhomogeneous convolution semigroup) on a Lie group and the associated stochastic processes with independent increments (not necessarily time homogeneous). The infinitesimal generator of a convolution hemigroup has been studied recently by Heyer-Paps [HP1], [HP2] and Kunita [Ku4] independently. In the latter, Condition (**D**) is assumed, which is stronger than the one assumed in the former. Then the infinitesimal generator is represented as an integro-differential operator of Hunt's type depending on time continuously. In the next section we reproduce main results in [Ku4] and then discuss Kolmogorov's forward and backward equations for convolution hemigroups

In Section 3, we consider a stochastic process on a Lie group with independent increments. We will represent it as a solution of an SDE with jumps driven by a Brownian motion on a Euclidean space and a Poisson random measure on the Lie group G. Then, by constructing the solution of a certain SDE, we show that for a given time dependent integro-differential operator, there exists a unique (in the sense of law) process with independent increments whose infinitesimal generator coincides with the given time dependent integro-differential operator. In particular we will show that Condition (**D**) is necessary and sufficient for our representation of the infinitesimal generator. See Theorem 3.5.

In section 4, we consider the case where the Lie group is simply connected and nilpotent. We represent the process with independent increments as a solution of an SDE driven by a process with independent increments on a Euclidean space. This fact will be applied to prove that

the process on the Lie group has the selfsimilar property if and only if the
driving process is selfsimilar.

The last section is devoted to the study of the selfdecomposable distribution. We introduce an Ornstein-Uhlenbeck type process on a simply
connected nilpotent Lie group and show that any selfdecomposable distribution can be obtained as a limit distribution of the Ornstein-Uhlenbeck
type process. See Sato-Yamazato [SY] for a similar result in Euclidean
space.

2. Convolution hemigroups of probability distributions on Lie groups

Let G be a connected Lie group of dimension d. Let $\{\mu_{s,t}\}_{0\leq s<t<\infty}$ be a
family of probability distributions on G. It is called a *convolution hemigroup*
if it satisfies the following two properties.

(i) (Hemigroup property) $\mu_{s,t}*\mu_{t,u} = \mu_{s,u}$ holds for any $0 \leq s < t < u < \infty$.

(ii) (Continuity) $\mu_{s,t}$ converges weakly to δ_e as $t - s \to 0$, where δ_e is the
Dirac measure concentrated at the unit e of the Lie group G.

A convolution hemigroup $\{\mu_{s,t}\}$ is called *homogeneous* if for any $s < t$,
$\mu_{s,t}$ depends only on $t - s$. We denote it by μ_{t-s}. Then one parameter
family of distributions $\{\mu_t\}_{t>0}$ satisfies $\mu_s * \mu_t = \mu_{s+t}$ for any $s, t > 0$ and
$\mu_h \to \delta_e$ weakly as $h \downarrow 0$. It is called a *convolution semigroup*.

We shall define a hemigroup of linear operators on the Banach space of
real continuous functions associated with the convolution hemigroup $\{\mu_{s,t}\}$.
Let $\mathcal{C} = \mathcal{C}(G)$ be the set of all bounded continuous functions f on G (such
that $\lim_{\sigma\to\infty} f(\sigma)$ exists if G is noncompact). It is a separable Banach
space with the supremum norm $\|\ \|$. We define

$$P_{s,t}f(\sigma) = \int f(\sigma\tau)\mu_{s,t}(d\tau), \quad f \in \mathcal{C}. \tag{2.1}$$

Then $\{P_{s,t}\}_{0\leq s<t<\infty}$ is a family of linear operators on $\mathcal{C}$ and satisfies
$P_{s,t}P_{t,u} = P_{s,u}$ for all $s < t < u$. Further $\lim_{t-s\to0} P_{s,t}f = f$ holds for
all $f \in \mathcal{C}$. The *infinitesimal generator* $\{\mathbf{A}(t)\}_{t\geq0}$ of $\{P_{s,t}\}$ or the *convolution hemigroup* $\{\mu_{s,t}\}$ is defined by

$$\mathbf{A}(t)f(\sigma) \equiv \lim_{h\downarrow0} \frac{P_{t,t+h}f(\sigma) - f(\sigma)}{h}, \tag{2.2}$$

if the right hand side converges uniformly as functions of $(\sigma, t) \in G \times [0, N]$
for any $N > 0$. Its domain $\mathcal{D}(\{\mathbf{A}(t)\})$ is the set of all $f \in \mathcal{C}$ such that the
above limit exists.

The representation of the infinitesimal generator of the convolution hemigroup was studied with details in Kunita [Ku4]. We shall reproduce it here for the convenience of the reader. We need some notations. Let $\mathcal{G}$ (or $\mathcal{G}'$) be the left invariant (or right invariant) Lie algebra of G. Elements of $\mathcal{G}$ (or $\mathcal{G}'$) are regarded as left invariant (or right invariant) first order differential operators (vector fields) and are denoted by X, Y, etc. (or X', Y', etc.) We fix its basis $\{X_1, ..., X_d\}$ (or $\{X_1', ..., X_d'\}$). We denote by $\mathcal{C}_2$ (or $\mathcal{C}_2'$) the totality of twice continuously differentiable functions f such that Xf and YZf (or $X'f$ and $Y'Z'f$) belong to $\mathcal{C}$ for any $X, Y, Z \in \mathcal{G}$ (or $X', Y', Z' \in \mathcal{G}'$). Set $\|f\|_2' = \|f\| + \sum_{i=1}^{d} \|X_i'f\| + \sum_{j,k=1}^{d} \|X_j'X_k'f\|$. Then $\mathcal{C}_2'$ is a Banach space with this norm. It holds $X'P_{s,t}f = P_{s,t}X'f$, etc., so that $P_{s,t}$ maps $\mathcal{C}_2'$ into itself and satisfies $\|P_{s,t}f\|_2' \leq \|f\|_2'$.

Now let $\tilde{G} = G \times [0, \infty)$ be the product manifold. Let $\tilde{\mathcal{C}} = \mathcal{C}(\tilde{G})$ be the set of all bounded continuous functions f on G such that $\lim_{\sigma \to \infty} f(\sigma, t)$ exists uniformly in $t \in [0, N]$ for any N if G is noncompact. It is a Frechet space. Let $\tilde{\mathcal{C}}_{2,1}'$ be the set of all $f(\sigma, t) \in \tilde{\mathcal{C}}$ which are twice continuously differentiable with respect to σ and continuously differentiable with respect to t such that $X'f, Y'Z'f, (\partial/\partial t)f$ belong to $\tilde{\mathcal{C}}$ for any $X', Y', Z' \in \mathcal{G}'$. For $f(\sigma, t) \in \tilde{\mathcal{C}}$, we set $f_t(\sigma) = f(\sigma, t)$. Then $f_t \in \mathcal{C}$ for any t. Now, for $\lambda > 0$ define

$$\tilde{R}_\lambda f(\sigma, t) = \int_0^\infty e^{-\lambda r} P_{t,t+r}(f_{t+r})(\sigma)dr. \tag{2.3}$$

We introduce a differentiability condition for the resolvent $\tilde{R}_\lambda$ with respect to t.

Condition (**D**). For any $f \in \tilde{\mathcal{C}}_{2,1}'$, $\tilde{R}_\lambda f(\sigma, t)$ is continuously differentiable with respect to $t \in [0, \infty)$ and $(\partial/\partial t)\tilde{R}_\lambda g(\sigma, t)$ belongs to $\tilde{\mathcal{C}}$.

In the following, we shall fix $x^1, ..., x^d, \phi$ of $\mathcal{C}_2$ satisfying

$$x^i(e) = 0, i = 1, ..., d, \quad X_i x^j(e) = \delta_{ij}, i, j = 1, ..., d, \tag{2.4}$$

$$\phi(e) = 0, \quad \phi(\sigma) > 0 \ (\sigma \neq e), \quad \lim_{\sigma \to \infty} \phi(\sigma) > 0 \ \text{if } G \text{ is noncompact},$$

$$\phi \cong \sum (x^i)^2 \ \text{near } e. \tag{2.5}$$

The following theorem is an extension of Hunt's theorem [Hu], where the infinitesimal generator of the convolution semigroup is obtained.

Theorem 2.1 *(Kunita [Ku4]) Assume Condition* (**D**). *Then* $\mathcal{C}_2 \subset \mathcal{D}(\{\mathbf{A}(t)\})$. *Further, for any* $f \in \mathcal{C}_2$, $\mathbf{A}(t)f(\sigma)$ *are represented by integro-*

differential operators, i.e., it holds $\mathbf{A}(t)f = \mathbf{L}(t)f$, *where*

$$\mathbf{L}(t)f(\sigma) = \frac{1}{2}\sum_{i,j}a^{ij}(t)X_iX_jf(\sigma) + \sum_i b^i(t)X_if(\sigma)$$

$$+ \int_G (f(\sigma\tau) - f(\sigma) - \sum_i x^i(\tau)X_if(\sigma))\nu_t(d\tau), \quad (2.6)$$

where $a(t) = (a^{ij}(t)), b(t) = (b^i(t))$ *and* ν_t *satisfy the following properties.*
(a) $a(t) = (a^{ij}(t)), t \geq 0$, *are symmetric nonnegative definite matrices continuous in* t,
(b) $b(t) = (b^i(t)), t \geq 0$, *are continuous in* t,
(c) $\nu_t, t \geq 0$, *are positive measures on* G *such that* $\nu_t(\{e\}) = 0$ *and the integrals* $\int \phi(\tau)f(\tau)\nu_t(d\tau)$ *are continuous in* t *for any* $f \in C$.
The matrices $a(t)$ *and the measures* ν_t *are uniquely determined from the convolution hemigroup* $\{\mu_{s,t}\}$, *but the vectors* $b(t)$ *may depend on the choice of the functions* $x^1, ..., x^d$.
If $\{\mu_{s,t}\} = \{\mu_t\}$ *is a convolution semigroup, then Condition* (D) *is always satisfied. In this case, the representation of* $\mathbf{L}(t)f$ *does not depend on* t, *i.e., the characteristics* (a, b, ν) *do not depend on* t.

As an application of the above theorem, we shall obtain Kolmogorov's forward and backward differential equation for the hemigroup $P_{s,t}f$.

Theorem 2.2 *Assume Condition* (D). *(1) Let* $f \in C_2$. *Then* $P_{s,t}f(\sigma)$ *is continuously differentiable in* $t \in (s, \infty)$ *for any fixed* $s > 0$ *and* σ. *It satisfies:*

$$\frac{\partial}{\partial t}P_{s,t}f(\sigma) = P_{s,t}\mathbf{L}(t)f(\sigma), \quad \forall t \in (s, \infty), \sigma \in G. \quad (2.7)$$

(2) Let $f \in C_2'$. *Then* $P_{s,t}f(\sigma)$ *is continuously differentiable in* $s \in (0, t)$ *for any fixed* $t > 0$ *and* σ. *Further, it satisfies*

$$\frac{\partial}{\partial s}P_{s,t}f(\sigma) = -\mathbf{L}(s)P_{s,t}f(\sigma), \quad \forall s \in (0, t), \sigma \in G. \quad (2.8)$$

Proof. The first assertion (1) is proved in [Ku4]. So we shall only prove the second assertion. Let $s \in (0, \infty)$ and $h > 0$. Set for $g \in C_2'$,

$$\tilde{\mathbf{A}}_h(s)g(\sigma) = \frac{1}{h}(P_{s-h,s}g(\sigma) - g(\sigma)). \quad (2.9)$$

Then we can prove similarly as in the proof of Lemma 3.2 and Lemma 3.3 of [Ku4], that the above converges as $h \downarrow 0$ for each fixed σ and the limit denoted by $\tilde{\mathbf{L}}(s)g(\sigma)$ is represented as an integro-differential operator like (2.6). We show that two operators $\mathbf{L}(s)$ and $\tilde{\mathbf{L}}(s)$ coincide each other. Indeed, if $g \in \mathcal{C}'_2$ is of compact support, we have by (2.7),

$$\tilde{\mathbf{A}}_h(s)g(\sigma) = \frac{1}{h} \int_{s-h}^{s} P_{s-h,u}\mathbf{L}(u)g(\sigma)du.$$

The right hand side converges to $\mathbf{L}(s)g(\sigma)$ as $h \downarrow 0$. Therefore $\mathbf{L}(s)g(\sigma) = \tilde{\mathbf{L}}(s)g(\sigma)$ holds for any $g \in \mathcal{C}'_2$ with compact supports. Consequently we have $\mathbf{L}(s) = \tilde{\mathbf{L}}(s)$.

Now set $g = P_{s,t}f$, where $f \in \mathcal{C}'_2$. Since it belongs to $\mathcal{C}'_2$, we have

$$\exists \lim_{h \downarrow 0} \frac{1}{h}(P_{s-h,t}f(\sigma) - P_{s,t}f(\sigma)) = \mathbf{L}(s)P_{s,t}f(\sigma).$$

Therefore $\psi(s) := P_{s,t}f(\sigma)$ is differentiable from the left and the left derivative $(d/ds)^{-}\psi(s)$ is continuous in s. We can show easily $\psi(t) - \psi(r) = \int_r^t (d/ds)^{-}\psi(s)ds$. Therefore we have $(f - P_{r,t}f)(\sigma) = -\int_r^t \mathbf{L}(s)P_{s,t}f(\sigma)ds$. This proves that $P_{s,t}f(\sigma)$ is continuously differentiable with respect to $s \in (0,t)$ for each fixed σ and the derivative satisfies equation (2.8). The proof is complete.

3. The associated stochastic differential equations

Let $\varphi_t, t \geq 0$ be a stochastic process with values in a Lie group G continuous in probability such that $\varphi_0 = e$. It is called a *process with independent increments* if G-valued random variables $\varphi_{t_{i-1}}^{-1}\varphi_{t_i}, i = 1, ..., n$ are independent for any $0 = t_0 < t_1 < \cdots < t_n < \infty$. If the law of $\varphi_h^{-1}\varphi_{t+h}$ does not depend on h for any $t > 0$, the process is called a Lévy process on the Lie group.

Given a process with independent increments φ_t, set

$$\mu_{s,t}(B) = P(\varphi_s^{-1}\varphi_t \in B), \quad 0 \leq s < t < \infty. \tag{3.1}$$

Then $\{\mu_{s,t}\}_{0 \leq s < t < \infty}$ is a convolution hemigroup of probability distributions on G. Conversely, given a convolution hemigroup of probability distributions $\{\mu_{s,t}\}_{0 \leq s < t < \infty}$, there exists a process with independent increments $\varphi_t, t \geq 0$ satisfying (3.1). Further, φ_t is a Lévy process, if and only if the associated convolution hemigroup is homogeneous.

We will show that any process with independent increments satisfying Condition (**D**) can be obtained by solving an SDE with jumps. The following theorem is proved in [AK] in the case where φ_t is a Lévy process.

Theorem 3.1 *Let $\varphi_t, t \geq 0$ be a stochastic process with independent increments on a Lie group G satisfying Condition (**D**), whose infinitesimal generator is represented by (2.6). Then the process $\varphi_t, t \geq 0$ has a modification such that it is right continuous with left hand limits. Further, there exists a nonhomogeneous Brownian motion $B(t) = (B^1(t), \dots, B^d(t))$ on $\mathbf{R}^d (\cong \mathcal{G})$ such that $Cov(B^i(t), B^j(t)) = \int_0^t a^{ij}(u)du$ and a nonhomogeneous Poisson random measure $N((s,t] \times E)$ on $[0, \infty) \times G$ with intensity $\int_s^t \nu_u(E)du$ satisfying the following properties:*

(1) $B(t)$ and $N((s,t] \times E)$ are independent for any s, t, E. Further

$$\sigma(\varphi_s; s \leq t) = \sigma(B(s), N((0,s] \times E); s \leq t, E \in \mathcal{B}(G)) \qquad (3.2)$$

holds for any t.

(2) Set $\varphi_{s,t}(p) = p\varphi_s^{-1}\varphi_t$. Then it satisfies a stochastic differential equation driven by the above $B(t)$ and $N((s,t] \times E)$, i.e., $\varphi_{s,t}(p)$ satisfies

$$
\begin{aligned}
f(\varphi_{s,t}(p)) \ = \ & f(p) + \sum_i \int_s^t X_i f(\varphi_{s,u}(p)) \circ dB^i(u) \\
& + \sum_i \int_s^t b^i(u) X_i f(\varphi_{s,u}(p)) du \\
& + \int_s^t \int_G (f(\varphi_{s,u-}(p)\sigma) - f(\varphi_{s,u-}(p)))(N(dud\sigma) - \nu_u(d\sigma)du) \\
& + \int_s^t \left(\int_G \Big(f(\varphi_{s,u-}(p)\sigma) - f(\varphi_{s,u-}(p)) \right. \\
& \qquad\qquad \left. - \sum_i x^i(\sigma) X_i f(\varphi_{s,u-}(p)) \Big) \nu_u(d\sigma) \right) du, \qquad (3.3)
\end{aligned}
$$

for any C^∞-function f over G, where $\circ dB^i(u)$ denotes the Stratonovich integral.

Proof. In the sequel, we fix a time s and a state p. Then the process $\varphi_{s,t}(p), t \geq s$ is a Markov process starting from p at time s. Its transition probabilities are $P_{s,t}(p, A) = \mu_{s,t}(p^{-1}A)$. For $f \in \mathcal{C}_2$, define

$$M_{s,t}f(p) = f(\varphi_{s,t}(p)) - f(p) - \int_s^t \mathbf{L}(u)f(\varphi_{s,u}(p))du. \qquad (3.4)$$

Then its mean is 0 for any t and p because of the formula (2.7). Then the Markov property implies that it is a martingale for any p. Therefore the real stochastic processes $M_{s,t}f(p), t \geq s$ and $f(\varphi_{s,t}(p)), t \geq s$ have modifications which are right continuous with left hand limits. Then the

G-valued stochastic process $\varphi_{s,t}(p), t \geq s$ has a modification which is right continuous with left hand limits.

Given $f, g \in C_2$ and $p, q \in G$, let $\langle M_{s,t}f(p), M_{s,t}g(q) \rangle, t \geq s$ be a continuous predictable process of bounded variation such that $M_{s,t}f(p)M_{s,t}g(q) - \langle M_{s,t}f(p), M_{s,t}g(q) \rangle$ is a martingale. The existence and uniqueness of a such process is known by Meyer's decomposition of semimartingales. It is called the bracket process of $M_{s,t}f(p)$ and $M_{s,t}g(q)$. It is represented by

$$\langle M_{s,t}f(p), M_{s,t}g(q) \rangle = \int_s^t B(f,g)_u(\varphi_{s,u}(p), \varphi_{s,u}(q))du, \qquad (3.5)$$

where,

$$
\begin{aligned}
B(f,g)_t(p,q) &= 2\sum_{ij} a^{ij}(t)X_if(p)X_jf(q) \\
&+ \int_G (f(p\sigma) - f(p))(g(q\sigma) - g(q))\nu_t(d\sigma). \qquad (3.6)
\end{aligned}
$$

c.f. Lemma 3.2 in [AK].

For a partition $\delta = \{0 = t_0 < t_1 < \cdots < t_n \cdots\}$ of $[0, \infty)$, set $Y_t^\delta f(p) = \sum_k M_{t \wedge t_k, t \wedge t_{k+1}}f(p)$. Then, $Y_t f(p) = \lim_{|\delta| \to 0} Y_t^\delta f(p)$ exists in $L^2(P)$ for any $f \in C_2, p \in G$ and $t > 0$. It is an L^2-martingale and is a real valued process with independent increments. Further we have $\langle Y_t f(p), Y_t g(q) \rangle = \int_0^t B_u(f,g)du$ and

$$M_{s,t}f(p) = \int_s^t dY_u f(\varphi_{s,u-}(p)). \qquad (3.7)$$

c.f. Lemma 3.3 in [AK].

The process $Y_t f(p)$ is decomposed uniquely as the sum of a continuous martingale $Y_t^c f(p)$ and a discontinuous one $Y_t^d f(p)$, both of which are processes with independent increments. For $x^1, ..., x^d$ of C_2 satisfying (2.4), define $B^i(t) = Y_t^c(x^i)(e), i = 1, ..., d$. Then $B(t) = (B^1(t), ..., B^d(t))$ is a Brownian motion with mean 0 and $\text{Cov}(B^i(t), B^j(t)) = \int_0^t a^{ij}(u)du$. Further, $Y_t^c f(p)$ is represented by $Y_t^c f(p) = \sum_i X_i f(p)B^i(t)$. Define

$$N((s,t] \times E) = \sharp\{u \in (s,t] : \varphi_{u-}^{-1}\varphi_u \in E\}. \qquad (3.8)$$

Then it is a Poisson random measure on $\mathbf{R}^+ \times G$ with intensity measure $dtd\nu_t$. Further,

$$Y_t^d f(p) = \int_0^t \int_G (f(p\sigma) - f(p))\tilde{N}(dud\sigma), \qquad (3.9)$$

where $\tilde{N}((s,t] \times E) = N((s,t] \times E) - \int_0^t \nu_u(E)du$. We have thus proved

$$Y_t f(p) = \sum_i X_i f(p) B^i(t) + \int_0^t \int_G (f(p\sigma) - f(p)) \tilde{N}(dud\sigma). \qquad (3.10)$$

c.f. Lemma 3.4 in [AK]. Therefore, we have by (3.7)

$$\begin{aligned}
M_{s,t} f(p) &= \sum_i \int_0^t X_i f(\varphi_{s,u-}(p)) dB^i(u) \\
&\quad + \int_0^t \int_G (f(\varphi_{s,u-}(p)\sigma) - f(\varphi_{s,u-}(p))) \tilde{N}(dud\sigma). \quad (3.11)
\end{aligned}$$

This proves that $\varphi_{s,t}(p)$ satisfies the SDE (3.3).

Corollary 3.2 *The following statements are equivalent in Theorem 3.1.*
1) The stochastic process $\varphi_t, t \geq 0$ is of diffusion type, i.e., its sample functions $\varphi_t, t \geq 0$ are continuous a.s.
2) The Lévy measures ν_t are identically 0 for all $t \geq 0$.

The distributions $\mu_{s,t}$ are called *Gaussian in the generalized sense* if one of 1) or 2) of the above corollary is satisfied. In particular if $\mu_{s,t}$ are temporally homogeneous, they are called *Gaussian distributions*. In the case where $\{\mu_{s,t}\}$ is a convolution hemigroup on a Euclidean space, "Gaussian" and "Gaussian in the generalized sense" are identical, since both are Gaussian distributions in the usual sense. However in the case of noncommutative Lie group, these two notions might be different.

We shall obtain a backward SDE for the inverse flow $\psi_{s,t}(p) = \varphi_{s,t}^{-1}(p)$.

Theorem 3.3 *(c.f. Kunita [Ku1]) Assume Condition (D). The inverse flow $\psi_{s,t}(p) = \varphi_{s,t}^{-1}(p)$ satisfies the following backward stochastic differential equation.*

$$\begin{aligned}
f(\psi_{s,t}(p)) &= f(p) - \sum_i \int_s^t X_i f(\psi_{u,t}(p)) \circ \hat{d}B^i(u) \\
&\quad - \sum_i \int_s^t b^i(u) X_i f(\psi_{u,t}(p)) du \\
&\quad + \int_s^t \int_G \Big(f(\tau\psi_{u,t}(p)) - f(\psi_{u,t}(p))\Big) (\hat{N}(dud\tau) - du\hat{\nu}_u(d\tau)) \\
&\quad + \int_s^t \Big(\int_G \Big(f(\tau\psi_{u,t}(p)) - f(\psi_{u,t}(p)) \\
&\qquad\qquad + \sum_i x^i(\tau^{-1}) X_i f(\psi_{u,t}(p))\Big) \hat{\nu}_u(d\tau)\Big) du. \quad (3.12)
\end{aligned}$$

Here, $\circ \hat{d}B^i(u)$ denotes the backward Stratonovich integral. $\hat{N}(dsd\sigma)$ is a Poisson random measure defined by $N(dsd\sigma^{-1})$ and $\hat{\nu}_s(d\sigma) = \nu_s(d\sigma^{-1})$.

Proof. Substitute $\psi_{s,t}(p)$ in place of p in each term of equation (3.3). We have $f(\varphi_{s,t}(\psi_{s,t}(p))) = f(p)$ and

$$\int_s^t X_i f(\varphi_{s,u}(p)) \circ dB^i(u) \Big|_{p=\psi_{s,t}(p)} = \int_s^t X_i f(\varphi_{u,t}(p)) \circ \hat{d}B^i(u). \quad (3.13)$$

A similar formula is valid for the integral $\int_s^t b^i(u) X_i f(\varphi_{s,u}(p)) du$.

Now the sum of the 4-th and 5-th terms of the right hand side of (3.3) is approximated by a sequence of integrals such that domains of these integrations are restricted to U_m^c instead of G, where $U_m, m = 1, 2, ...$ are open neighborhoods of e decreasing to $\{e\}$. Since $\nu_u(U_m^c) < \infty$ holds for any m, the sum of these two integrals are written as

$$\int_s^t \int_{U_m^c} (f(\varphi_{s,u-}(p)\sigma) - f(\varphi_{s,u-}(p))) N(dud\sigma)$$

$$- \int_s^t \int_{U_m^c} \sum_i x^i(\sigma) X_i f(\varphi_{s,u-}(p)) \nu_u(d\sigma) du.$$

We have,

$$\int_s^t \int_{U_m^c} (f(\varphi_{s,u-}(p)\sigma) - f(\varphi_{s,u-}(p))) N(dud\sigma) \Big|_{p=\psi_{s,t}(p)}$$

$$= - \int_s^t \int_{\hat{U}_m^c} (f(\tau \psi_{u,t}(p)) - f(\psi_{u,t}(p))) \hat{N}(dud\tau), \quad (3.14)$$

$$\int_s^t \int_{U_m^c} x^i(\sigma) X_i f(\varphi_{s,u}(p)) \nu_u(d\sigma) du \Big|_{p=\psi_{s,t}(p)}$$

$$= \int_s^t \int_{\hat{U}_m^c} x^i(\tau^{-1}) X_i f(\varphi_{u,t}(p)) \hat{\nu}_u(d\tau) du, \quad (3.15)$$

where $\hat{U}_m^c = \{\sigma; \sigma^{-1} \in U_m^c\}$. Summing up all these formulas and letting $m \to \infty$, we get (3.12).

Corollary 3.4 Set $\hat{P}_{s,t} f(\sigma) = E[f(\psi_{s,t}(p))]$. Then,

$$\hat{P}_{s,t} f(p) = f(p) + \int_s^t \hat{P}_{u,t} \hat{L}(u) f(p) du, \quad (3.16)$$

where

$$\hat{L}(t)f(\sigma) = \frac{1}{2}\sum_{i,j} a^{ij}(t)X_iX_jf(\sigma) - \sum_i b^i(t)X_if(\sigma)$$

$$+ \int_G (f(\tau\sigma) - f(\sigma) + \sum_i x^i(\tau^{-1})X_if(\sigma))\hat{\nu}_t(d\tau).(3.17)$$

Proof. We shall rewrite the term in (3.12) involving the Stratonovich backward integral. It holds

$$\int_s^t X_if(\psi_{u,t}(p)) \circ \hat{d}B^i(u) = \int_s^t X_if(\psi_{u,t}(p))\hat{d}B^i(u) \qquad (3.18)$$

$$- \frac{1}{2}\sum_j \int_s^t a^{ij}(u)X_iX_jf(\psi_{u,t}(p))du.$$

The first term of the right hand side of (3.18) is the backward Ito integral and it is a backward martingale with mean is 0. The intensity measure of $\hat{N}(dud\tau)$ is $du\hat{\nu}_u(d\tau)$. The term in (3.12) involving the stochastic integral based on $\hat{N}(dud\tau) - du\hat{\nu}_u(d\tau)$ is a backward martingale with mean 0. Therefore, taking expectations of each term of (3.12), we get the formula (3.16).

Given a time dependent operator $\mathbf{L}(t)$ of (2.6), we shall study the existence and uniqueness problem of a convolution hemigroup whose infinitesimal generator is represented by $\mathbf{L}(t)$. It will be done through solving a certain stochastic differential equation.

Theorem 3.5 *Suppose that we are given a Brownian motion $B(t)$ on $\mathbf{R}^d$ ($\cong \mathcal{G}$) with mean 0 and covariance $\int_0^t a^{ij}(u)du$, and a Poisson random measure $N((s,t] \times E)$ on $[0,\infty) \times G$ with intensity $dtd\nu_t$, which are mutually independent. Then for any $s > 0$ and $p \in G$ the stochastic differential equation (3.3) has a unique solution $\varphi_{s,t}(p), t \geq 0$. Further $\varphi_t := \varphi_{0,t}(e)$ is a process with independent increments.*

*Let $\{\mu_{s,t}\}$ be the associated convolution hemigroup. Then it satisfies Condition (**D**). Its infinitesimal generator is represented by $\mathbf{L}(t)f$ of (2.6).*

Proof. The existence and the uniqueness of the global solution of equation (3.3) can be proved similarly as in [AK], where the cases of time homogeneous Brownian motion and time homogeneous Poisson random measure are discussed. Further φ_t has independent increments as in shown in [AK].

We will prove that the process satisfies Condition (**D**). Note that Corollary 3.4 tells us that $\hat{P}_{s,t}f$ is continuously differentiable with respect to s if

$f \in \mathcal{C}_2$. We have $P_{s,t}f(\sigma) = \hat{P}_{s,t}\hat{f}(\sigma^{-1})$, where $\hat{f}(\sigma) = f(\sigma^{-1})$. Note that $f \in \mathcal{C}'_2$ if and only if $\hat{f} \in \mathcal{C}_2$. Then $P_{s,t}f(\sigma)$ is continuously differentiable with respect to s if $f \in \mathcal{C}'_2$. We can now prove the differentiability of $P_{t,t+r}f$ with respect to t if $f \in \mathcal{C}'_2$. In fact, we have

$$\exists \lim_{h \to 0} \frac{1}{h}(P_{t+h,t+r+h} - P_{t,t+r})f = \left.\frac{\partial P_{s,t}f}{\partial s}\right|_{s=t,t=t+r} + \left.\frac{\partial P_{s,t}f}{\partial t}\right|_{s=t,t=t+r},$$

which is continuous in t. It is easy to see that $P_{t,t+r}(f_{t+r})$ is also continuously differentiable with respect to t if $f \in \tilde{\mathcal{C}}'_{21}$. Then the resolvent $\tilde{R}_\lambda f(\sigma, t)$ of (2.3) is continuously differentiable with respect to t and the derivative belongs to $\tilde{\mathcal{C}}$. The last assertion of the theorem is obvious. The proof is complete.

4. Processes with independent increments with selfsimilar properties

In [Ku4], we studied with details the case where the underlying Lie group G is simply connected and nilpotent. In this case the exponential map $\exp : \mathcal{G} \to G$ is a diffeomorphism. Then any $\sigma \in G$ can be represented uniquely by $\sigma = \exp \sum_{i=1}^{d} x_i X_i$, where $(x_1, ..., x_d) \in \mathbf{R}^d$. We set $\xi(x) = \exp \sum_i^d x_i X_i$. Then (2.6) can be represented in another form. Indeed, we have

$$\mathbf{L}(t)f(\sigma) = \frac{1}{2}\sum_{i,j} a^{ij}(t)X_i X_j f(\sigma) + \sum_i b^i(t)X_i f(\sigma)$$

$$+ \int_{\mathbf{R}^d}\{f(\sigma\xi(x)) - f(\sigma) - \sum_i \frac{x_i}{1+|x|^2}X_i f(\sigma)\}\hat{\nu}_t(dx) \quad (4.1)$$

where $\hat{\nu}_t$ are Lévy measures on the Euclidean space $\mathbf{R}^d$.

The following theorem is proved in [Ku2] in the case where $\varphi_t, t \geq 0$ is a Lévy process.

Theorem 4.1 *Suppose that G is simply connected and nilpotent. Let $\varphi_t, t \geq 0$ be a G-valued process with independent increments satisfying Condition (D). Then there exists a unique $\mathbf{R}^d$-valued process with independent increments $Z(t) \equiv Z_t, t \geq 0$ and φ_t is represented by SDE:*

$$f(\varphi_t) = f(\varphi_s) + \sum_j \int_s^t X_j f(\varphi_{u-})dZ_u^j$$

$$+ \quad \frac{1}{2} \sum_{i,j} \int_s^t a^{ij}(u) X_i X_j f(\varphi_u) du$$

$$+ \quad \sum_{s \leq u \leq t} \{ f(\varphi_{u-}\xi(\Delta Z_u)) - f(\varphi_{u-}) - \sum_j X_j f(\varphi_{u-}) \Delta Z_u^j \} \quad (4.2)$$

Proof. We shall rewrite equation (3.3). Since $\exp : \mathcal{G} \to G$ is a diffeomorphism, it holds $\varphi_{s-}\varphi_s = \exp \sum_j p_s^j X_j = \xi(p_s)$, where $p_s = (p_s^1, \ldots, p_s^d)$ is a Poisson point process on $\mathbf{R}^d$. Define

$$J((s,t] \times E) = \sharp\{ u \in (s,t]; p_u \in E \}. \quad (4.3)$$

It is a Poisson random measure on $\mathbf{R}^d \times [0, \infty)$. Let $\hat{\nu}_s ds$ be the intensity measure of J. Then ν_t is the pullback of $\hat{\nu}_t$ by ξ, i.e., $\nu_t = \xi^* \hat{\nu}_t$. Further, $N((s,t] \times E) = J((s,t] \times \xi^{-1}(E))$. Set

$$Z(t) \quad = \quad B(t) + \int_0^t b(u) du + \int_0^t \int_{0 < |x| < \delta_0} x \left(J(dudx) - du \hat{\nu}_u(dx) \right)$$

$$+ \int_0^t \int_{|x| \geq \delta_0} x J(dudx). \quad (4.4)$$

Then (3.3) is written as (4.2). The uniqueness of the process $Z(t), t \geq 0$ follows from the uniqueness of $B(t), t \geq 0$ and $N((s,t] \times E)$ in Theorem 3.1. See [Ku2].

The process $Z(t), t \geq 0$ of Theorem 4.1 is called the *generating process* of $\varphi_t, t \geq 0$.

In [Ku4] we characterized the selfsimilar convolution hemigroup through its infinitesimal generator. Here we shall characterize it through the stochastic differential equation. Let us first recall the definition of a selfsimilar convolution hemigroup.

Suppose that we are given a one parameter group of automorphisms $\{\gamma_r\}_{r>0}$ of the Lie group G, i.e., (i) For each $r > 0$, γ_r is an automorphism of G, (ii) $\gamma_r \gamma_s = \gamma_{rs}$ holds for any $r, s > 0$, (iii) γ_r is continuous in $r \in (0, \infty)$. It is called a *dilation* if it satisfies (iv) $\gamma_r(\sigma) \to e$ uniformly on compact sets as $r \to 0$.

For a distribution μ on G, we denote by μ^{γ_r} the distribution such that $\mu^{\gamma_r}(A) = \mu(\gamma_r^{-1}(A))$ holds for any Borel set A of G. Then it holds $\mu^{\gamma_r}(f) = \mu(f \circ \gamma_r)$, where $f \circ \gamma_r$ is the composition of the function f and the map γ_r.

A convolution hemigroup $\{\mu_{s,t}\}_{0 \leq s < t < \infty}$ is called *selfsimilar with respect to a dilation* $\{\gamma_r\}_{r>0}$ if $\mu_{s,t}^{\gamma_r} = \mu_{rs,rt}$ holds for any $r > 0$ and

$0 \leq s < t < \infty$. Let $\varphi_t, t \geq 0$ be a G-valued stochastic process with independent increments associated with $\{\mu_{s,t}\}$. Then it is selfsimilar with respect to $\{\gamma_r\}$, if and only if the law of the process $\varphi_t^{(\gamma_r)} \equiv \gamma_r(\varphi_t), t \geq 0$ is equal to the law of the process $\varphi_t^{(r)} \equiv \varphi_{rt}, t \geq 0$, for any $r > 0$.

Proposition 4.2 *Let $\{\mu_{s,t}\}$ be a convolution hemigroup. It is selfsimilar with respect to the dilation $\{\gamma_r\}$ if and only if*

$$P_{s,t}(f \circ \gamma_r)(\sigma) = P_{rt,rt}f \circ \gamma_r(\sigma), \qquad (4.5)$$

*holds for all $f \in \mathcal{C}$ and $0 \leq s < t, r > 0$. Further any selfsimilar convolution hemigroup satisfies Condition (**D**).*

Proof. The first assertion is verified easily. The second assertion is shown in [Ku4].

Let $d\gamma_r$ be the differential of the automorphism γ_r. Then $\{d\gamma_r\}_{r>0}$ is a one parameter group of automorphisms of $\mathcal{G}$. It satisfies $d\gamma_r X \to 0$ as $r \to 0$ for any $X \in \mathcal{G}$. The linear map $d\gamma_r$ is represented by $d\gamma_r = \exp(\log r)Q = r^Q$, where Q is a linear map of $\mathcal{G}$ such that all of its eigenvalues have positive real parts. The linear map Q is called the *exponent* of the dilation $\{\gamma_r\}_{r>0}$.

Theorem 4.3 *Let $\varphi_t, t \geq 0$ be a process with independent increments on a simply connected nilpotent Lie group G equipped with a dilation $\{\gamma_r\}$, satisfying Condition (**D**). It is selfsimilar with respect to the dilation $\{\gamma_r\}$, if and only if the generating $\mathbf{R}^d$-valued process $Z(t), t \geq 0$ is selfsimilar with respect to the dilation $\{r^Q\}$.*

Proof. Let $Z_t \equiv Z(t)$ be the generating process of φ_t. Then it satisfies the stochastic differential equation (4.2). Now, apply (4.2) for the function $f \circ \gamma_r(\sigma)$ and set $\varphi_t^{(\gamma_r)} = \gamma_r(\varphi_t)$. Then we have

$$
\begin{aligned}
f(\varphi_t^{(\gamma_r)}) &= f(\varphi_s^{(\gamma_r)}) + \sum_{j=1}^d \int_s^t d\gamma_r X_j f(\varphi_{u-}^{(\gamma_r)}) dZ_u^j \\
&\quad + \frac{1}{2} \sum_{i,j=1}^d \int_s^t a^{ij}(u) d\gamma_r X_i d\gamma_r X_j f(\varphi_{u-}^{(\gamma_r)}) du \\
&\quad + \sum_{u \leq t} \{ f(\varphi_{u-}^{(\gamma_r)} \xi^{(\gamma_r)}(\Delta Z_u)) - f(\varphi_{u-}^{(\gamma_r)}) - \sum_j d\gamma_r X_j f(\varphi_{u-}^{(\gamma_r)}) \Delta Z_u^j \},
\end{aligned}
$$

where $\xi^{(\gamma_r)}(x) = \exp \sum x_i d\gamma_r X_i$. Therefore, the generating process of $\varphi_t^{(\gamma_r)}, t \geq 0$ is $r^Q Z(t), t \geq 0$, since $r^Q = d\gamma_r$. On the other hand, the generating process of $\varphi_t^{(r)} \equiv \varphi_{rt}, t \geq 0$ is $Z^{(r)}(t) \equiv Z(rt), t \geq 0$. Consequently, $\{\varphi_t^{(r)}, t \geq 0\} \cong \{\varphi_t^{(\gamma_r)}, t \geq 0\}$ (equivalent in the sense of the law) holds if and only if $\{r^Q Z(t), t \geq 0\} \cong \{Z^{(r)}(t), t \geq 0\}$. This shows that $\varphi_t, t \geq 0$ is selfsimilar with respect to $\{\gamma_r\}$ if and only if $Z(t), t \geq 0$ is selfsimilar with respect to $\{r^Q\}$. The proof is complete.

Corollary 4.4 *Let $\{\mu_{s,t}\}$ be a convolution hemigroup on a simply connected nilpotent Lie group G equipped with a dilation $\{\gamma_r\}$ satisfying Condition (**D**). Let $\{\tilde{\mu}_{s,t}\}$ be its generating convolution hemigroup on $\mathcal{G}$. Then $\{\mu_{s,t}\}$ is selfsimilar with respect to the dilation $\{\gamma_r\}$ if and only if $\{\tilde{\mu}_{s,t}\}$ is selfsimilar with respect to $\{r^Q\}$, where $r^Q = d\gamma_r$.*

5. Ornstein-Uhlenbeck type processes and their limit distributions

Let μ be a distribution on a Lie group G. It is called *infinitely divisible in the generalized sense* if for any $\epsilon > 0$, there exist distributions $\mu_1, ..., \mu_n$ such that $\mu_k(U_\epsilon^c) < \epsilon, k = 1, ..., n$ and $\mu = \mu_1 * \cdots * \mu_n$, where U_ϵ is an ϵ-neighborhood of e. In particular if we can choose $\mu_1 = \cdots = \mu_n$, the distribution μ is called *infinitely divisible* as usual. Any distribution $\mu_{s,t}$ of a convolution hemigroup $\{\mu_{s,t}\}_{0 \leq s < t < \infty}$ is infinitely divisible in the generalized sense and any distribution μ_t of a convolution semigroup $\{\mu_t\}_{0 < t < \infty}$ is infinitely divisible.

Let β be an automorphism of the Lie group G. A distribution μ on G is called *β-decomposable* if there exists a distribution μ_β such that $\mu = \beta\mu * \mu_\beta$. Let $\{\gamma_r\}$ be a dilation on G. A distribution μ is called *$\{\gamma_r\}$-selfdecomposable* if it is γ_r-decomposable for any $0 < r < 1$. If $\{\mu_{s,t}\}$ is selfsimilar with respect to $\{\gamma_r\}$, $\mu_{0,t}$ is $\{\gamma_r\}$-selfdecomposable for any $t > 0$. In [Ku4] we show that a selfdecomposable distribution μ can be embedded in a convolution hemigroup $\{\mu_{s,t}\}$ such that $\mu = \mu_{0,1}$, provided that μ is infinitely divisible in the generalized sense.

Sato-Yamazato [SY] introduces a certain Ornstein-Uhlenbeck type process on a Euclidean space and shows that any operator-selfdecomposable distribution can be obtained as a limit distribution of the transition probabilities of the Ornstein-Uhlenbeck type process. In this section we will introduce a similar Ornstein-Uhlenbeck type process on a simply connected nilpotent Lie group G equipped with a dilation $\{\gamma_r\}$. It will be shown that any transition probabilities of the Ornstein-Uhlenbeck type process can be obtained from a selfsimilar convolution hemigroup $\{\mu_{s,t}\}$ and vice versa. It will turn out that the limit distribution of the Ornstein-Uhlenbeck type process is nothing but the distribution $\mu_{0,1}$.

Given a dilation $\{\gamma_r\}$ on the Lie group G, we set $\delta_t = \gamma_{e^t}$, $-\infty < t < \infty$. Then $\{\delta_t\}$ is a one parameter group of diffeomorphisms of G. Let $\mathbf{Y}$ be its infinitesimal generator ($=$ a complete C^∞-vector field):

$$\mathbf{Y}f(\sigma) = \lim_{t \to 0} \frac{f(\delta_t(\sigma)) - f(\sigma)}{t} = \lim_{t \to 1} \frac{f(\gamma_t(\sigma)) - f(\sigma)}{t - 1}. \tag{5.1}$$

Note that $\mathbf{Y}$ is not an element of $\mathcal{G}$. Indeed, it is represented by

$$\mathbf{Y}f(\sigma) = \sum_{j,k} q_{jk} x_j(\sigma) X_k f(\sigma), \tag{5.2}$$

where (q_{ij}) is the exponent of the dilation $\{\gamma_r\}$ and $\sigma = \exp \sum x_j(\sigma) X_j$. It satisfies

$$\mathbf{Y}f(\sigma\tau) = \mathbf{Y}(f \circ L_\sigma)(\tau) + \mathbf{Y}(f \circ R_\tau)(\sigma), \tag{5.3}$$

where L_σ (R_σ) is the left (right) translation by σ, i.e., $f \circ L_\sigma(\tau) = f(\sigma\tau)$ ($f \circ R_\sigma(\tau) = f(\tau\sigma)$). Set $\mathcal{C}_2^0 = \{f \in \mathcal{C}_2; \mathbf{Y}f \in \mathcal{C}\}$.

Let $\mathbf{L}$ be an integro-differential operator such that

$$
\begin{aligned}
\mathbf{L}f(\sigma) &= \frac{1}{2}\sum_{i,j} a^{ij} X_i X_j f(\sigma) + \sum_i b^i X_i f(\sigma) \\
&+ \int_{\mathbf{R}^d} \{f(\sigma\xi(x)) - f(\sigma) - \sum_i \frac{x_i}{1 + |x|^2} X_i f(\sigma)\} \hat{\nu}(dx),
\end{aligned} \tag{5.4}
$$

where $a = (a^{ij})$ is a symmetric nonnegative definite matrix, $b = (b^i)$ is a vector and $\hat{\nu}$ is a Levy measure on $\mathbf{R}^d$. It is invariant with respect to L_σ, i.e., for any $f \in \mathcal{C}_2$ it holds $\mathbf{L}(f \circ L_\sigma)(\tau) = \mathbf{L}f(\sigma\tau)$.

Let $\{p_t(\sigma, A), t \geq 0, \sigma \in G, A \in \mathcal{B}(G)\}$ be a family of transition probabilities over the Lie group G such that $\{Q_t, t > 0\}$ defined by

$$Q_t f(\sigma) = \int_G f(\tau) p_t(\sigma, d\tau), \tag{5.5}$$

makes a strongly continuous semigroup of linear operators on $\mathcal{C}$. Suppose that the domain $\mathcal{D}(\mathbf{G})$ of the infinitesimal generator $\mathbf{G}$ includes $\mathcal{C}_2^0$ and $\mathbf{G}f, f \in \mathcal{C}_2^0$ is represented by

$$\mathbf{G}f = (\mathbf{L} - \mathbf{Y})f. \tag{5.6}$$

Then the family of transition probabilities $\{p_t(\sigma, \cdot)\}$ is called an *Ornstein-Uhlenbeck type process* associated with the operator (5.6) or associated with $\mathbf{L}$ and $\{\gamma_r\}$.

In the remainder of this paper, we shall consider convolution hemigroups under a slightly more general setting. Let $\{\mu_{s,t}\}_{0<s<t<\infty}$ be a two parameter family of probability distributions on a Lie group G satisfying the hemigroup property and the continuity property mentioned in Section 2. Note that $\mu_{0,t}$ is not defined. We shall call it a *convolution hemigroup, open at 0*. If $\lim_{s\downarrow 0} \mu_{s,t} =: \mu_{0,t}$ exists as a probability distribution, then $\{\mu_{s,t}\}_{0\leq s<t<\infty}$ is clearly a convolution hemigroup. In such a case we will call it *closable*. The selfsimilar property can be defined for convolution hemigroup, open at 0, too.

Theorem 5.1 *Let $\{\mu_{s,t}\}_{0<s<t<\infty}$ be a convolution hemigroup, open at 0, selfsimilar with respect to a dilation $\{\gamma_r\}$. Define*

$$p_t(\sigma, A) = \mu_{e^{-t},1}(\delta_{-t}(\sigma)^{-1}A). \tag{5.7}$$

Then it is an Ornstein-Uhlenbeck type process associated with $\mathbf{L}(1)$ and $\{\gamma_r\}$, where $\mathbf{L}(1)f$ is the representation formula (2.6) of the infinitesimal generator of $\{\mu_{s,t}\}$ at time 1.

Proof. Define $Q_t f(\sigma)$ using the above $p_t(\sigma, A)$. Then we have

$$Q_t f(\sigma) = \mu_{e^{-t},1}(f \circ L_{\delta_{-t}(\sigma)}) = (P_{e^{-t},1}f) \circ \delta_{-t}(\sigma), \tag{5.8}$$

where $P_{s,t}$ is the linear operator defined by (2.1). We shall prove the semigroup property $Q_s Q_t = Q_{s+t}$. Note that $Q_s Q_t f = P_{e^{-s},1}(P_{e^{-t},1}f \circ \delta_{-t}) \circ \delta_{-s}$. Since $\{\mu_{s,t}\}$ is selfsimilar with respect to $\{\gamma_r\}$, $P_{e^{-s},1}(g \circ \delta_{-t}) = P_{e^{-(s+t)},e^{-t}}(g) \circ \delta_{-t}$ holds by Proposition 4.1. Then $Q_s Q_t f(\sigma)$ is equal to

$$P_{e^{-(s+t)},e^{-t}}(P_{e^{-t},1}f) \circ \delta_{-(s+t)} = (P_{e^{-(s+t)},1}f) \circ \delta_{-(s+t)} = Q_{s+t}f.$$

We shall next compute its infinitesimal generator $\mathbf{G}$. Suppose $f \in C_2^0$. We have by (5.8)

$$Q_h f(\sigma) - f(\sigma) = \int_{e^{-h}}^1 P_{e^{-h},r}(\mathbf{L}(r)f)(\delta_{-h}(\sigma))dr + \{f(\delta_{-h}(\sigma)) - f(\sigma)\}.$$

Divide each term of the above by h and let $h \to 0$. Then we find that $C_2^0 \subset \mathcal{D}(\mathbf{G})$ and $\mathbf{G}f = (\mathbf{L}(1) - \mathbf{Y})f$. The proof is complete.

Theorem 5.2 *Let $\{p_t(\sigma, \cdot)\}$ be an Ornstein-Uhlenbeck type process associated with the operator $\mathbf{L}$ of (5.4) and the dilation $\{\gamma_r\}$. Define $\mu_{s,t}$ by*

$$\mu_{s,t}(A) = p_{\log t/s}(e, \gamma_t^{-1}(A)). \tag{5.9}$$

Then $\{\mu_{s,t}\}_{0<s<t<\infty}$ is a convolution hemigroup open at 0, whose infinitesimal generator $\mathbf{A}(1)$ (at time 1) is represented by (5.4). It is selfsimilar with respect to the dilation $\{\gamma_r\}$.

Proof. Since the operator $\mathbf{Y}$ satisfies (5.3) and $\mathbf{L}$ is invariant with respect to the left translation L_σ, the infinitesimal generator $\mathbf{G}$ of the Ornstein-Uhlenbeck type process satisfies

$$\mathbf{G}f(\sigma\tau) = \mathbf{G}(f \circ L_\sigma)(\tau) - \mathbf{Y}(f \circ R_\tau)(\sigma). \tag{5.10}$$

Then its semigroup Q_t should satisfy $Q_t f(\sigma\tau) = Q_t(f \circ L_{\gamma_{e-t}(\sigma)})(\tau)$. In particular, the transition probabilities satisfy $p_t(\sigma, A) = p_t(e, \gamma_{e-t}(\sigma)^{-1}A)$.

Now we shall consider the distribution $\mu_{s,t}$ defined by (5.9). We have

$$\mu_{s,t}(\sigma^{-1}A) = p_{\log t/s}(e, \gamma_t^{-1}(\sigma^{-1})\gamma_t^{-1}(A)) = p_{\log t/s}(\gamma_s^{-1}(\sigma), \gamma_t^{-1}(A)).$$

Therefore,

$$P_{s,t}f(\sigma) = \int f(\tau)\mu_{s,t}(\sigma^{-1}d\tau) = Q_{\log t/s}(f \circ \gamma_t)(\gamma_s^{-1}(\sigma)).$$

This implies for $0 < s < t < u < \infty$,

$$
\begin{aligned}
P_{s,t}P_{t,u}f(\sigma) &= Q_{\log t/s}(Q_{\log u/t}(f \circ \gamma_u) \circ \gamma_t^{-1} \circ \gamma_t)(\gamma_s^{-1}(\sigma)) \\
&= Q_{\log t/s}Q_{\log u/t}(f \circ \gamma_u)(\gamma_s^{-1}(\sigma)) \\
&= Q_{\log u/s}(f \circ \gamma_u)(\gamma_s^{-1}(\sigma)) = P_{s,u}f(\sigma),
\end{aligned}
$$

proving the hemigroup property $\mu_{s,t} * \mu_{t,u} = \mu_{s,u}$.

We shall differentiate $P_{s,t}f$ for $t > s$. Suppose $f \in \mathcal{C}_2^0$. Since

$$\frac{\partial}{\partial t}(f \circ \gamma_t) = \frac{\partial}{\partial t}(f \circ \delta_{\log t}) = \frac{1}{t}\frac{\partial}{\partial u}(f \circ \delta_u)\Big|_{u=\log t} = \frac{1}{t}\mathbf{Y}(f \circ \gamma_t),$$

we have

$$
\begin{aligned}
\frac{\partial}{\partial t}P_{s,t}f(\sigma) &= \frac{\partial}{\partial t}Q_{\log t/s}(f \circ \gamma_t)(\gamma_s^{-1}(\sigma)) \\
&= \frac{1}{t}Q_{\log t/s}\mathbf{G}(f \circ \gamma_t)(\gamma_s^{-1}(\sigma)) + \frac{1}{t}Q_{\log t/s}\mathbf{Y}(f \circ \gamma_t)(\gamma_s^{-1}(\sigma)).
\end{aligned}
$$

Then the infinitesimal generator $\mathbf{A}(t)$ of $\{\mu_{s,t}\}$ is represented by

$$
\begin{aligned}
\mathbf{L}(t)f(\sigma) &= \lim_{s\uparrow t}\frac{\partial}{\partial t}P_{s,t}f(\sigma) = \frac{1}{t}\mathbf{G}(f \circ \gamma_t)(\gamma_t^{-1}(\sigma)) + \frac{1}{t}\mathbf{Y}(f \circ \gamma_t)(\gamma_t^{-1}(\sigma)) \\
&= \frac{1}{t}\mathbf{L}(f \circ \gamma_t)(\gamma_t^{-1}(\sigma)).
\end{aligned}
$$

Therefore $\{\mu_{s,t}\}$ is selfsimilar with respect to the dilation $\{\gamma_r\}$. See [Ku4]. The proof is complete.

Corollary 5.3 *Transition probabilities $p_t(\sigma, \cdot)$ of Ornstein-Uhlenbeck type process are infinitely divisible in the generalized sense. Further, if $\nu = 0$ holds in the operator $\mathbf{L}$ of (5.4), transition probabilities are Gaussian distributions in the generalized sense.*

The equality (4.5) implies that if $f \in \mathcal{D}(\{\mathbf{A}(t)\})$, then $f \circ \gamma_r \in \mathcal{D}(\{\mathbf{A}(t)\})$ and it satisfies

$$\mathbf{A}(t)f(\sigma) = \frac{1}{t}\mathbf{A}(1)(f \circ \gamma_t)(\gamma_t^{-1}(\sigma)). \tag{5.11}$$

Therefore $\mathbf{A}(1)$ or its representation $\mathbf{L}$ of (5.4) determines the selfsimilar convolution hemigroup $\mu_{s,t}$, open at 0. It is shown in [Ku4] that the self-similar convolution hemigroup, open at 0, is closable if and only if the Lévy measure $\hat{\nu}$ in (5.4) satisfies the integrability condition:

$$\int_{\mathbf{R}^d} \log(1 + |x|^2)\hat{\nu}(dx) < \infty. \tag{5.12}$$

The next corollary indicates that the above condition is again necessary and sufficient that the Ornstein-Uhlenbeck type process is positive recurrent.

Theorem 5.4 *Let $\{p_t(\sigma, A)\}$ be the transition probabilities of the Ornstein-Uhlenbeck type process associated with the operator $\mathbf{L}$ of (5.4) and the dilation $\{\gamma_r\}$. Then for any σ, $\{p_t(\sigma, \cdot)\}$ converges weakly to a unique distribution μ as $t \to \infty$ if and only if Condition (5.12) is satisfied. Further the limit distribution is selfdecomposable with respect to the dilation $\{\gamma_r\}$.*

Proof. Consider the convolution hemigroup $\{\mu_{s,t}\}$, open at 0 defined by (5.9). Then it satisfies (5.7). If Condition (5.12) is satisfied, then $\mu_{e^{-t},1} \to \mu_{0,1}$ and $\gamma_{e^{-t}}(\sigma) \to e$ as $t \to \infty$. Therefore $p_t(\sigma, \cdot)$ converges weakly to $\mu_{0,1}$ for any σ. Further $\mu = \mu_{0,1}$ is selfdecomposable with respect to $\{\gamma_r\}$.

Conversely, if Condition (5.12) is not satisfied, then $\mu_{e^{-t},1}$ does not converge. Therefore, $p_t(\sigma, A)$ does not converge. The proof is complete.

References

[AK] D. Applebaum, H. Kunita, Lévy flows on manifolds and Lévy processes on Lie groups, *Kyoto J. Math.*, **33** (1993), 1103-1123.

[HP1] H. Heyer, G. Pap, Convergence of noncommutative triangular arrays of probability measures on a Lie group, submitted to *J. Theor. Probab.* (1996).

[HP2] H. Heyer, G. Pap, Convolution hemigroups of bounded variation on a Lie projective group, preprint.

[Hu] G.A. Hunt, Semigroups of measures on Lie groups, *Trans. Amer. Math. Soc.*, **81** (1956), 264-293.

[JM] Z.J. Jurek, J.D. Mason, *Operator-limit distributions in probability theory*, J. Wiley & Sons, 1993.

[Ku1] H. Kunita, Convergence of stochastic flows with jumps and Lévy processes in diffeomorphisms group, *Ann. Inst. Henri Poincaré*, **22** (1986), 297-321.

[Ku2] H. Kunita, Stable Lévy processes on Lie groups, *Stochastic analysis on infinite dim. space*, ed. by Kunita-Kuo, Pitman Research Notes in Math. 310 (1994), 167-182.

[Ku3] H. Kunita, Convolution semigroups of stable distributions over a nilpotent Lie group, *Proc. Japan Acad.* **79**. Ser. A (1994), 305-310.

[Ku4] H. Kunita, Infinitesimal generators of nonhomogeneous convolution semigroups on Lie groups, *Osaka Journal of Mathematics*, to appear.

[Sa] K. Sato, Self-similar processes with independent increments, *Probab. Th. Rel. Fields*, **89** (1991), 285-300.

[SY] K. Sato, M. Yamazato, Operator-selfdecomposable distributions as limit distributions of processes of Ornstein-Uhlenbeck type, *Stochastic Process Appl.*, **17** (1984), 73-100.

Kyushu University
Fukuoka 812, Japan

Received October 1, 1996

Optimal Damping of Forced Oscillations in Discrete-time Systems by Output Feedback [1]

Anders Lindquist and Vladimir A. Yakubovich

Abstract

In this paper we consider optimal control by output feedback of a linear discrete-time system corrupted by an additive harmonic vector disturbance with known frequencies but unknown amplitudes and phases. We consider both a deterministic and a stochastic version of the problem. The object is to design a robust optimal regulator which is universal in the sense that it does not depend on the unknown amplitudes and phases and is optimal for all choices of these values. We show that, under certain natural technical conditions, an optimal universal regulator (OUR) exists in a suitable class of stabilizing and realizable linear regulators, provided the dimension of the output is no smaller than the dimension of the harmonic disturbance. When this dimensionality condition is not satisfied, the existence of an OUR is not a generic property, and consequently it does not exist from a practical point of view. For the deterministic problem we also show that, under slightly stronger technical conditions, any linear OUR is also optimal in a very wide class of nonlinear regulators. In the stochastic case we are only able to show optimality in the linear class of regulators.

1. Introduction

Many important engineering problems can be formulated mathematically as a linear-quadratic regulator problem with the added complication of an unobserved harmonic additive disturbance, for which only the frequencies are known. Some examples, among many others, are vibration damping in industrial machines and helicopters [FF, Fr, GEI, Gu, BLS, WKK], noise reduction in vehicles and transformers [SBKB], control of aircraft in the presence of wind shear [Le, Mi, ZB], and control of the roll motion of a ship

[1]This research was supported in part by grants from the Royal Swedish Academy of Sciences, INTAS, NUTEK and the Swedish Foundation for Strategic Research.

[KO]. Such a harmonic disturbance adds critically stable dynamics which is unobservable and unstabilizable, and therefore traditional linear-quadratic methods cannot be used.   Nor can one in general use a discrete-time version of the methods proposed in [Fr, FW]. In [LY] this problem was solved in the case of complete state information. In the present paper, the same methodology is extended to take care of the case of output feedback.

Specifically, we consider a discrete-time linear system

$$x_{t+1} = Ax_t + Bu_t + Ew_t \tag{1.1}$$
$$y_t = Cx_t \tag{1.2}$$

($t = 0, 1, 2, \ldots$) with state $x_t \in \mathbb{R}^n$, output $y_t \in \mathbb{R}^m$ and two vector inputs, namely a control $u_t \in \mathbb{R}^k$ an unobserved disturbance $w_t \in \mathbb{R}^\ell$ which we shall take to be harmonic with known frequencies but unknown amplitudes and phases. More precisely,

$$w_t = \sum_{j=1}^{N} w^{(j)} e^{i\theta_j t}, \tag{1.3}$$

where the frequencies

$$-\pi < \theta_1 < \theta_2 < \ldots < \theta_N \leq \pi \tag{1.4}$$

are known, but the complex vector amplitudes $w^{(1)}, w^{(2)}, \ldots, w^{(N)}$, in which the phases have been absorbed, are not. Moreover, A, B, C, E are constant real matrices of appropriate dimensions such that (A, B) is stabilizable and (C, A) is detectable, and without loss of generality

$$rank\ C = m \quad \text{and} \quad rank\ E = \ell. \tag{1.5}$$

In fact, if the first condition is not satisfied, some components of y_t could be eliminated. Moreover, if $\hat{\ell} := rank\ E < \ell$, Ew_t may be exchanged by $\hat{E}\hat{w}_t$, where $\hat{w}_t \in \mathbb{R}^{\hat{\ell}}$ by an obvious reformulation. Of course, (1.5) implies that $m \leq n$ and that $\ell \leq n$.

The deterministic problem to be considered in this paper is to damp the forced oscillation in the system (1.1) by output feedback. This is to be done so as to minimize a cost functional

$$\Phi = \limsup_{T \to \infty} \frac{1}{T} \sum_{t=0}^{T} \Lambda(x_t, u_t), \tag{1.6}$$

where $\Lambda(x, u)$ is a real quadratic form

$$\Lambda(x, u) = \begin{pmatrix} x \\ u \end{pmatrix}^* \begin{pmatrix} Q & S \\ S^* & R \end{pmatrix} \begin{pmatrix} x \\ u \end{pmatrix} \tag{1.7}$$

with properties to be specified in Section 2. This functional is a measure of the forced oscillations in the closed-loop system and, for the classes of admissible regulators to be defined next, it does not depend on initial conditions.

What we want to construct is a regulator which is optimal in some suitable class and which does not depend on the unknown complex vector amplitudes $w^{(1)}, w^{(2)}, \ldots, w^{(N)}$ and consequently is *universal* in the sense that it simultaneously solves the complete family of optimization problems corresponding to different choices of these complex amplitudes. Such a regulator will be referred to as an *optimal universal regulator* (OUR). Moreover, this optimal regulator must be robust with respect to possible estimation errors in the known frequencies $\theta_1, \theta_2, \ldots, \theta_N$ in the sense that the cost Φ is continuous in the estimation errors and tends to its true optimal value as the errors tend to zero. It is not hard to see that there are optimal regulators which depend on $w^{(1)}, w^{(2)}, \ldots, w^{(N)}$, but that there actually exist universal ones is perhaps surprising.

At first sight it might be tempting to try to apply standard linear-quadratic regulator theory to an extended control system obtained by amending to (1.1) the critically stable autonomous system

$$\begin{cases} z_{t+1} = F z_t \\ w_t = H z_t \end{cases} \tag{1.8}$$

where F and H are matrices of dimensions $N\ell \times N\ell$ and $\ell \times N\ell$ respectively given by

$$F = \begin{bmatrix} e^{i\theta_1} I_\ell & 0 & \cdots & 0 \\ 0 & e^{i\theta_2} I_\ell & \cdots & 0 \\ \vdots & \vdots & \ddots & \vdots \\ 0 & 0 & \cdots & e^{i\theta_N} I_\ell \end{bmatrix} \qquad H = \begin{bmatrix} I_\ell & I_\ell & \cdots & I_\ell \end{bmatrix} \tag{1.9}$$

and $z_0 := \mathrm{col}(w^{(1)}, w^{(2)}, \ldots, w^{(N)}) \in \mathbb{C}^{N\ell}$. In this context, universality of a regulator would imply that it does not depend on z_0 and that it is optimal for all z_0. However, since this would add uncontrollable, critically stable modes to the system, standard linear-quadratic regulator theory does not apply.

If we were to consider the simple optimization problem to find a *process* (x_t, u_t) minimizing Φ subject to the constraints (1.1a), it would, as pointed out in [LY], be necessary to assume that

$$\frac{1}{\sqrt{t}} |x_t| \to 0 \quad \text{as } t \to \infty. \tag{1.10}$$

This is a weak stability condition for the closed-loop system, and it guarantees that the cost is finite. We shall denote by $\mathcal{A}$ the class of all processes (x_t, u_t) satisfying (1.1a) and this stability condition. (To insure that the infimum of Φ over it is not $-\infty$, we must of course introduce some condition on the quadratic form (1.7). This will be done in Section 2.)

However, this class of admissible processes is much too large since we would like to optimize by feedback and in such a way that the optimal regulator is universal. Therefore we shall consider two classes of regulators, one that is linear and one that allows for nonlinear regulators.

Let $\mathcal{L}$ be the class of all linear realizable and stabilizing regulators

$$\mathcal{L}: \quad D(\sigma)u_t = N(\sigma)y_t, \qquad (1.11)$$

where $D(\lambda)$ and $N(\lambda)$ are real matrix polynomials of dimensions $k \times k$ and $k \times m$ respectively, and σ is the forward shift $\sigma y_t = y_{t+1}$. Here *realizable* means that the leading coefficient matrix of $D(\lambda)$ is nonsingular and $\deg N \le \deg D$ so that $D(\lambda)^{-1}N(\lambda)$ is a proper rational matrix function. By *stabilizing* we mean that the coefficient matrix of the closed-loop system is asymptotically stable, i.e., the matrix polynomial

$$\Xi(\lambda) = \begin{bmatrix} \lambda I_n - A & -B \\ -N(\lambda)C & D(\lambda) \end{bmatrix} \qquad (1.12)$$

is such that $\det \Xi(\lambda) \ne 0$ for $|\lambda| \ge 1$. (Here, of course, I_n is the $n \times n$ identity matrix.)

Secondly, we consider a class $\mathcal{N}$ of in general nonlinear regulators

$$\mathcal{N}: \quad u_t = f_t(y_t, y_{t-1}, \dots, y_0, u_{t-1}, \dots, u_0), \qquad (1.13)$$

which are stabilizing in the sense that the processes (x_t, u_t), generated by such feedback controls all belong to $\mathcal{A}$. Clearly, $\mathcal{L} \subset \mathcal{N}$. It is trivial but useful for the subsequent analysis to note that, if $\mathcal{A}_\mathcal{L}$ and $\mathcal{A}_\mathcal{N}$ are the classes of processes (x_t, u_t) which are generated by the regulators in $\mathcal{L}$ and $\mathcal{N}$ respectively, then

$$\mathcal{A}_\mathcal{L} \subset \mathcal{A}_\mathcal{N} \subset \mathcal{A}. \qquad (1.14)$$

In this paper we show that, under suitable technical conditions, there exists an optimal universal regulator in the linear class $\mathcal{L}$, provided $\ell \le m$, and that this regulator is also OUR in the class $\mathcal{N}$ under slightly stronger conditions. The optimal universal regulator is not unique so a general description of all such regulators is obtained. If $\ell > m$, an OUR will exist only when the system parameters satisfy certain equations, making the existence of an OUR a nongeneric property. From a practical point of view this implies that there is no OUR if $\ell > m$. Nonuniversal optimal regulators are given in Section 6 in the case of nonexistence of an OUR.

We stress that our solutions are optimal in the sense stated in this paper only, and that other desirable design specifications may not be satisfied for an arbitrary universal optimal regulator. Therefore it is an important property of our procedure that the OUR is not uniquely defined, allowing for a considerable degree of design freedom to satisfy these additional specifications..

Next, let us consider a stochastic version of this problem. Merely replacing the amplitudes $w^{(1)}, w^{(2)}, \ldots, w^{(N)}$ by (jointly distributed) random vectors and the cost function (1.6) by

$$\Phi = \lim_{T \to \infty} \mathrm{E}\{\frac{1}{T} \sum_{t=0}^{T} \Lambda(x_t, u_t)\}, \tag{1.15}$$

where $\mathrm{E}\{\cdot\}$ denotes mathematical expectation, would, as explained in Section 8 , amount to a trivial extension of the deterministic problem formulation described above. It turns out that the appropriate stochastic problem is obtained by replacing (1.8) by the stochastic system

$$\begin{cases} z_{t+1} = Fz_t + Gv_t \\ w_t = Hz_t + Kv_t \end{cases} \tag{1.16}$$

where z_0 a random vector and where $v_0, v_1, v_2, \ldots$ is a white noise process with an intensity V_t tending sufficiently quickly to zero to insure that $\mathrm{E}\{|w_t|^2\}$ remains bounded. An optimal regulator is *universal* for this problem if it does not depend on z_0, G, K and V_t, and it is optimal for all values of these quantities. In Section 8 we show that any optimal universal regulator for the deterministic control problem is an optimal universal regulator for the stochastic problem, at least in the class $\mathcal{L}$. Whether it is also optimal in the class $\mathcal{N}$ is still an open question.

The outline of our paper is as follows. Section 2 is a preliminary section in which we introduce some technical conditions to be used in different contexts later. Sections 3–5 are devoted to the underlying optimization problem over $\mathcal{A}_{\mathcal{L}}$. This is done in a simpler way, but somewhat more limited context, than in [LY], and therefore we shall need the more general result of [LY] later in Section 7 for the case of nonlinear regulators. In Section 4 a parameterization of all regulators in $\mathcal{L}$, akin to that of Youla but especially adapted to our present problem, is introduced. Section 5 presents the design of the OUR in $\mathcal{L}$ in the case that $\ell \leq m$, and Section

6 considers the case $\ell > m$. In Section 7 the existence of optimal universal regulators in the class $\mathcal{N}$ is studied. The theorems stated there could be regarded as our main result. Finally, Section 8 is devoted to the stochastic case. The results of this paper were announced, without proofs, in [LY2].

2. Assumptions and definitions

In this section we introduce some technical conditions to be referred to later in this paper.

First we need to specify required properties of the quadratic form (1.7). It could be indefinite, but in order to insure that the cost function (1.6) is bounded from below we must introduce some positivity condition.

Strong frequency domain condition (SFDC). There is a $\delta > 0$ such that

$$\Lambda(\tilde{x}, \tilde{u}) \geq \delta(|\tilde{x}|^2 + |\tilde{u}|^2) \qquad (2.1)$$

for all $\tilde{x} \in \mathbb{C}^n$, $\tilde{u} \in \mathbb{C}^k$, $\lambda \in \mathbb{C}$ such that $|\lambda| = 1$ and

$$\lambda\tilde{x} = A\tilde{x} + B\tilde{u}. \qquad (2.2)$$

Since $\tilde{x}$ and $\tilde{u}$ are complex, $*$ in (1.7) will of course have to be taken as Hermitian conjugation instead of merely transposition.

For the linear case the following weaker condition will suffice.

Weak frequency domain condition (WFDC). There is a $\delta > 0$ such that (2.1) holds for all $\tilde{x} \in \mathbb{C}^n$, $\tilde{u} \in \mathbb{C}^k$ and

$$\lambda = e^{i\theta_j} \quad j = 1, 2, \ldots, N \qquad (2.3)$$

satisfying (2.2).

Note that both of these conditions are invariant under the action of the feedback group

$$(A, B) \rightarrow (TAT^{-1} + TBK, TB), \qquad (2.4)$$

where T is a nonsingular matrix and K is an arbitrary matrix of appropriate dimensions. Moreover, if A does not have any eigenvalues on the unit circle, SFDC is equivalent to

$$\Lambda(\tilde{x}, \tilde{u}) > 0 \quad \text{for all } \tilde{u} \neq 0, \quad \tilde{x} = (\lambda I - A)^{-1} B\tilde{u} \qquad (2.5)$$

and λ on the unit circle, and WFDC is equivalent to (2.5) for all $\lambda = e^{i\theta_j}, j = 1, 2, \ldots, N$. Therefore, writing

$$\Lambda(\tilde{x}, \tilde{u}) = \tilde{u}^*\Pi(\lambda)\tilde{u} \quad \text{where } \tilde{x} = (\lambda I - A)^{-1} B\tilde{u} \qquad (2.6)$$

and where the Hermitian $k \times k$ matrix function

$$\Pi(\lambda) = B^*(\bar{\lambda}I - A^*)^{-1}Q(\lambda I - A)^{-1}B + B^*(\bar{\lambda}I - A^*)^{-1}S + S^*(\lambda I - A)^{-1}B + R, \tag{2.7}$$

SFDC may be written

$$\Pi(\lambda) > 0 \quad \text{for all } \lambda \text{ on the unit circle} \tag{2.8}$$

and WFDC as

$$\Pi(e^{i\theta_j}) > 0 \quad \text{for } j = 1, 2, \ldots, N. \tag{2.9}$$

Secondly, without loss of generality, we also assume that the matrix A is stable, i.e.,

$$\det(\lambda I - A) \neq 0 \quad \text{for } |\lambda| \geq 1. \tag{2.10}$$

In fact, if it is not, we can always stabilize by dynamic feedback, in general at the price of increased dimension of the system. Under very special conditions (see, for example, [Ki]), there is a matrix K such that $\Gamma := A + BKC$ is a stable matrix, and then the feedback law

$$u_t = KCx_t + v_t \tag{2.11}$$

allows us to exchange (1.1a) for a similar system where A and u_t are exchanged for $A + BKC$ and v_t respectively. In general, however, an observer must be used. As is well-known, one can always use the controller

$$\begin{cases} \hat{x}_{t+1} = A\hat{x}_t + Bu_t + L(y_t - C\hat{x}_t) \\ u_t = K\hat{x}_t + v_t \end{cases} \tag{2.12}$$

leading to a closed-loop system

$$\begin{bmatrix} x_{t+1} \\ \hat{x}_{t+1} \end{bmatrix} = \begin{bmatrix} A & BK \\ LC & A + BK - LC \end{bmatrix} \begin{bmatrix} x_t \\ \hat{x}_t \end{bmatrix}$$
$$+ \begin{bmatrix} B \\ B \end{bmatrix} v_t + \begin{bmatrix} E \\ 0 \end{bmatrix} w_t \tag{2.13}$$

$$y_t = \begin{bmatrix} C & 0 \end{bmatrix} \begin{bmatrix} x_t \\ \hat{x}_t \end{bmatrix} \tag{2.14}$$

which has precisely the form (1.1). Pre- and postmultiplying the new "A-matrix" with $\begin{bmatrix} I & 0 \\ I & -I \end{bmatrix}$ we see that it has the characteristic polynomial

$$\det(\lambda I - A - BK)\det(\lambda I - A + LC), \tag{2.15}$$

as is well-known. Due to stabilizability of (A, B) and detectability of (C, A), by the Pole Assignment Theorem, K and L can be chosen so that (2.15) has all its roots in the open unit disc as required. The dynamic regulator (2.12) is, however, never minimal, and observers of lower dimension can be found in any standard textbook on the subject (see, e.g., [AM, KS]).

3. The auxiliary optimization problem

We now consider the problem of minimizing the cost function (1.6) subject to (1.1) over the class $\mathcal{A}_\mathcal{L}$ of admissible processes (x_t, u_t) corresponding to regulators (1.11) in the linear class $\mathcal{L}$.

To this end, let Ψ_x, Ψ_u and Ψ_y be the transfer functions from the harmonic input Ew_t to x_t, u_t and y_t respectively. Clearly, these rational matrix functions are determined by

$$(\lambda I - A)\Psi_x = B\Psi_u + I \tag{3.1}$$

$$D(\lambda)\Psi_u = N(\lambda)\Psi_y \tag{3.2}$$

$$\Psi_y = C\Psi_x \tag{3.3}$$

Since a regulator in $\mathcal{L}$ is realizable by definition, $W(\lambda) := D(\lambda)^{-1}N(\lambda)$ is a proper rational matrix function, i.e., $W(\infty)$ is finite. But it follows from (3.1), (3.2) and (3.3) that

$$[I - \lambda^{-1}A - \lambda^{-1}BW(\lambda)C]\Psi_x(\lambda) = \lambda^{-1}I,$$

and consequently $\Psi_x(\infty) = 0$. i.e., Ψ_x is strictly proper. Then, by (3.3) the same is true for Ψ_y, and by (3.2) for Ψ_u, i.e., $\Psi_y(\infty) = 0$ and $\Psi_u(\infty) = 0$.

Due to the stability of (1.12), the process (x_t, u_t, y_t) tends asymptotically to the unique harmonic solution

$$x_t = \sum_{j=1}^{N} x^{(j)} e^{i\theta_j t}, \quad u_t = \sum_{j=1}^{N} u^{(j)} e^{i\theta_j t}, \quad y_t = \sum_{j=1}^{N} y^{(j)} e^{i\theta_j t}, \tag{3.4}$$

where

$$x^{(j)} = \Psi_x(\lambda_j)Ew^{(j)}, \quad u^{(j)} = \Psi_u(\lambda_j)Ew^{(j)}, \quad y^{(j)} = \Psi_y(\lambda_j)Ew^{(j)} \tag{3.5}$$

and

$$\lambda_j = e^{i\theta_j}, \quad j = 1, 2, \dots, N. \tag{3.6}$$

Therefore, the usual limit (rather than just limsup) does exist in (1.6), and it is given by

$$\Phi = \sum_{j=1}^{N} \Lambda(x^{(j)}, u^{(j)}). \tag{3.7}$$

To see this, observe that, if f_t and g_t are two harmonic sequences

$$f_t = \sum_{j=1}^{N} f^{(j)} e^{i\theta_j t} \quad \text{and} \quad g_t = \sum_{j=1}^{N} g^{(j)} e^{i\theta_j t}, \tag{3.8}$$

and M is an arbitrary matrix of appropriate dimensions, then

$$\limsup_{T\to\infty} \frac{1}{T} \sum_{j=1}^{T} f_t^* M g_t = \sum_{j=1}^{N} \sum_{k=1}^{N} f^{(j)*} M g^{(k)} \varphi_{jk}, \tag{3.9}$$

where

$$\varphi_{jk} = \lim_{T\to\infty} \frac{1}{T} \sum_{t=0}^{T} e^{i(\theta_k - \theta_j)t}. \tag{3.10}$$

The limit (3.10) does exist and equals one if $j = k$ and zero otherwise. Consequently,

$$\limsup_{T\to\infty} \frac{1}{T} \sum_{t=1}^{T} f_t^* M g_t = \lim_{T\to\infty} \frac{1}{T} \sum_{t=1}^{T} f_t^* M g_t = \sum_{j=1}^{N} f^{(j)*} M g^{(j)}. \tag{3.11}$$

Using this formula (3.7) follows.

Now, in view of the constraint (1.1),

$$x^{(j)} = (\lambda_j I - A)^{-1} (Bu^{(j)} + Ew^{(j)}), \tag{3.12}$$

and therefore

$$\Lambda(x^{(j)}, u^{(j)}) = u^{(j)*} \Pi(\lambda_j) u^{(j)} + p_j^* u^{(j)} + u^{(j)*} p_j + q_j, \tag{3.13}$$

where Π is given by (2.7),

$$p_j = [Q(\lambda_j I - A)^{-1} B + S]^* (\lambda_j I - A)^{-1} E w^{(j)} \tag{3.14}$$

and

$$q_j = w^{(j)*} E^* (\bar{\lambda}_j I - A^*)^{-1} Q(\lambda_j I - A)^{-1} E w^{(j)}. \tag{3.15}$$

Next, consider the auxiliary optimization problem to minimize Φ, given by (3.7) and (3.13)–(3.15), where $u^{(1)}, u^{(2)}, \ldots, u^{(N)}$ are regarded as independent variables in $\mathbb{C}^k$. Since A is stable, $(\lambda_j I - A)^{-1}$ does exist, and WFDC implies that

$$\Pi(\lambda_j) > 0 \quad \text{for } j = 1, 2, \ldots, N. \tag{3.16}$$

Therefore

$$\Lambda(x^{(j)}, u^{(j)}) = (u^{(j)} - \hat{u}^{(j)})^* \Pi(\lambda_j)(u^{(j)} - \hat{u}^{(j)}) + \Phi^{(j)}_{\min}, \qquad (3.17)$$

where

$$\hat{u}^{(j)} = -\Pi(\lambda_j)^{-1} p_j \quad \text{and} \quad \Phi^{(j)}_{\min} = q_j - p_j^* \Pi(\lambda_j)^{-1} p_j, \qquad (3.18)$$

so $u^{(j)} = \hat{u}^{(j)}$ is the solution of the problem to minimize $\Lambda(x^{(j)}, u^{(j)})$ given (3.12). More precisely, $\hat{u}^{(j)}$ has the form

$$\hat{u}^{(j)} = U(\lambda_j) w^{(j)}, \qquad (3.19)$$

where

$$U(\lambda) = -\Pi(\lambda)^{-1} [Q(\lambda I - A)^{-1} B + S]^* (\lambda I - A)^{-1} E. \qquad (3.20)$$

Consequently,

$$\Phi = \sum_{j=1}^{N} (u^{(j)} - \hat{u}^{(j)})^* \Pi(\lambda_j)(u^{(j)} - \hat{u}^{(j)}) + \sum_{j=1}^{N} \Phi^{(j)}_{\min}, \qquad (3.21)$$

and therefore the solution to the auxiliary optimization problem is given by $u^{(j)} = \hat{u}^{(j)}, j = 1, 2, \ldots, N$.

However, the variables $u^{(1)}, u^{(2)}, \ldots, u^{(N)}$ are really not independent since the optimization should be done over the class $\mathcal{A}_{\mathcal{L}}$ of admissible processes and hence tied together via (3.5) and a regulator (1.11) in $\mathcal{L}$ which must be universal and thus not depend on the unknown vector amplitudes $w^{(1)}, w^{(2)}, \ldots, w^{(N)}$. Therefore, we next proceed to characterizing the class of all regulators in $\mathcal{L}$.

First, however, let us observe for later reference that, in view of (3.5), (3.19), (3.18), (3.14) and (3.15), the cost function (3.21) takes the form

$$\Phi = w^* \Omega w, \quad \text{where } w := \begin{bmatrix} w^{(1)} \\ w^{(2)} \\ \vdots \\ w^{(N)} \end{bmatrix} \qquad (3.22)$$

for some symmetric Hermitian matrix Ω which depends on the choice of regulator in $\mathcal{L}$.

4. Parameterization of the class $\mathcal{L}$

Next, we shall present a parameterization of all regulators in $\mathcal{L}$, akin to the Youla parameterization but more suitable for our purposes. To this end, we first need a definition of equivalence.

Definition 4.1 *We shall say that two regulators*

$$D_1(\sigma)u_t = N_1(\sigma)y_t \quad and \quad D_2(\sigma)u_t = N_2(\sigma)y_t$$

are equivalent if there exist matrix polynomials D_0 and N_0, of dimensions $k \times k$ and $k \times m$ respectively, such that

$$D_1 = M_1 D_0, \quad N_1 = M_1 N_0 \quad D_2 = M_2 D_0, \quad N_2 = M_2 N_0$$

for some stable $k \times k$ matrix polynomials M_1 and M_2. We recall that a square matrix polynomial is stable if $\det M(\lambda) \neq 0$ for $|\lambda| \geq 1$.

Clearly, as can be seen from (3.1), Ψ_x, Ψ_u and Ψ_y are invariant under this equivalence.

Lemma 4.2 *(i) Let A be a stable matrix with characteristic polynomial $\chi(\lambda)$, and let $V(\lambda)$ be the matrix polynomial*

$$V(\lambda) = \chi(\lambda)C(\lambda I_n - A)^{-1}. \tag{4.1}$$

Let $\rho(\lambda)$ be an arbitrary stable scalar polynomial and let $R(\lambda)$ be an arbitrary $k \times m$ polynomial such that

$$\deg(RV) < \deg \rho. \tag{4.2}$$

Then the regulator

$$D(\sigma)u_t = N(\sigma)y_t \tag{4.3}$$

with

$$D(\lambda) = \rho(\lambda)I_k + R(\lambda)V(\lambda)B, \quad N(\lambda) = \chi(\lambda)R(\lambda) \tag{4.4}$$

is realizable and stabilizing, and for this regulator

$$\Psi_u(\lambda) = \frac{R(\lambda)}{\rho(\lambda)}V(\lambda), \quad \det \Xi(\lambda) = \chi(\lambda)\left[\rho(\lambda)\right]^k \tag{4.5}$$

where Ξ is given by (1.12).

(ii) Conversely, any realizable and stabilizing regulator (4.3) belongs to the class of regulators (4.3)–(4.4) in the sense that it is equivalent to one in this class.

This is a overparameterization of $\mathcal{L}$ which is an advantage in our application. We can for example chose $\rho = \chi \rho_1$ for some stable scalar polynomial ρ_1 and take the degree of R to be at most $\deg \rho_1$.

Proof.   (i) Let $D(\lambda)$ and $N(\lambda)$ be defined by (4.4). It is evident that the realizability condition holds. For stabilizability we need to show that $\Xi(\lambda)$, as defined by (1.12), is a stable matrix polynomial. We have

$$\det \Xi(\lambda) = \det(\lambda - A) \det[D - NC(\lambda I - A)^{-1}B]. \tag{4.6}$$

But, in view of (4.1) and (4.4), $D - NC(\lambda I - A)^{-1}B = \rho I_k$, and hence the second of equations (4.5) follows. To prove the first of equations (4.5), note that, in view of (3.1) and (4.4),

$$\begin{aligned} N\Psi_y &= NC(\lambda I - A)^{-1}(B\Psi_u + I) \\ &= RVB\Psi_u + RV \end{aligned}$$

and

$$D\Psi_u = \rho\Psi_u + RVB\Psi_u.$$

Therefore, the first of equations (4.5) follows from the second of equations (3.1).

(ii) Next, let (4.3) be an arbitrary regulator in $\mathcal{L}$. Then, Ξ, defined by (1.12), is stable, and, by (4.6),

$$\det \Xi = \chi^{1-k} \det P,$$

where P is the $k \times k$ matrix polynomial

$$P = \chi D - NVB, \tag{4.7}$$

which is stable and nontrivial since $\det P = \chi^{k-1} \det \Xi$ is stable. In view of (3.1) and (3.2),

$$D\Psi_u = NC(\lambda I - A)^{-1}(B\Psi_u + I). \tag{4.8}$$

Solving (4.8), taking (4.1) and (4.7) into account, yields $P\Psi_u = NV$. Then,

$$\Psi_u = P^{-1}NV = \frac{P_a NV}{\det P}, \tag{4.9}$$

where P_a is the adjoint matrix polynomial $P_a := P^{-1} \det P$. Now, choose $\rho := \det P$, which has just been shown to be stable, and $R := P_a N$. Then the first of equations (4.5) holds. Since $\Psi_u(\infty) = 0$ for all regulators in $\mathcal{L}$, this in turn implies that (4.2) holds. Moreover, Ψ_x is given by (3.1), from which it also follows that $\det \Psi_x \not\equiv 0$. Next, define the matrix polynomials

$$\hat{D} = \rho I_k + RVB \quad \text{and} \quad \hat{N} = \chi R.$$

Then, by the first part of the lemma, the linear regulator

$$\hat{D}(\sigma)u_t = \hat{N}(\sigma)y_t \tag{4.10}$$

belongs to $\mathcal{L}$, and the corresponding transfer functions, $\hat{\Psi}_u$ and $\hat{\Psi}_x$, have the properties $\hat{\Psi}_u = \Psi_u$ and $\hat{\Psi}_x = \Psi_x$. Consequently, (3.2) and (3.3) yield

$$D^{-1}NC = \Psi_u\Psi_x^{-1} = \hat{\Psi}_u\hat{\Psi}_x^{-1} = \hat{D}^{-1}\hat{N}C. \tag{4.11}$$

Since $\det CC^* \neq 0$, it follows that

$$D^{-1}N = \hat{D}^{-1}\hat{N}. \tag{4.12}$$

Let the $k \times k$ matrix polynomial $\hat{M}$ be the greatest common left divisor of $\hat{D}$ and $\hat{N}$, i.e.

$$\hat{D} = \hat{M}D_0, \qquad \hat{N} = \hat{M}N_0, \tag{4.13}$$

where D_0 and N_0 are left coprime matrix polynomials. Since (4.10) belongs to $\mathcal{L}$, $\hat{M}$ is stable, and, since $\det\hat{D} \not\equiv 0$, we see that $\det\hat{M} \not\equiv 0$ and $\det D_0 \not\equiv 0$. From (4.11) we have $N = DD_0^{-1}N_0$, so setting $M := DD_0^{-1}$, we obtain

$$D = MD_0, \qquad N = MN_0. \tag{4.14}$$

Since D_0 and N_0 are left coprime, there exist matrix polynomials Π_1 and Π_2 such that

$$D_0\Pi_1 + N_0\Pi_2 = I.$$

(See, e.g., [Fu].) Therefore

$$M = M(D_0\Pi_1 + N_0\Pi_2) = D\Pi_1 + N\Pi_2$$

is a matrix polynomial. Since $Dy_t = Ny_t$ belongs to $\mathcal{L}$, M is stable. From (4.13) and (4.14) we now see that the regulators $Dy_t = Nx_t$ and $\hat{D}y_t = \hat{N}x_t$ are equivalent, as claimed.

5. Design of a linear optimal universal regulator: The case $m \geq \ell$

Let ρ and R be real polynomials defined as in Lemma 4.2. Then, by (4.5), the harmonic component of the control u_t, as defined in (3.4) and (3.5), is given by

$$u^{(j)} = \frac{R(\lambda_j)}{\rho(\lambda_j)}V(\lambda_j)Ew^{(j)}. \tag{5.1}$$

We recall that the harmonic components of x_t and u_t are the only parts that contribute to the cost functional (1.6), and, as explained in Section 3, optimality is achieved if

$$u^{(j)} = \hat{u}^{(j)} \quad \text{for } j = 1, 2, \ldots, N, \tag{5.2}$$

where $\hat{u}^{(j)} = U(\lambda_j)w^{(j)}$, as seen from (3.21) and (3.19).

The question now is whether there are real polynomials ρ and R, satisfying the conditions of Lemma 4.2, such that (5.2) holds for all choices of $w^{(1)}, w^{(2)}, \ldots, w^{(N)}$. If this is so, there does exist an optimal universal regulator in $\mathcal{L}$, and it is given by Lemma 4.2 in terms of R and ρ. If not, an optimal universal regulator may not exist, and we shall see in Section 6 that it does not exist as a rule, but an optimal regulator which is not universal may exist.

Consequently, an optimal universal regulator does exist, if

$$R(\lambda_j)F_j = \rho(\lambda_j)U(\lambda_j), \quad j = 1, 2, \ldots, N, \tag{5.3}$$

where $F_1, F_2, \ldots, F_N$ are $m \times \ell$ complex matrices defined by

$$F_j := V(\lambda_j)E = \chi(\lambda_j)C(\lambda_j I - A)^{-1}E. \tag{5.4}$$

Now, if $m \geq \ell$ and

$$\det F_j^* F_j \neq 0, \quad j = 1, 2, \ldots, N, \tag{5.5}$$

then it is easy to see that

$$R(\lambda_j) = \rho(\lambda_j)U(\lambda_j)(F_j^* F_j)^{-1}F_j^* + \tilde{R}_j, \quad j = 1, 2, \ldots, N \tag{5.6}$$

is a solution of (5.3) for all $\tilde{R}_1, \tilde{R}_2, \ldots, \tilde{R}_N$ such that $\tilde{R}_j F_j = 0$, for $j = 1, 2, \ldots, N$, and that these are precisely all solutions to (5.3).

Therefore, for any stable scalar polynomial ρ, of sufficiently large degree, there is an R satisfying (5.6) such that the corresponding regulator (4.3) is an optimal universal regulator. In fact, for this regulator, (5.2) holds so that (3.21) implies that

$$\Phi = \Phi_{\min} := \sum_{j=1}^{N} \Phi_{\min}^{(j)} \tag{5.7}$$

and that $\Phi \geq \Phi_{\min}$ for all other regulators in $\mathcal{L}$, proving optimality. Since, in addition, D and N do not depend on $w^{(1)}, w^{(2)}, \ldots, w^{(N)}$, the optimal regulator is universal.

Let us summarize the results obtained so far. We consider the problem of finding an optimal universal regulator in the linear class $\mathcal{L}$. This corresponds to the optimization problem to minimize the cost function Φ, defined by (1.6), over the class $\mathcal{A}_{\mathcal{L}}$ of admissible processes, subject to the constraint (1.1), under the condition that $w^{(1)}, w^{(2)}, \ldots, w^{(N)}$ are unknown and hence must not affect the regulator or the optimality of it.

Theorem 5.1 *Suppose that*

1. *$m := \dim y_t \geq \ell := \dim w_t$,*

2. *A is a stable matrix with characteristic polynomial χ,*

3. *$\det F_j^* F_j \neq 0, \quad j = 1, 2, \ldots, N$,*

4. *WFDC holds, i.e., $\Pi(\lambda_j) > 0$ for $j = 1, 2, \ldots, N$.*

Then:

(i) There exists an optimal universal regulator in the class $\mathcal{L}$, and it is defined by formulas (4.4) where ρ is a stable scalar polynomial and R is a $k \times m$ matrix polynomial such that $\deg(RV) < \deg \rho$ which satisfies the interpolation conditions (5.6).

(ii) Any optimal universal regulator in $\mathcal{L}$ is equivalent to one of the type mentioned in (i).

(iii) The optimal value of the cost function Φ is

$$\Phi_{min} := \sum_{j=1}^{N} \Phi_{min}^{(j)}, \tag{5.8}$$

where $\Phi_{min}^{(1)}, \Phi_{min}^{(2)}, \ldots, \Phi_{min}^{(N)}$ are defined by (3.18).

Proof. Statements (i) and (iii) have already been proven above, so it only remains to prove (ii). To this end, let

$$\hat{D}(\sigma)u_t = \hat{N}(\sigma)y_t \tag{5.9}$$

be an optimal universal regulator. Then (5.9) is optimal for any choice of the vector amplitudes $w^{(1)}, w^{(2)}, \ldots, w^{(N)}$, for example, for the choice that all these vector amplitudes are zero except the j:th one so that $w_t = w^{(j)} e^{i\theta_j t}$. Then

$$\min_{\mathcal{A}_{\mathcal{L}}} \Phi = \Phi_{min}^{(j)},$$

and the optimal control is $u^{(j)} = \hat{u}^{(j)}$, where $\hat{u}^{(j)} = U(\lambda_j)w^{(j)}$. Since this should hold for all values of $w^{(1)}, w^{(2)}, \ldots, w^{(N)}$, this implies that

$$\Psi_u(\lambda_j)E = U(\lambda_j). \tag{5.10}$$

Since j is arbitrary, this holds for each $j = 1, 2, \ldots, N$. By Lemma 4.2, the regulator (5.9) is equivalent to some regulator (4.3) with D and N given by (4.4), and thus it has the same transfer function Ψ_u so that (5.10) holds for $j = 1, 2, \ldots, N$. This is equivalent to the interpolation condition (5.6). This proves statement (ii).

6. Design of linear optimal regulators: The case $m < \ell$

In this section we show that an optimal universal regulator in $\mathcal{L}$ does not in general exist in the case that

$$m := \dim y_t < \ell := \dim w_t \tag{6.1}$$

unless certain algebraic relations on the system parameters are satisfied, and therefore it does not exist in practice.

Suppose for the moment that $w^{(1)}, w^{(2)}, \ldots, w^{(N)}$ are fixed, and let us determine all optimal, possibly nonuniversal, regulators in the class $\mathcal{L}$. To this end, recall that the cost function is

$$\Phi = \sum_{j=1}^{N} (u^{(j)} - \hat{u}^{(j)})^* \Pi(\lambda_j)(u^{(j)} - \hat{u}^{(j)}) + \sum_{j=1}^{N} \Phi_{\min}^{(j)}, \tag{6.2}$$

for any regulator in $\mathcal{L}$, where $\hat{u}^{(j)} := U(\lambda_j)w^{(j)}$ and $u^{(j)} = \Psi_u(\lambda_j)Ew^{(j)}$ are given by (3.19) and (3.5) respectively. Consider first those regulators in $\mathcal{L}$ which are defined by Lemma 4.2(i) and formulas (4.4) via appropriate polynomials R and ρ. For such a choice of polynomials

$$u^{(j)} = X_j F_j w^{(j)}, \tag{6.3}$$

where the $m \times \ell$ matrices $F_1, F_2, \ldots, F_N$ are given by (5.4) and

$$X_j = \frac{R(\lambda_j)}{\rho(\lambda_j)}. \tag{6.4}$$

Let $\mathcal{J}' := \{j \mid F_j w^{(j)} \neq 0\}$ and $\mathcal{J}'' := \{j \mid F_j w^{(j)} = 0\}$ so that $\mathcal{J}' \cup \mathcal{J}'' = \{1, 2, \ldots, N\}$. Then (6.2) implies that

$$\Phi = \sum_{j \in \mathcal{J}'} (u^{(j)} - \hat{u}^{(j)})^* \Pi(\lambda_j)(u^{(j)} - \hat{u}^{(j)}) + \Phi_{\min}, \tag{6.5}$$

where

$$\Phi_{\min} = \sum_{j \in \mathcal{I}''} (\hat{u}^{(j)})^* \Pi(\lambda_j) \hat{u}^{(j)} + \sum_{j=1}^{N} \Phi_{\min}^{(j)}. \tag{6.6}$$

Recall that $\Pi(\lambda_j) > 0$. It follows from (6.5) that if we find a regulator $D(\sigma)u_t = N(\sigma)y_t$ with R, ρ such that

$$u^{(j)} = \hat{u}^{(j)} \quad \text{for } j \in \mathcal{I}', \tag{6.7}$$

then it is optimal. The corresponding polynomials ρ and R must satisfy relations

$$X_j F_j w^{(j)} = U(\lambda_j) w^{(j)}, \quad j \in \mathcal{I}'. \tag{6.8}$$

In this case the infimum of Φ is attained in $\mathcal{L}$ and

$$\inf_{\mathcal{L}} \Phi = \Phi_{\min}. \tag{6.9}$$

It is easy to see that all solutions X_j of (6.8) are given by

$$X_j = \frac{U(\lambda_j) w^{(j)} (w^{(j)})^* F_j^*}{|F_j w^{(j)}|^2} + \tilde{X}_j \quad \text{where } \tilde{X}_j F_j w^{(j)} = 0 \text{ and } j \in \mathcal{I}'. \tag{6.10}$$

Obviously there exist a matrix polynomial $R(\lambda)$ and a scalar stable polynomial $\rho(\lambda)$ such that

$$R(\lambda_j) = \rho(\lambda_j) X_j, \quad j \in \mathcal{I}' \tag{6.11}$$

and condition $\deg(RV) < \deg \rho$ of Lemma 4.2 holds. Here ρ may be any stable scalar polynomial of sufficiently high degree. The corresponding regulator is optimal, and we have also proved relation (6.9). (Note that in the case $m \geq \ell$ considered in Section 5 the conditions $F_j w^{(j)} = 0$ imply that $w^{(j)} = 0$ under the assumption that $\det F_j^* F_j \neq 0$. Therefore $\hat{u}^{(j)} = 0$ for $j \in \mathcal{I}''$ and (6.6) coincides with (5.8).)

Now, consider an arbitrary optimal regulator in $\mathcal{L}$ obtained via formulas (4.5) of Lemma 4.2. Then $\Phi = \Phi_{\min}$, and, because of (6.5), we obtain first (6.7) and then (6.8), (6.10) and (6.11). Hence the regulator is determined in the way mentioned above.

Consider next an arbitrary optimal regulator

$$\hat{D}(\sigma)u_t = \hat{N}(\sigma)y_t$$

in $\mathcal{L}$, not necessarily obtained via formulas (4.5). By Lemma (4.2) it is equivalent to a regulator $Du_t = Ny_t$ determined via formulas (4.5). Since Φ

depends only on Ψ_u, which is the same for these regulators, and $\Phi = \Phi_{\min}$ for the regulator $\hat{D}u_t = \hat{N}y_t$, we have $\Phi = \Phi_{\min}$ also for the regulator $Du_t = Ny_t$. Therefore, $Du_t = Ny_t$ is optimal also, and consequently it is of the kind discussed above.

Let us formulate the results obtained so far. Suppose $m < \ell$, and let the complex amplitudes $w^{(1)}, w^{(2)}, \ldots, w^{(N)}$ be fixed. Then an optimal regulator exists. Any regulator defined by formulas (4.5) with ρ, R satisfying the conditions of Lemma 4.2 and the interpolation conditions (6.11) is optimal. Conversely, any regulator which is optimal in $\mathcal{L}$ is equivalent to one obtained in this way.

Now suppose that a *universal* optimal regulator exists and that it is defined by formulas (4.5) in Lemma 4.2. Then (6.8) must hold, i.e.,

$$X_j F_j w^{(j)} = U(\lambda_j) w^{(j)} \quad \text{if } F_j w^{(j)} \neq 0.$$

But universality implies that this must hold for all values of $w^{(1)}, w^{(2)}, \ldots, w^{(N)}$, and consequently, as seen from a continuity argument, we must have

$$X_j F_j = U(\lambda_j), \quad j = 1, 2, \ldots, N \tag{6.12}$$

precisely as in Section 5 . The difference from the situation in Section 5 is that, in the case $m < \ell$, (6.12) does not in general have a solution since it is an overdetermined system of $k\ell N$ linear equations with kmN unknown variables, the components of the $k \times m$ matrices $X_1, X_2, \ldots, X_N$. Let us find the conditions for the existence of such a solution.

Suppose that the rows of the $m \times \ell$ matrices $F_1, F_2, \ldots, F_N$ are linearly independent, i.e., that

$$\det F_j F_j^* \neq 0, \quad j = 1, 2, \ldots, N. \tag{6.13}$$

Then postmultiplying (6.12) by $F_j^*(F_j F_j^*)^{-1}$ for $j = 1, 2, \ldots, N$, we obtain

$$X_j = U(\lambda_j) F_j^*(F_j F_j^*)^{-1}, \quad j = 1, 2, \ldots, N. \tag{6.14}$$

It follows from (6.12) that

$$U(\lambda_j) \left[F_j^*(F_j F_j^*)^{-1} F_j - I \right] = 0, \quad j = 1, 2, \ldots, N. \tag{6.15}$$

Conversely, (6.14) and (6.15) imply (6.12), and hence (6.15) is equivalent to the existence of a solution $X_1, X_2, \ldots, X_N$ in (6.12). This is of course a very strict condition, showing that the existence of a universal regulator in the case $m < \ell$ is nongeneric. For example, if $m = k = 1$, this condition implies that the ℓ-dimensional row vectors F_j and $U(\lambda_j)$ are proportional, i.e.,

$$U(\lambda_j) = \kappa_j F_j, \quad j = 1, 2, \ldots, N,$$

where $\kappa_1, \kappa_2, \ldots, \kappa_N$ are scalars.

We have thus established that (6.15) is a necessary condition for the existence of an optimal universal regulator. It is also sufficient, because, under this condition, equations

$$R(\lambda_j) = \rho(\lambda_j) U(\lambda_j) F_j^* (F_j F_j^*)^{-1}, \quad j = 1, 2, \ldots, N, \tag{6.16}$$

i.e., equations (6.14), satisfy (6.8) for all choices of $w^{(1)}, w^{(2)}, \ldots, w^{(N)}$, and consequently appropriate ρ and R do exist so that the corresponding regulator (4.3), determined by (4.4), is an optimal universal regulator. Any other optimal universal regulator in $\mathcal{L}$ is equivalent to one constructed in this way. In fact, by Lemma 4.2(ii), any regulator in $\mathcal{L}$ is equivalent to one constructed via (4.2)–(4.4) and thus has the same closed-loop transfer functions Ψ_u, Ψ_x, Ψ_y, and hence the same $x^{(j)}, u^{(j)}, y^{(j)}$ in (3.1), and consequently the same value of the cost function Φ. Moreover, we just showed that any optimal universal regulator constructed via (4.2)–(4.4) must satisfy (6.16).

We summarize the results of this section in the following theorem. First, however, let us recall the problem under consideration. Find a regulator (1.11) in the class $\mathcal{L}$ such that the overall closed-loop system consisting of (1.1) and (1.11) generates a process (x_t, u_t) minimizing the cost function (1.6). The external harmonic disturbance (1.3) is such that (1.4) holds. We say that the regulator (1.11) is an optimal universal regulator (OUR) in $\mathcal{L}$ if D and N do not depend on $w^{(1)}, w^{(2)}, \ldots, w^{(N)}$ and it is optimal for all values of $w^{(1)}, w^{(2)}, \ldots, w^{(N)}$.

Theorem 6.1 *Suppose that*

1. $m := \dim y_t < \ell := \dim w_t$,

2. A *is a stable matrix with characteristic polynomial* $\chi(\lambda)$,

3. *WFDC holds, i.e.,* $\Pi(\lambda_j) > 0$ *for* $j = 1, 2, \ldots, N$.

Then:

(i) *Let* $\det F_j F_j^* \neq 0$, $j = 1, 2, \ldots, N$. *Then there exists an optimal universal regulator in the class* $\mathcal{L}$ *if and only if conditions (6.15) hold, where* $U(\lambda)$ *is defined by (3.20),* $\lambda_1, \lambda_2, \ldots, \lambda_N$ *by (3.6) and* $F_1, F_2, \ldots, F_N$ *by (5.4). In this case, the OUR is defined by formulas (4.4) where* ρ *is* an arbitrary stable scalar polynomial of sufficiently high degree, and R is a $k \times m$ matrix polynomial such that $\deg(RV) < \deg \rho$ which satisfies the interpolation conditions (6.16). Any other OUR in $\mathcal{L}$ is equivalent to some regulator of this type. The transfer function Ψ_u from Ew_t to u_t for the corresponding optimal closed-loop system is

$$\Psi_u(\lambda) = \frac{R(\lambda)}{\rho(\lambda)} V(\lambda). \tag{6.17}$$

(ii) If (6.15) fails for some $j = 1, 2, \ldots, N$, then there is no OUR. Then there is an optimal nonuniversal regulator $D(\sigma)u_t = N(\sigma)y_t$ in $\mathcal{L}$ defined via (4.4), where ρ is an arbitrary stable (real) scalar polynomial of sufficiently high degree, and R is a $k \times m$ matrix polynomial such that $\deg(RV) < \deg \rho$ which satisfies the interpolation conditions

$$R(\lambda_j) = \frac{\rho(\lambda_j)}{|F_j w^{(j)}|^2} U(\lambda_j) w^{(j)} (w^{(j)})^* F_j^* + \tilde{R}_j \quad \text{where } \tilde{R}_j F_j w^{(j)} = 0, \tag{6.18}$$

for all $j \in \mathcal{J}'$, i.e., for all j for which $F_j w^{(j)} \neq 0$. Any other optimal regulator is equivalent to one constructed in this way. The transfer matrix Ψ_u from Ew_t to u_t is given by (6.17).

Consequently, the existence of an optimal universal regulator is a highly nongeneric property when $m < \ell$, so, from a practical point of view, OUR does not exist in this case.

Remark 6.2 If (6.15) holds and $F_j w^{(j)} \neq 0$ for some j, then we may replace (6.18) by

$$R(\lambda_j) = \rho(\lambda_j) U(\lambda_j) F_j^* (F_j F_j^*)^{-1}. \tag{6.19}$$

For this j we have $u_j = \hat{u}_j$ for all $w^{(j)} \in \mathbb{C}^\ell$.

7. Optimal universal regulators in the class $\mathcal{N}$

In this section we show that an optimal universal regulator in $\mathcal{L}$ is also an OUR in the larger class $\mathcal{N}$, defined in Section 1, under conditions which are only slightly stronger than those in Theorems 5.1 and 6.1. This fact is a corollary of the following main lemma.

Lemma 7.1 Suppose that the strong frequency domain condition (SFDC) holds, i.e.,

$$\Pi(\lambda) > 0 \quad \text{for all } \lambda \text{ on the unit circle.} \tag{7.1}$$

Let $\mathcal{A}$ and $\mathcal{A}_{\mathcal{L}}$ be the classes of admissible processes (x_t, u_t) defined in Section 1. Then, if the problem to minimize the cost function (1.6) over all processes in $\mathcal{A}_{\mathcal{L}}$ has an optimal solution satisfying the interpolation condition

$$u^{(j)} = \hat{u}^{(j)} \quad \text{for } j \in \mathfrak{I}', \tag{7.2}$$

this solution is also optimal for the problem to minimize (1.6) over $\mathcal{A}$.

Proof. Define $\hat{\mathcal{L}}$ to be the class of all linear stabilizing regulators

$$\hat{\mathcal{L}}: \quad \hat{D}(\sigma)u_t = \hat{N}(\sigma)y_t,$$

obtained by setting $C := I$ in the definition of $\mathcal{L}$. Then, any D and N corresponding to a regulator in $\mathcal{L}$ define a regulator in $\hat{\mathcal{L}}$ by setting $\hat{D} = D$ and $\hat{N} = NC$, and consequently

$$\mathcal{L} \subset \hat{\mathcal{L}}. \tag{7.3}$$

Now, the quantities $\Pi(\lambda_j)$, $\hat{u}^{(j)}$ and $\Phi_{\min}^{(j)}$ do not depend on C (see (2.7), (3.14), (3.15) and (3.18)–(3.20)) and therefore $\Phi_{\min}$ in (6.6) does not depend on C either, although $u^{(j)}$ does. Consequently, it follows from (6.5) that

$$\Phi \geq \Phi_{\min} \tag{7.4}$$

for all processes (x_t, u_t) in $\mathcal{A}_{\hat{\mathcal{L}}}$. But, since a process which is optimal in $\mathcal{A}_{\mathcal{L}}$ satisfies (7.2) so that $\Phi = \Phi_{\min}$, it is optimal also in $\mathcal{A}_{\hat{\mathcal{L}}}$. Moreover, under the strong frequency domain condition (SFDC), it was proven in [LY, Theorem 5.1 and Remark 5.2] that a minimum of Φ over all $(x_t, u_t) \in \mathcal{A}$ can be obtained by choosing a process in $\mathcal{A}_{\hat{\mathcal{L}}}$. Consequently, a process which is optimal in $\mathcal{A}_{\mathcal{L}}$ is optimal in $\mathcal{A}$ also, as claimed.

Consequently, we have established the main result of this paper, namely the following extension of Theorem 5.1 to the larger class $\mathcal{N}$ of nonlinear regulators. We recall that $\mathcal{L} \subset \mathcal{N}$.

Theorem 7.2 *Suppose that*

1. $m := \dim y_t \geq \ell := \dim w_t$,

2. *A is a stable matrix,*

3. $\det F_j^* F_j \neq 0, \quad j = 1, 2, \ldots, N$,

4. *SFDC holds, i.e., $\Pi(\lambda) > 0$ for all λ on the unit circle.*

Then there is an optimal universal regulator in the class N, which actually belongs to L ⊂ N, and it can be determined as in Theorem 5.1.

Proof.   By Theorem 5.1, there exists an optimal universal regulator in L under the stated conditions, and it follows from the proof of Theorem 5.1 that the interpolation conditions (5.2), and consequently also (7.2), holds for this regulator. Consequently, by Lemma 7.1, the corresponding process (x_t, u_t) is also optimal in A and hence in the class of all admissible processes generated by N. Therefore the optimal universal regulator in L is an OUR also in N.

Similarly we also have the following extension of Theorem 6.1.

Theorem 7.3 *Suppose that*

1. $m := \dim y_t < \ell := \dim w_t$,

2. *A is a stable matrix,*

3. $\det F_j F_j^* \neq 0$, $j = 1, 2, \ldots, N$,

4. *SFDC holds, i.e.,* $\Pi(\lambda) > 0$ *for all λ on the unit circle.*

Then, provided condition (6.15) holds, there is an optimal universal regulator in the class N, which actually belongs to L ⊂ N, and it can be determined as in point (i) of Theorem 6.1. If condition (6.15) fails, there is a nonuniversal optimal regulator in N which belongs to L and is given in point (ii) of Theorem 6.1.

The proof of Theorem 7.3 follows the same principles as that of Theorem 7.2.

8. The stochastic case

It is interesting to note that an optimal universal regulator in L for the (deterministic) control problem discussed in the previous sections is optimal in L for the stochastic control problem obtained by taking $w^{(1)}, w^{(2)}, \ldots, w^{(N)}$ to be (jointly distributed) random vectors and replacing the cost function (1.6) by

$$\Phi = \lim_{T \to \infty} \mathrm{E}\{\frac{1}{T} \sum_{t=0}^{T} \Lambda(x_t, u_t)\}, \tag{8.1}$$

where $\mathrm{E}\{\cdot\}$ denotes mathematical expectation.   In fact, universality implies that the same regulator (1.12) is optimal for all values of

$w^{(1)}, w^{(2)}, \ldots, w^{(N)}$, so summing with respect to the appropriate probability measure shows that this regulator is optimal also for the stochastic problem (Lemma 8.2).

Therefore it is natural to ask whether there is a more nontrivial stochastic version of the problem previously discussed in this paper for which the optimal universal regulator is optimal. As pointed out in Section 1, the appropriate stochastic problem is obtained by replacing (1.8) by the stochastic system

$$\begin{cases} z_{t+1} = Fz_t + Gv_t \\ w_t = Hz_t + Kv_t \end{cases} \tag{8.2}$$

where z_0 now is a random $N\ell$-vector and where $v_0, v_1, v_2, \ldots$ is a zero-mean vector-valued white noise, independent of z_0, i.e.,

$$E\{v_s v_t^*\} = V_t \delta_{st}, \quad E\{v_t\} = 0, \tag{8.3}$$

with $\{|V_t|\}_{t=0}^{\infty}$ being an ℓ_1 sequence, i.e.,

$$\sum_{t=0}^{\infty} |V_t| < \infty. \tag{8.4}$$

The condition (8.4) insures that $E\{w_s w_t^*\}$ is bounded for all $s, t \in \mathbb{Z}_+$. We say that an optimal regulator is *universal* (for the stochastic problem) if it does not depend on z_0, G, K and $\{V_t\}_{t \in \mathbb{Z}_+}$, and it is optimal for all values of these quantities.

Theorem 8.1 *Consider the stochastic control system*

$$\begin{aligned} x_{t+1} &= Ax_t + Bu_t + Ew_t \tag{8.5} \\ y_t &= Cx_t \tag{8.6} \end{aligned}$$

with w_t generated by the critically stable stochastic system (8.2). Then any optimal universal regulator in $\mathcal{L}$ for the deterministic problem to control (1.1), with w_t given by (1.3), so as to minimize (1.6) is also an optimal universal regulator in the class $\mathcal{L}$ for the stochastic problem to control (8.5) so as to minimize (8.1).

We first prove the statement of Theorem 8.1 in the special case when $v_t \equiv 0$ so that all stochastics is generated by the initial condition z_0.

Lemma 8.2 *Let $v_t \equiv 0$. Then the limit (8.1) exists for all (x_t, u_t) in $\mathcal{A}_{\mathcal{L}}$ and*

$$\lim_{T \to \infty} E\{\frac{1}{T} \sum_{t=0}^{T} \Lambda(x_t, u_t)\} = E\{\lim_{T \to \infty} \frac{1}{T} \sum_{t=0}^{T} \Lambda(x_t, u_t)\}, \tag{8.7}$$

where the second limit is in the sense of almost sure convergence. Consequently, any optimal regulator in $\mathcal{L}$ is an optimal regulator for the stochastic
problem and does not depend on z_0.

Proof. Let us first consider the deterministic problem with $(x_t, u_t) \in$
$\mathcal{A}_{\mathcal{L}}$. A straightforward calculation shows that

$$\frac{1}{T} \sum_{t=0}^{T} \Lambda(x_t, u_t) = w^* \Omega_T w + \omega_T^* w + \eta_T, \tag{8.8}$$

where w is defined as in (3.22) and Ω_T is a symmetric Hermitian matrix
which depends on the choice of regulator in $\mathcal{L}$. Since the regulator is stabilizing, $\omega_T \to 0$ and $\eta_T \to 0$ as $T \to \infty$, and, as seen from (3.22), (8.8)
tends $w^* \Omega w$ for any choice of w. Consequently, $\Omega_T \to \Omega$ as $T \to \infty$. Next,
let w be the stochastic vector z_0 as required in the present problem. Then

$$\mathrm{E}\{\frac{1}{T} \sum_{t=0}^{T} \Lambda(x_t, u_t)\} = \mathrm{E}\{z_0^* \Omega_T z_0\} + \omega_T^* \mathrm{E}\{z_0\} + \eta_T.$$

Consequently, since $\omega_T \to 0$ as $T \to \infty$, (8.7) follows if

$$\lim_{T \to \infty} \mathrm{E}\{z_0^* \Omega_T z_0\} = \mathrm{E}\{\lim_{T \to \infty} z_0^* \Omega_T z_0\}. \tag{8.9}$$

But, $\Omega_T \to \Omega$, and therefore there is a matrix $M > \Omega$ and a $T_0 > 0$ such
that $\Omega_T \leq M$ for all $T \geq T_0$ so that

$$z_0^* \Omega_T z_0 \leq z_0^* M z_0 \quad \text{for } T \geq T_0$$

and for each value of z_0. Consequently, (8.9) follows by dominated convergence; see, e.g., [Kr, Theorem 2.4].

Proof of Theorem 8.1 Let us first normalize the white noise sequence
$v_0, v_1, v_2, \ldots$ by setting

$$v_t := L_t \eta_t \tag{8.10}$$

so that η_t is a zero-mean, p-dimensional, normalized white noise, i.e.,

$$\mathrm{E}\{\eta_s \eta_t^*\} = I_p \delta_{st}, \quad \mathrm{E}\{\eta_t\} = 0, \tag{8.11}$$

implying that L_t is a matrix-valued function such that $L_t L_t^* = V_t$. Then

$$w_t = \bar{w}_t + \sum_{k=1}^{p} \sum_{s=0}^{t} w_t(s, k)(\eta_s)_k, \tag{8.12}$$

where

$$w_t(s, k) = HF^t g_{sk} \quad \text{where } g_{sk} = \begin{cases} F^{-s-1}GL_s e_k & \text{for } s < t \\ KL_s e_k & \text{for } s = t \end{cases} \tag{8.13}$$

e_k being the k:th axis unit vector, and where

$$\bar{w}_t = HF^t z_0. \tag{8.14}$$

Now, any regulator in $\mathcal{L}$ applied to (8.5) yields a closed-loop system (8.2), (8.5), (1.11), driven by the white noise η_t so that

$$x_t = \bar{x}_t + \sum_{k=1}^{p} \sum_{s=0}^{t-1} x_t(s, k)(\eta_s)_k \tag{8.15}$$

$$u_t = \bar{u}_t + \sum_{k=1}^{p} \sum_{s=0}^{t-1} u_t(s, k)(\eta_s)_k, \tag{8.16}$$

$$y_t = \bar{y}_t + \sum_{k=1}^{p} \sum_{s=0}^{t-1} y_t(s, k)(\eta_s)_k \tag{8.17}$$

where $\bar{x}_t, \bar{u}_t, \bar{y}_t$ are stochastic vector sequences generated by the initial condition z_0, and thus independent of $\{\eta_t\}$, and $x_t(s, k), u_t(s, k), y_t(s, k)$ are deterministic vector sequences. All these sequences of course depend on the particular choice of regulator. It follows from (8.15)–(8.17) that $\bar{x}_t, \bar{u}_t, \bar{y}_t$ are the conditional expected values of x_t, u_t, y_t given z_0, and therefore

$$\bar{x}_{t+1} = A\bar{x}_t + B\bar{u}_t + E\bar{w}_t \tag{8.18}$$

$$\bar{y}_t = C\bar{x}_t \tag{8.19}$$

Since $x_t(s, k) = \mathrm{E}\{x_t(\eta_s)_k\}$ for $t \geq k + 1$, and the corresponding relations hold for $u_{t+1}(s, k)$ and $y_{t+1}(s, k)$,

$$x_{t+1}(s, k) = Ax_t(s, k) + Bu_t(s, k) + Ew_t(s, k), \tag{8.20}$$

$$x_{s+1}(s, k) = Ew_s(s, k)$$

$$y_t(s, k) = Cx_t(s, k) \tag{8.21}$$

for $t = s+1, s+2, \ldots$, it follows from (8.13) and (8.14) that $w_t(s, k)$ and $\bar{w}_t$ satisfy the autonomous system (1.8), differing only in the initial conditions, which correspond to $w^{(1)}, w^{(2)}, \ldots, w^{(N)}$ in (1.3), and therefore (8.18) and (8.20) have the same structure as the deterministic system (1.1)–(1.3).

Next, we show that the cost function can be decomposed accordingly. In fact, it is easy to check that

$$\mathrm{E}\{\Lambda(x_t, u_t)\} = \mathrm{E}\{\Lambda(\bar{x}_t, \bar{u}_t)\} + \sum_{k=1}^{p} \sum_{s=0}^{t-1} \Lambda(x_t(s, k), u_t(s, k)) \tag{8.22}$$

so, if we agree to define $x_t(s, k)$ and $u_t(s, k)$ to be zero for $k \geq t$, we have

$$
\begin{aligned}
&\mathrm{E}\{\tfrac{1}{T}\sum_{t=0}^{T}\Lambda(x_t, u_t)\} = \mathrm{E}\{\tfrac{1}{T}\sum_{t=0}^{T}\Lambda(\bar{x}_t, \bar{u}_t)\}\\
&+\sum_{k=1}^{p}\sum_{s=0}^{\infty}\left[\tfrac{1}{T}\sum_{t=s+1}^{T}\Lambda(x_t(s, k), u_t(s, k))\right].
\end{aligned}
\tag{8.23}
$$

By Lemma 8.2, the limit

$$
\bar{\Phi} = \mathrm{E}\{\lim_{T\to\infty}\frac{1}{T}\sum_{t=0}^{T}\Lambda(\bar{x}_t, \bar{u}_t)\}
\tag{8.24}
$$

exists. Therefore, provided the limits

$$
\Phi_{sk} = \lim_{T\to\infty}\frac{1}{T}\sum_{t=s+1}^{T}\Lambda(x_t(s, k), u_t(s, k))
\tag{8.25}
$$

exist, and provided

$$
\lim_{T\to\infty}\sum_{k=0}^{\infty}\left[\frac{1}{T}\sum_{t=s+1}^{T}\Lambda(x_t(s, k), u_t(s, k))\right] = \sum_{s=0}^{\infty}\Phi_{sk},
\tag{8.26}
$$

the limit does also exist in the cost function (8.1) and

$$
\Phi = \bar{\Phi} + \sum_{k=1}^{p}\sum_{s=0}^{\infty}\Phi_{sk}.
\tag{8.27}
$$

Under these condition, which must be verified, the stochastic control problem thus decomposes into separate decoupled control problems, all having the structure of the one considered earlier in this paper, namely the problem $\bar{P}$ of Lemma 8.2 to minimize $\bar{\Phi}$ given (8.18) and the deterministic problems P_{sk} to minimize Φ_{sk} given (8.20). The latter problems differ only in the vector amplitudes $w^{(1)}, w^{(2)}, \ldots, w^{(N)}$ in (1.3) and in the initial time (which does not affect the steady-state behavior measured by the cost function). Consequently, if there is a universal optimal regulator $Du_t = Ny_t$ in $\mathcal{L}$ for the deterministic problem to control (1.1) so as to minimize (1.6), then

$$
D(\sigma)u_t(s, k) = N(\sigma)y_t(s, k)
\tag{8.28}
$$

is optimal in $\mathcal{L}$ for P_{sk}, for all $s = 0, 1, 2, \ldots$ and $k = 1, 2, \ldots, p$. Moreover, by Lemma 8.2,

$$
D(\sigma)\bar{u}_t = N(\sigma)\bar{y}_t
\tag{8.29}
$$

is optimal in $\mathcal{L}$ for the problem $\bar{P}$. Hence the cost function (8.27) for the stochastic system is minimized by these control actions. But, in view of

(8.27) and (8.28), the stochastic processes u_t and y_t, given by (8.16) and (8.17), satisfy

$$D(\sigma)u_t = N(\sigma)y_t, \qquad (8.30)$$

and therefore the regulator (8.29) is optimal in $\mathcal{L}$ for the stochastic problem of Theorem 8.1. Clearly this regulator does not depend on z_0, G, K and $\{V_t\}_{t\in\mathbb{Z}_+}$, and it is optimal for all values of these quantities. Hence it is a universal regulator, as claimed.

It remains to show that the limits (8.23) and (8.24) exist and that (8.25) holds under the feedback conditions (8.28) and (8.27). It was established in Section 3 that the limits do exist under linear stabilizing feedback, so we only need to verify (8.25). To this end, recall that, for any regulator in $\mathcal{L}$, the cost function takes the form (3.22) in the deterministic problem. Moreover, for the problems P_{sk}, it follows from (8.13) and $V_s = L_s L_s^*$ that the norm of $w = \mathrm{col}(w^{(1)}, w^{(2)}, \ldots, w^{(N)})$ is bounded by $\kappa |V_s|^{\frac{1}{2}}$ for some positive constant κ which depends on the regulator. Hence (3.22) is bounded by $\kappa^2 |V_s|$ so, in view of (8.4),

$$\sum_{k=0}^{\infty} \Phi_{sk} < \infty.$$

Moreover,

$$\left| \frac{1}{T} \sum_{t=s+1}^{T} \Lambda(x_t(s,k), u_t(s,k)) \right| < \kappa^2 |V_s|,$$

which is an ℓ_1 sequence. Consequently (8.25) follows from the dominated convergence theorem.

We note that the decomposition (8.18), (8.20) and (8.26) is analogous to the one used in [Li], so a natural question is whether the admissible class of regulators could be extended to include nonlinear control laws as in [Li]. However, this leads to technical difficulties related to the existence of the limits (8.23) and (8.24) and the validity of (8.25).

References

[AM] B. D. O. Anderson and J. B. Moore, *Optimal Control: Linear Quadratic Methods*, Prentice-Hall, London,1989.

[BLS] S. Bittanti, F. Lorito and S. Strada, *An LQ approach to active control of vibrations in helicopters*, Trans ASME, J. Dynamical Systems, Measurement and Control **118** (1996), 482–488.

[Fr]   B. A. Francis, *The linear multivariable regulator problem*, SIAM
        J. Control and Optimization **15** (1977), 486–505.

[FW]   B. A. Francis and W. M. Wonham, *The internal model principle
        of control theory*, Automatica **12** (1977), 457–465.

[FF]   K. V. Frolov and F. A. Furman, *Applied theory of vibration pro-
        tected systems*, Mashinostroenie, 1980 (in Russian).

[Fr]   K. V. Frolov, *Vibration in Engineering*, Mashinostroenie, 1981 (in
        Russian).

[Fu]   P. A. Fuhrmann, *Linear Systems and Operators in Hilbert Space*,
        McGraw-Hill, New York,1981.

[GEI]   M. D. Genkin, V. G. Elezov and V. D. Iablonski, *Methods of
        Controlled Vibration Protection of Engines*, Moscow, Nauka,1985
        (in Russian).

[Gu]   D. Guicking, *Active Noise and Vibration Control*, Reference bib-
        liography, Drittes Physical Institute. Univ. of Goettingen, Jan.,
        1990.

[KO]   C. G. Källström and P. Ottosson, *The generation and control of
        roll motion of ships in closed turns*, pp. 1–12, 1982.

[Ki]   H. Kimura, *Pole assignment by output feedback: A longstanding
        open problem*, Proc. 33rd Conference on Decision and Control,
        Lake Buena Vista, Florida, december 1994.

[Kr]   K. Krickeberg, *Probability Theory*, Addison-Wesley, 1965.

[KS]   H. Kwakernaak and R. Sivan, *Modern Signals and Systems*, Pren-
        tice Hall, 1991.

[Le]   G. Leitmann and S. Pandey, *Aircraft control under conditions
        of windshear*, Proc. 29th Conf. Decision and Control, Honululu,
        1990, 747–752.

[Li]   A. Lindquist, *On feedback control of linear stochastic systems*,
        SIAM J. Control **11** (1973), 323–343.

[LY]   A. Lindquist and V. A. Yakubovich, *Optimal damping of forced os-
        cillations in discrete-time systems*, IEEE Trans. Automatic Con-
        trol, to be published.

[LY2]   A. Lindquist and V. A. Yakubovich, *Universal controllers for op-timal damping of forced oscillations in linear discrete systems*, Doklady Mathematics SS (1997), 156–159. (Translated from Doklady Akademii Nauk 352 (1997), 314–417.

[Mi]   A. Miele, *Optimal trajectories and guidance trajectories for aircraft flight through windshears*, Proc. 29th Conf. Decision and Control, Honululu, 1990, 737–746.

[SBKB]   R. Shoureshi, L. Brackney, N. Kubota and G. Batta, *A modern control approach to active noise control*, Trans ASME, J. Dynamical Systems, Measurement and Control **115** (1993), 673–678.

[WKK]   V. Z. Weytz, M. Z. Kolovski and A. E. Koguza, *Dynamics of controlled machine units*, Moscow, Nauka, 1984 (in Russian).

[Ya]   V. A. Yakubovich, *A frequency theorem in control theory*, Sibirskij Mat. Zh. **4**(1973), 386–419 (in Russian). English translation in Sibirian Mathem. Journal.

[ZB]   Y. Zhao and A. E. Bryson, *Aircraft control in a downburst on take-off and landing*, Proc. 29th Conf. Decision and Control, Honululu, 1990, 753–757.

Anders Linquist
Division of Optimization and Systems Theory
Royal Institute of Technology
100 44 Stockholm
Sweden

Vladimir Yakubovitch
Department of Mathematics and Mechanics
St. Petersburg University
St. Petersburg 198904
Russia

Received November 28, 1996

The Variation of the Forecast of Lévy's Brownian Motion as the Observation Domain Undergoes Deformation [1]

László Márkus

1. Introduction, the description of the problem

While one observes a random phenomenon evolving in an area or in space it may well happen, that the supposed observation domain undergoes a deformation under the influence of heat, pressure, etc. Therefore, in this paper we do not regard the observation domain as usual, something known, immovable and unchangeable. On the contrary, we wish to investigate the dependence of the forecast on the observation domain. In a simple case we wish to give the variation of the forecast of Lévy's Brownian Motion (LBM) for infinitesimal deformations of the observation domain.

Suppose we are given a random field *X(t)* , that means a multivariate random function over the d dimensional Euclidean space R^d, or more precisely a mapping:

$$X : R^d \rightarrow \mathcal{L}_2 \left(\Omega, \mathcal{A}, P \right)$$

We observe the field X in a smooth, bounded domain D, and based on this observation we consider the forecast of the field at a distant point t. We prefer to write t in its spherical coordinates as $t = (T, \theta)$, where T is the distance of t from the origin, and θ is the vector of angles. Beside the point t, the best forecast, $Y_D(t)$, that is the conditional expectation of $X(t)$ given X in D, depend obviously on the domain D. Since we are interested in this latter dependence, we consider Y_D at a fixed point t as random function of domains.

Imagine $D = D_0$ smoothly embedded in a one parameter family of smooth bounded domains $\{D_S : S \in (-\varepsilon, \varepsilon)\}$ the deformed ones of D_0. Let us be more precise about the *infinitesimal* variation of domains $\{D_S : |S| < \varepsilon\}$. In a neighborhood U of the boundary ∂D we can establish coordinates $(d, S) \in \partial D \times (-\varepsilon, \varepsilon)$ in which a point is written $u = (d, S)$ if it lies at a distance S along the line determined by the outward pointing normal

[1]Research supported by the Nat. Sci. Research Fund OTKA, grants No. T014116, T166665

$n_D(d)$ to ∂D at $d \in \partial D$. Let us have a one parameter set of smooth transformations $\{\Gamma_S : S \in (-\varepsilon, \varepsilon)\}$, which we call deformations, described by a smooth function $\Gamma(d, S) \in C^\infty (\partial D \times (-\varepsilon, \varepsilon))$, such that $|\Gamma(d, S)| < \varepsilon$ and $\Gamma(d, 0) = 0$. Now D_S is determined by

$$\partial D_S = \big\{ u \ : \ u = (d, \Gamma(d, s)) \quad d \in \partial D \big\}.$$

Thus *infinitesimal* variations of the domain $D = D_0$ along deformations Γ_S are put in one to one correspondence with smooth normal vector fields on ∂D via the relationship:

$$\gamma(d) n_D(d), \ \gamma \in C^\infty(\partial D) \sim \{D_S : \ |S| < \varepsilon, \ S \to 0 \}$$
$$\iff \qquad \gamma(d) = \frac{\partial}{\partial S} \Gamma(d, S) \bigg|_{S=0}$$

Let us consider Y_D, or a function G_D, which depend on D beside of its arguments, (e.g. Green's function $G_D(t, s)$). It is natural to conceive of Y_D or G_D as (random or deterministic) functions on some kind of *manifold* whose points are smooth, bounded domains $D_S \subset R^d$. This motivates to compute variations $\frac{\delta}{\delta\gamma} Y_D$ or $\frac{\delta}{\delta\gamma} G_D$ by differentiating the functions Y_{D_S}, G_{D_S} of the real variable S.

We aim to describe in this set-up the variation of the forecast $\frac{\delta}{\delta\gamma} Y_D$ as D changes infinitesimally along deformations Γ_S.

The forecast problem was studied in this spirit for LBM, when observation took place on a closed plane curve. Using white noise analysis, Si Si obtained a white noise integral representation for the variation [Si].

2. The case of dilating (pulsating) balls in three dimension

We consider here the 3 dimensional parameter case, when, like in all odd dimensions, the germ Markov property of LBM holds true [McK]. Thus - unlike in Si Si's case - the observation on the whole domain can be reduced to an observation in the infinitesimal environment of the boundary, meaning the germ field. Furthermore, we can take advantage of partial differential equation description of LBM and its forecast (see [McK], [Ro], [Go]). Restricting ourselves to domains of R^3 with connected smooth boundaries, the germ of it has long been known to be generated by the actual values of the field, and its *generalised* normal derivatives [McK] [Pi].

In order to avoid here detailed technicalities we take dilations as deformations, and we restrict our considerations to the case of the ball $\mathbf{B}_R$ of radius R with its boundary, the sphere $\mathbf{S}_R$, though most of what follows can be transmitted to more general domains. In our notations $\Gamma(d, S) = S$ and instead of $\frac{\delta}{\delta\gamma}$ we will denote variations along dilations by δ. As we know [Ci], [Ro] LBM satisfies in 3 dimension the following partial differential equation:

$$\Delta X = W_\sigma$$

with the Laplacian Δ, the (nonstandard) Gaussian white noise W_σ and the initial condition $X(0) = 0$. Accordingly, [McK], [Ro], its best forecast $Y = Y_{B_R}$ satisfies the following Lauricella type problem [La], the study of which dates back to the beginning of the century (see [Ni] and references therein):

$$\Delta^2 Y = 0 \quad in \ \mathbf{B}_R, \tag{2.1}$$

$$Y = X \quad in \ \mathbf{S}_R, \tag{2.2}$$

$$\frac{\partial}{\partial n}Y = \frac{\partial}{\partial n}X \quad in \ \mathbf{S}_R. \tag{2.3}$$

The solution can be given by means of the biharmonic Green's function $G(t,r)$ in the form:

$$Y(t) = \int_{S_R} \Delta_r G(t,r)\frac{\partial}{\partial n}X(r) - X(r)\frac{\partial}{\partial n}\Delta_r G(t,r)d\sigma(r) \tag{2.4}$$

substituting here the actual value of Green's function for the ball $\mathbf{B}_R$ we have the forecast as in [McK]:

$$Y(R,T,\theta) = \frac{1}{2}\int_{S_R} \frac{(T-R)^2}{R}\cdot\frac{\partial}{\partial R}\frac{RX(r)}{T} - \frac{(T^2-R^2)^2}{R}\cdot\frac{\partial}{\partial R}\frac{RX(r)}{\rho^3}d\sigma(r)$$

where ρ is the distance between $r \in \mathbf{B}_R$ and t . Introducing spherical harmonics $h_{n,k}$, let us consider the series expansion of LBM by them. As is known [McK] the coefficient motions $X_{n,k}$ are double Markov processes and so they have a canonical representation by the canonical white noise processes. These latter add up to a white noise W over the whole R^3 by spherical harmonics series expansion. According to what have been said in the introduction the variation along dilations of the ball $\mathbf{B}_R$ turns out to be the derivative in R. Nevertheless some cautiousness is needed since these derivatives must be taken in the generalised sense. From the spherical harmonics series expansion for the forecast [McK] we have after the generalised derivation for every $f(R) \in C_0^\infty(R-\varepsilon, R+\varepsilon)$:

$$\int_{R-\varepsilon}^{R+\varepsilon} f(R) \cdot \delta Y(R,T,\theta) dR = \int_{R-\varepsilon}^{R+\varepsilon} f(R) \frac{1}{2} \sum_{n,k} h_{n,k}(\theta) T^{-(n+1)} R^{n-2} (T^2 - R^2)$$
$$\times \left[R^2 X_{n,k}(R) + (2n+1) R X'_{n,k}(R) + (n^2-1) X''_{n,k}(R) \right] dR$$

The canonical representation shows that the term in between the brackets is exactly R times the canonical white noise. Estimation shows that the sum is convergent, and the convergence is uniform in θ. Adding up the sum we arrive at the following conclusion.

Proposition 2.1 *The variation δY of the forecast of LBM exists as a generalised process over the open interval $(R-\varepsilon, R+\varepsilon)$ and it has the white noise integral representation:*

$$\int_{R-\varepsilon}^{R+\varepsilon} f(R) \cdot \delta Y(R,T,\Theta) dR = \frac{1}{2} \int_{R-\varepsilon}^{R+\varepsilon} f(R) \int_{S_R} \frac{(T-R)^2}{RT} \frac{(T^2-R^2)^2}{R\rho^3} W(dR, d\phi)$$

for every $f(R) \in C_0^\infty(R-\varepsilon, R+\varepsilon)$, where (R, ϕ) denotes a point of the sphere and ρ its distance from t.

3. The PDE method

Now let us focus our attention to the PDE description. Introducing polar coordinates $r = (R, \phi)$, $\phi = (\phi_1, \phi_2)$ we denote by $H_S(R, \phi)$ the function $\Delta_r G_{D_S}(t, r) = \Delta_r G_{D_S}(t, R, \phi)$, the Laplacian of the biharmonic Green's function [Ni], corresponding to the domain D_S. For simplicity put $H_0(R, \phi) = H(R, \phi)$. We will omit to write the first argument if it is clear, how far we are from the origin (e.g. on the ball $\mathbf{B}_R$ or on $\mathbf{B}_{R+S}$). With these notations the solution (2.4) of the Lauricella problem (2.1)-(2.3) writes as:

$$Y_{D_0}(t) = \int_{S_R} H(\phi) \frac{\partial}{\partial n} X(R, \phi) - X(R, \phi) \frac{\partial}{\partial n} H(\phi) d\sigma(\phi) \qquad (3.1)$$

or for the dilated ball

$$Y_{D_S}(t) = \int_{S_{R+S}} H_S(\phi) \frac{\partial}{\partial n} X(R+S, \phi) - X(R+S, \phi) \frac{\partial}{\partial n} H_S(\phi) d\sigma_S(\phi) \qquad (3.2)$$

with σ and σ_S being the reference measures on the angles corresponding to the surface areas of the spheres $\mathbf{S}_R$ and $\mathbf{S}_{R+S}$ respectively. Consider the difference of $Y_{D_S} - Y_{D_0}$, in the form of $(3.1) - (3.2)$. We change first the reference measure σ_S in (3.2) to $\left(1 + \frac{R}{S}\right)^2 d\sigma(\phi)$ then compute the difference of the right hand sides of equations (3.1), (3.2)

$$Y_{D_S} - Y_{D_0} =$$

$$\left(1 + \frac{R}{S}\right)^2 \int_{S_R} H_S(\phi)\frac{\partial}{\partial n}X(R+S,\phi) - X(R+S,\phi)\frac{\partial}{\partial n}H_S(\phi)d\sigma(\phi) \quad (3.3)$$

$$-\int_{S_R} H(\phi)\frac{\partial}{\partial n}X(R,\phi) - X(R,\phi)\frac{\partial}{\partial n}H(\phi)d\sigma(\phi)$$

In order to determine the variation we evaluate the difference only up to first order terms in S. For this purpose we write from the definition of the variation of the biharmonic Green's function

$$H_S(\phi) \;=\; H(\phi) + S\cdot\delta H(\phi) + o(S) \qquad (3.4)$$

$$\frac{\partial}{\partial n}H_S(\phi) \;=\; \frac{\partial}{\partial n}H(\phi) + S\cdot\delta\frac{\partial}{\partial n}H(\phi) + o(S) \qquad (3.5)$$

Transform the term $X(R+S,\phi)$ as

$$X(R+S,\phi) = S\cdot\frac{X(R+S,\phi) - X(R,\phi)}{S} + X(R,\phi) \qquad (3.6)$$

and similarly

$$\frac{\partial}{\partial n}X(R+S,\phi) = S\cdot\frac{\frac{\partial}{\partial n}X(R+S,\phi) - \frac{\partial}{\partial n}X(R,\phi)}{S} + \frac{\partial}{\partial n}X(R,\phi) \qquad (3.7)$$

Substituting (3.4)-(3.7) in (3.3) we drop all the terms of the order of magnitude $o(S^2)$. Since the dispersion $\mathcal{D}^2\left(X(R+S,\phi) - X(R,\phi)\right) = S$, a bounded term$\times S\cdot(X(R+S,\phi) - X(R,\phi))$ is of order $o(S)$ in $\mathcal{L}_2\left(\Omega, \mathcal{A}, P\right)$, therefore it, too, can be dropped. However, the term

$$\int_{S_R} \frac{\partial}{\partial n}H(\phi)\frac{X(R+S,\phi) - X(R,\phi)}{S}d\sigma(\phi)$$

just tend to the evaluation at $\frac{\partial}{\partial n}H(\phi)$ of the generalised normal derivative $\frac{\partial}{\partial n}X(R,\phi)$. Special care require the term

$$\int\limits_{S_R} H(\phi)\frac{\frac{\partial}{\partial n}X(R+S,\phi)-\frac{\partial}{\partial n}X(R,\phi)}{S}d\sigma(\phi) \qquad (3.8)$$

because this is in fact the difference of evaluations of the generalised normal derivatives, taken on different balls. It is well known for LBM however [Pi], that in d dimensions only its generalised normal derivatives of order $\frac{d-1}{2}$ exist as generalised random fields over the sphere. It means for three dimension, that *only the first* normal derivative of LBM is well defined as a *generalised field over the sphere*. Nevertheless as we turn to spherical coordinates, the normal derivatives on balls centered at the origin turn out to be derivatives in the coordinate R, the distance from the origin. This way derivatives of any order can be taken in the generalised sense. Of course this derivation will result in a *generalised field in a neighbourhood* of the sphere. It means that though (3.8) will not converge by itself, but it will, if we evaluate it at a smooth function f(R) of the radius by integration, as it is usual for generalised functions! To avoid a clumsy integral formula we shall not write this evaluation in detail, and we still shall use the notation $\frac{\partial^2}{(\partial n)^2}X(R,\phi)$ keeping in mind its exact meaning. To keep R to denote the radius of the original ball $\mathbf{B}_R$ we shall rather denote in the integral the argument of f by Z. Thus we summarize what has been told in the following.

Theorem 3.1 *The variation δY of the forecast of LBM exists and it is well defined as a generalised process over the open interval $(R-\varepsilon, R+\varepsilon)$:*

$$\int\limits_{R-\varepsilon}^{R+\varepsilon} f(Z)\cdot\delta Y(Z,t)dZ$$

Using the partial differential equation description of the forecast in our notations $\delta Y(R,t)$ can be written as:

$$\delta Y(R,t) = \frac{2}{R}\int\limits_{S_R} H(\phi)\frac{\partial}{\partial n}X(R,\phi)-X(R,\phi)\frac{\partial}{\partial n}H(\phi)d\sigma(\phi)+$$

$$\int\limits_{S_R} \delta H(\phi)\frac{\partial}{\partial n}X(R,\phi)-X(R,\phi)\delta\frac{\partial}{\partial n}H(\phi)d\sigma(\phi)+ \quad(3.9)$$

$$\int\limits_{S_R} H(\phi)\frac{\partial^2}{(\partial n)^2}X(R,\phi)-\frac{\partial}{\partial n}X(R,\phi)\frac{\partial}{\partial n}H(\phi)d\sigma(\phi)$$

Remark 3.2 *Heuristically speaking, the first row corresponds to the changes as if we would integrate on the dilated ball* $\mathbf{B}_{R+S}$ *in (2.3) absolutely the same function as the one on* $\mathbf{B}_R$. *The second row corresponds the change of Green's function, while the third row covers the change in the observation of the field.*

The method of partial differential equations, we used in this section, can easily be applied in case of deformations, other than dilation. Because of the lost rotational symmetry it does not seem clear, whether the same is true for the spherical harmonics decomposition method.

Acknowledgment. I wish to thank three excellent scholars, M. Arató, Yu.A. Rozanov and T. Hida, who were my teachers in the field of stochastic processes and fields. T. Hida also inspired this research and paid attention to my work, for which I am especially grateful to him.

References

[Ci]   Z. Ciesielski, Stochastic Equation for Realization of the Lévy-Brownian Motion with a Several Dimensional Time, *Bull. Acad. Polonaise,* **11** (1975) 1193–1198.

[Go]   A. Goldman, Techniques biharmoniques pour l'étude du mouvement brownien de P.Lévy à trois paramètres, *Ann. Inst. Henri Poincaré* **25** (1989) No. 4., 351–381.

[La]   G. Lauricella, Integrazione dell'equazione $\Delta(\Delta u) = 0$ in un campo di forma circolare, *Torino, Atti,* **31** (1896), 610–618.

[McK] H. P. McKean, Brownian Motion with Several Dimensional Time *Theory Probab. & Appl.* **8** (1963), 334–354.

[Ni]   M. Nicolesco, *Les Fonctions Polyharmoniques,* Hermann, 1936.

[Pi]   L. I. Piterbarg, On Prediction of a Class of Random Fields, *Theory Probab. Appl.* **28** (1983), 184–194.

[Ro]   Yu. A. Rozanov, Markov Random Fields and Boundary Value Problems for Stochastic Partial Differential Equations, *Theory Probab. & Appl.* **32** (1987), 3–34.

[Si] Si Si, A note on Lévy's Brownian Motion, *Nagoya Math J.* **114** (1989), 165–172.

Eötvös Loránd University
Dept. of Probability Theory and Statistics,
Budapest, Múzeum krt. 6-8. H-1088 Hungary

Received February 28, 1997

A Maximal Inequality
for the Skorohod Integral

Elisa Alòs and David Nualart[1]

Abstract

In this paper we establish a maximal inequality for the Skorohod integral of stochastic processes belonging to the space $\mathbb{L}^F$ and satisfying an integrability condition. The space $\mathbb{L}^F$ contains both the square integrable adapted processes and the processes in the Sobolev space $\mathbb{L}^{2,2}$. Processes in $\mathbb{L}^F$ are required to be twice weakly differentiable in the sense of the stochastic calculus of variations in points (r, s) such that $r \vee s \geq t$.

1. Introduction

A stochastic integral for processes which are not necessarily adapted to the Brownian motion was introduced by Skorohod in [Sk]. The Skorohod integral turns out to be a generalization of the classical Itô integral, and on the other hand, it coincides with the adjoint of the derivative operator on the Wiener space. The techniques of the stochastic calculus of variations, introduced by Malliavin in [Ma], have allowed to develop a stochastic calculus for the Skorohod integral of processes in the Sobolev space $\mathbb{L}^{2,2}$ (see [NP]).

In a recent paper ([AN]) we have introduced a space of square integrable processes, denoted by $\mathbb{L}^F$, which is included in the domain of the Skorohod integral, and contains both the space of adapted processes and the Sobolev space $\mathbb{L}^{2,2}$. A process $u = \{u_t, t \in [0, T]\}$ in $\mathbb{L}^F$ is required to have square integrable derivatives $D_s u_t$ and $D^2_{r,s} u_t$ in the regions $\{s \geq t\}$ and $\{s \vee r \geq t\}$, respectively. We have proved in [AN] that the Skorohod integral of processes in the space $\mathbb{L}^F$ verifies the usual properties (quadratic variation, continuity, local property) and a change-of-variable formula can also be established.

The purpose of this paper is to prove a maximal inequality for processes in the space $\mathbb{L}^F$. Section 2 is devoted to recall some preliminaries on the

---

[1]Supported by the DGICYT grant number PB93–0052

stochastic calculus for the Skorohod integral. In Section 3 we show the maximal inequality (Theorems 3.1 and 3.2). The main ingredients of the proof are the factorization method used to deduce maximal inequalities for stochastic convolutions (see [dPT]) and the Itô formula for the Skorohod integral following the ideas introduced by Hu and Nualart in [HN].

2. A class of Skorohod integrable processes

Let $(\Omega, \mathcal{F}, P)$ be the canonical probability space of the one-dimensional Brownian motion $W = \{W_t, t \in [0, T]\}$. Let H be the Hilbert space $L^2([0, T])$. For any $h \in H$ we denote by $W(h)$ the Wiener integral $W(h) = \int_0^T h(t) dW_t$. Let $\mathcal{S}$ be the set of smooth and cylindrical random variables of the form:

$$F = f(W(h_1), ..., W(h_n)), \tag{2.1}$$

where $n \geq 1$, $f \in C_b^\infty(\mathbb{R}^n)$ (f and all its derivatives are bounded), and $h_1, ..., h_n \in H$. Given a random variable F of the form (2.1), we define its derivative as the stochastic process $\{D_t F, t \in [0, T]\}$ given by

$$D_t F = \sum_{j=1}^n \frac{\partial f}{\partial x_j}(W(h_1), ..., W(h_n)) h_j(t), \quad t \in [0, T].$$

In this way the derivative DF is an element of $L^2([0, T] \times \Omega) \cong L^2(\Omega; H)$. More generally, we can define the iterated derivative operator on a cylindrical random variable by setting

$$D_{t_1, ..., t_n}^n F = D_{t_1} \cdots D_{t_n} F.$$

The iterated derivative operator D^n is a closable unbounded operator from $L^2(\Omega)$ into $L^2([0, T]^n \times \Omega)$ for each $n \geq 1$. We denote by $\mathbb{D}^{n,2}$ the closure of $\mathcal{S}$ with respect to the norm defined by

$$\| F \|_{n,2}^2 = \| F \|_{L^2(\Omega)}^2 + \sum_{l=1}^n \| D^l F \|_{L^2([0,T]^l \times \Omega)}^2 .$$

We denote by δ the adjoint of the derivative operator D that is also called the Skorohod integral with respect to the Brownian motion $\{W_t\}$. That is, the domain of δ (denoted by $\mathrm{Dom}\,\delta$) is the set of elements $u \in L^2([0, T] \times \Omega)$ such that there exists a constant c verifying

$$\left| E \int_0^T D_t F u_t dt \right| \leq c \| F \|_2,$$

for all $F \in \mathcal{S}$. If $u \in \text{Dom }\delta$, $\delta(u)$ is the element in $L^2(\Omega)$ defined by the duality relationship

$$E(\delta(u)F) = E \int_0^T D_t F u_t dt, \quad F \in \mathcal{S}.$$

We will make use of the following notation: $\int_0^T u_t dW_t = \delta(u)$.

The Skorohod integral is an extension of the Itô integral in the sense that the set $L_a^2([0,T] \times \Omega)$ of square integrable and adapted processes is included into $\text{Dom }\delta$ and the operator δ restricted to $L_a^2([0,T] \times \Omega)$ coincides with the Itô stochastic integral (see [NP]).

Let $\mathbb{L}^{n,2} = L^2([0,T]; \mathbb{D}^{n,2})$ equipped with the norm

$$\| v \|_{n,2}^2 = \| v \|_{L^2([0,T] \times \Omega)}^2 + \sum_{j=1}^n \| D^j v \|_{L^2([0,T]^{j+1} \times \Omega)}^2.$$

We recall that $\mathbb{L}^{1,2}$ is included in the domain of δ, and for a process u in $\mathbb{L}^{1,2}$ we can compute the variance of the Skorohod integral of u as follows:

$$E(\delta(u)^2) = E \int_0^T u_t^2 dt + E \int_0^T \int_0^T D_s u_t D_t u_s ds dt. \tag{2.2}$$

Let $\mathcal{S}_T$ be the set of processes of the form $u_t = \sum_{j=1}^q F_j h_j(t)$, where $F_j \in \mathcal{S}$ and $h_j \in H$. We will denote by $\mathbb{L}^F$ the closure of $\mathcal{S}_T$ by the norm:

$$\| u \|_F^2 = E \int_0^T u_t^2 dt + E \int_{\{s \geq t\}} (D_s u_t)^2 ds dt + E \int_{\{r \vee s \geq t\}} (D_r D_s u_t)^2 dr ds dt. \tag{2.3}$$

That is, $\mathbb{L}^F$ is the class of stochastic processes $\{u_t, t \in [0,T]\}$ such that for each time t, the random variable u_t is twice weakly differentiable with respect to the Wiener process in the two-dimensional future $\{(r,s), r \vee s \geq t\}$.

The space $L_a^2([0,T] \times \Omega)$ is contained in $\mathbb{L}^F$. Furthermore, for all $u \in L_a^2([0,T] \times \Omega)$ we have $D_s u_t = 0$ for almost all $s \geq t$, and, hence,

$$\| u \|_F = \| u \|_{L^2([0,T] \times \Omega)}. \tag{2.4}$$

The next proposition provides an estimate for the L^2 norm of the Skorohod integral of processes in the space $\mathbb{L}^F$.

Proposition 2.1 $\mathbb{L}^F \subset \text{Dom }\delta$ *and we have that, for all u in $\mathbb{L}^F$,*

$$E|\delta(u)|^2 \leq 2 \| u \|_F^2. \tag{2.5}$$

Proof. Suppose first that u has a finite Wiener chaos expansion. In this case we can write:

$$E\left|\int_0^t u_s dW_s\right|^2 = E\int_0^t u_s^2 ds + E\int_0^t \int_0^t D_s u_\theta D_\theta u_s d\theta ds$$

$$= E\int_0^t u_s^2 ds + 2E\int_0^t \int_0^\theta D_s u_\theta D_\theta u_s d\theta ds$$

$$= E\int_0^t u_s^2 ds + 2E\int_0^t u_\theta \left(\int_0^\theta D_\theta u_s dW_s\right) d\theta.$$

Using now the inequality $2\langle a, b\rangle \le |a|^2 + |b|^2$ we obtain

$$E|\delta(u)|^2 \le 2E\int_0^T u_s^2 ds + E\int_0^T |\int_0^\theta D_\theta u_s dW_s|^2 d\theta. \qquad (2.6)$$

Because u has a finite chaos decomposition we have that $\{D_\theta u_s \mathbf{1}_{[0,\theta]}(s), s \in [0, T]\}$ belongs to $\mathbb{L}^{1,2} \subset \mathrm{Dom}\,\delta$ for each $\theta \in [0, T]$, and furthermore we have

$$E\int_0^T |\int_0^\theta D_\theta u_s dW_s|^2 d\theta \le E\int_0^T \int_0^\theta (D_\theta u_s)^2 ds d\theta$$

$$+ E\int_0^T \int_0^\theta \int_0^\theta (D_\sigma D_\theta u_s)^2 d\sigma ds d\theta. \qquad (2.7)$$

Now substituting (2.7) into (2.6) we obtain

$$E|\delta(u)|^2 \le 2E\int_0^T u_s^2 ds + E\int_0^T \int_0^\theta (D_\theta u_s)^2 ds d\theta$$

$$+ E\int_0^T \int_0^\theta \int_0^\theta (D_\sigma D_\theta u_s)^2 d\sigma ds d\theta$$

$$\le 2\| u \|_F^2,$$

which proves (2.5) in the case that u has a finite chaos decomposition. The general case follows easily from a limit argument. ∎

Note that $u \in \mathbb{L}^F$ implies $u\mathbf{1}_{[r,t]} \in \mathbb{L}^F$ for any interval $[r, t] \subset [0, T]$, and, by Proposition 2.1 we have that $u\mathbf{1}_{[r,t]} \in \mathrm{Dom}\,\delta$.

The following results, which are proved in [AN] are some basic properties for the Skorohod integral of processes u in $\mathbb{L}^F$.

(1) *(Local property for the operator δ)* Let $u \in \mathbb{L}^F$ and $A \in \mathcal{F}$ be such that $u_t(\omega) = 0$, a.e. on the product space $[0, T] \times A$. Then $\delta(u) = 0$ a.e. on A.

(2) *(Quadratic variation)* Let $u \in \mathbb{L}^F$. Then

$$\sum_{i=1}^{n-1} \left(\int_{t_i}^{t_{i+1}} u_s dW_s \right)^2 \rightarrow \int_0^T u_s^2 ds, \qquad (2.8)$$

in $L^1(\Omega)$, as $|\pi| \rightarrow 0$, where π runs over all finite partitions $\{0 = t_0 < t_1 < \cdots < t_n = T\}$ of $[0, T]$.

The local property allows to extend the Skorohod integral to processes in the space $\mathbb{L}^F_{loc}$. That is, $u \in \mathbb{L}^F_{loc}$ if there exists a sequence $\{(\Omega_n, u^n), n \geq 1\} \subset \mathcal{F} \times \mathbb{L}^F$ such that $u = u^n$ on Ω_n for each n, and $\Omega_n \uparrow \Omega$, a.s. Then we define $\delta(u)$ by

$$\delta(u)|_{\Omega_n} = \delta(u^n)|_{\Omega_n}.$$

Suppose that u is an adapted process verifying $\int_0^T u_s^2 ds < \infty$ a.s. Then one can show that u belongs to $\mathbb{L}^F_{loc}$ and $\delta(u)$ coincides with the Itô integral of u.

Let $\mathbb{L}^F_b$ denote the space of processes $u \in \mathbb{L}^F$ such that $\| \int_0^T u_s^2 ds \|_\infty < \infty$. We have proved in [AN] the following Itô's formula for the Skorohod integral:

Theorem 2.2 *Consider a process of the form $X_t = \int_0^t u_s dW_s$, where $u \in (\mathbb{L}^F_b)_{loc}$. Assume also that the indefinite Skorohod integral $\int_0^t u_s dW_s$ has a continuous version. Let $F : \mathbb{R} \rightarrow \mathbb{R}$ be a twice continuously differentiable function. Then we have*

$$F(X_t) = F(0) + \int_0^t F'(X_s) u_s dW_s + \frac{1}{2} \int_0^t F''(X_s) u_s^2 ds$$

$$+ \int_0^t F''(X_s) \left(\int_0^s D_s u_r dW_r \right) u_s ds. \qquad (2.9)$$

3. Maximal inequality for the Skorohod integral process

The purpose of this section is to prove a maximal inequality for the Skorohod integral process where its integrand belongs to the space $\mathbb{L}^F$, using the ideas of [HN].

Theorem 3.1 *Let $2 < p < \infty$, $q > \frac{p}{2}$, $q \geq 2$ and $\frac{1}{q} + \frac{1}{r} = \frac{2}{p}$. Let $u = \{u_\theta, \theta \in [0, T]\}$ be a stochastic process in the space $\mathbb{L}^F$ such that*

(i) $\int_0^T E|u_s|^{p\vee r}ds < \infty,$

(ii) $\int_{\{s\geq\theta\}} |E(D_s u_\theta)|^q dsd\theta < \infty,$

(iii) $\int_{\{r\vee s\geq\theta\}} |E(D_r D_s u_\theta)|^q drdsd\theta < \infty,$

Then $\int_0^t u_s dW_s$ is in L^p for all $t \in [0,T]$ and

$$E(\sup_{0\leq t\leq T} |\int_0^t u_s dW_s|^p) \leq K_{p,q}\{\int_0^T E|u_s|^{p\vee r}ds$$

$$+ \int_{\{s\geq\theta\}} |E(D_s u_\theta)|^q dsd\theta + \int_{\{r\vee s\geq\theta\}} E|D_r D_s u_\theta|^q drdsd\theta\}, (3.1)$$

where $K_{p,q}$ is a constant depending only on T, p and q.

Proof.   We will assume that $u \in \mathcal{S}_T$. The general case will follow using a density argument similar to the one in [AN], pg. 8. Let $\alpha \in (\frac{1}{p}, \frac{1}{2})$. Using the fact that

$$\int_0^1 (1-u)^{\alpha-1} u^{-\alpha} du = \frac{\pi}{\sin(\alpha\pi)}$$

and applying Fubini's stochastic theorem and Hölder's inequality we obtain that

$$E(\sup_{0\leq t\leq T} |\int_0^t u_s dW_s|^p)$$

$$= \frac{\sin(\alpha\pi)}{\pi} E(\sup_{0\leq t\leq T} |\int_0^t (\int_s^t (t-\sigma)^{\alpha-1}(\sigma-s)^{-\alpha}d\sigma)u_s dW_s|^p)$$

$$= \frac{\sin(\alpha\pi)}{\pi} E(\sup_{0\leq t\leq T} |\int_0^t (\int_0^\sigma (\sigma-s)^{-\alpha}u_s dW_s)(t-\sigma)^{\alpha-1}d\sigma|^p)$$

$$\leq \frac{\sin(\alpha\pi)}{\pi} E(\sup_{0\leq t\leq T} \{(\int_0^t |\int_0^\sigma (\sigma-s)^{-\alpha}u_s dW_s|^p d\sigma)$$

$$\times|\int_0^t (t-\sigma)^{\frac{(\alpha-1)p}{(p-1)}} d\sigma|^{p-1}\})$$

$$= \frac{\sin(\alpha\pi)}{\pi}(\frac{p-1}{\alpha p-1})^{p-1}T^{\alpha p-1}E(\int_0^T |\int_0^\sigma (\sigma-s)^{-\alpha}u_s dW_s|^p d\sigma).$$

Let us now define for any $\sigma \in [0,T]$ the process

$$V_t^\sigma := \int_0^t (\sigma-s)^{-\alpha}u_s dW_s, \qquad t \in [0,\sigma],$$

and denote

$$C_{p,\alpha} = \frac{\sin(\alpha\pi)}{\pi}(\frac{p-1}{\alpha p-1})^{p-1}T^{\alpha p-1}.$$

We have proved that

$$E(\sup_{0\leq t\leq T}|\int_0^t u_s dW_s|^p) \leq C_{p,\alpha}E(\int_0^T |V_\sigma^\sigma|^p d\sigma). \tag{3.2}$$

Now we are going to use the same ideas as in [HN]. Applying Theorem 2.2 to $F(x) = |x|^p$ and taking the expectation, we obtain:

$$E|V_t^\sigma|^p = \frac{p(p-1)}{2}\int_0^t E[|V_s^\sigma|^{p-2}(\sigma-s)^{-2\alpha}u_s^2]ds$$
$$+p(p-1)\int_0^t E[|V_s^\sigma|^{p-2}(\sigma-s)^{-\alpha}u_s\int_0^s D_s u_\theta(\sigma-\theta)^{-\alpha}dW_\theta]ds$$
$$=: I_1 + I_2.$$

Applying Hölder's inequality we get

$$I_1 \leq \frac{p(p-1)}{2}\int_0^t (E|V_s^\sigma|^p)^{\frac{p-2}{p}}(E|u_s|^p)^{\frac{2}{p}}(\sigma-s)^{-2\alpha}ds$$

and

$$I_2 \leq p(p-1)\int_0^t (E|V_s^\sigma|^p)^{\frac{p-2}{p}}(E|u_s\int_0^s D_s u_\theta(\sigma-\theta)^{-\alpha}dW_\theta|^{\frac{p}{2}})^{\frac{2}{p}}(\sigma-s)^{-\alpha}ds.$$

Denote

$$A_s := \frac{p(p-1)}{2}(E|u_s|^p)^{\frac{2}{p}}(\sigma-s)^{-2\alpha}$$
$$+p(p-1)(E|u_s\int_0^s D_s u_\theta(\sigma-\theta)^{-\alpha}dW_\theta|^{\frac{p}{2}})^{\frac{2}{p}}(\sigma-s)^{-\alpha}$$

and $G_s = E|V_s^\sigma|^p$. Then we have that, for every $t \leq \sigma$

$$G_t \leq \int_0^t G_s^{\frac{p-2}{p}}A_s ds.$$

Using the lemma of [Za], p.171 we obtain

$$G_t \leq (\frac{2}{p}\int_0^t A_s ds)^{\frac{p}{2}}.$$

Therefore

$$
\begin{aligned}
E|V_t^\sigma|^p \;\leq\; & \{(p-1)\int_0^t (E|u_s|^p)^{\frac{2}{p}}(\sigma-s)^{-2\alpha}ds \\
& +2(p-1)\int_0^t (E|u_s\int_0^s D_s u_\theta(\sigma-\theta)^{-\alpha}dW_\theta|^{\frac{p}{2}})^{\frac{2}{p}}(\sigma-s)^{-\alpha}ds\}^{\frac{p}{2}} \\
\;\leq\; & (p-1)^{\frac{p}{2}}\Big[2^{\frac{p}{2}-1}\{\int_0^t (E|u_s|^p)^{\frac{2}{p}}(\sigma-s)^{-2\alpha}ds\}^{\frac{p}{2}} \\
& +2^{p-1}\{\int_0^t (E|u_s\int_0^s D_s u_\theta(\sigma-\theta)^{-\alpha}dW_\theta|^{\frac{p}{2}})^{\frac{2}{p}}(\sigma-s)^{-\alpha}ds\}^{\frac{p}{2}}\Big].
\end{aligned}
$$

By Hölder's inequality we have:

$$
\begin{aligned}
I_3 \;:=\; & \{\int_0^t (E|u_s\int_0^s D_s u_\theta(\sigma-\theta)^{-\alpha}dW_\theta|^{\frac{p}{2}})^{\frac{2}{p}}(\sigma-s)^{-\alpha}ds\}^{\frac{p}{2}} \\
\;\leq\; & \{\int_0^t (E|u_s|^r)^{\frac{1}{r}}(E|\int_0^s D_s u_\theta(\sigma-\theta)^{-\alpha}dW_\theta|^q)^{\frac{1}{q}}(\sigma-s)^{-\alpha}ds\}^{\frac{p}{2}} \\
\;\leq\; & \{\int_0^t (E|u_s|^r)^{\frac{q}{(q-1)r}}(\sigma-s)^{\frac{\alpha q}{1-q}}ds\}^{\frac{p(q-1)}{2q}} \\
& \times\{\int_0^t E|\int_0^s D_s u_\theta(\sigma-\theta)^{-\alpha}dW_\theta|^q ds\}^{\frac{p}{2q}} \\
\;\leq\; & c_1\{\int_0^t (E|u_s|^r)(\sigma-s)^{\frac{\alpha q}{1-q}}ds\}^{\frac{2q-p}{2q}} \\
& \times\{\int_0^t E|\int_0^s D_s u_\theta(\sigma-\theta)^{-\alpha}dW_\theta|^q ds\}^{\frac{p}{2q}},
\end{aligned}
$$

for some constant c_1 depending only on p, q, α and T. Since $\frac{2q-p}{2q}+\frac{p}{2q}=1$, using the inequality $ab\leq\frac{2q-p}{2q}a^{\frac{2q}{2q-p}}+\frac{p}{2q}b^{\frac{2q}{p}}$ for $a,b\geq 0$, we have

$$
I_3\leq c_1\{\int_0^t (E|u_s|^r)(\sigma-s)^{\frac{\alpha q}{1-q}}ds+\int_0^t E|\int_0^s D_s u_\theta(\sigma-\theta)^{-\alpha}dW_\theta|^q ds\}.
$$

Now we can estimate the Skorohod integral using Meyer's inequalities (see [Nu], Section 3.2) and we obtain

$$
\begin{aligned}
I_4 \;:=\; & \int_0^t E|\int_0^s D_s u_\theta(\sigma-\theta)^{-\alpha}dW_\theta|^q ds \\
\;\leq\; & c_2\Big\{\int_0^t (\int_0^s (\sigma-\theta)^{-2\alpha}|E(D_s u_\theta)|^2 d\theta)^{\frac{q}{2}}ds \\
& +\int_0^t E(\int_0^T\int_0^s (\sigma-\theta)^{-2\alpha}|D_r D_s u_\theta|^2 d\theta dr)^{\frac{q}{2}}ds\Big\},
\end{aligned}
$$

for some constant c_2. Hence, taking into account (3.2), we get

$$E(\sup_{0 \le t \le T} |\int_0^t u_s dW_s|^p) \le c_3(I_5 + I_6 + I_7 + I_8),$$

where c_3 is a constant depending on p, q, α and T, and

$$I_5 := \int_0^T (\int_0^\sigma (E|u_s|^p)^{\frac{2}{p}}(\sigma - s)^{-2\alpha}ds)^{\frac{p}{2}}d\sigma,$$

$$I_6 := \int_0^T \int_0^\sigma (\sigma - s)^{\frac{\alpha q}{1-q}} E|u_s|^r ds d\sigma,$$

$$I_7 := \int_0^T \int_0^\sigma (\int_0^s (\sigma - \theta)^{-2\alpha}|E(D_s u_\theta)|^2 d\theta)^{\frac{q}{2}} ds d\sigma,$$

$$I_8 := \int_0^T \int_0^\sigma E(\int_0^T \int_0^s (\sigma - \theta)^{-2\alpha}|D_r D_s u_\theta|^2 d\theta dr)^{\frac{q}{2}} ds d\sigma.$$

Now using Hölder's inequality and Fubini's theorem we obtain that

$$\begin{aligned}
I_5 &\le \frac{T^{1-2\alpha}}{1-2\alpha} \int_0^T \int_0^\sigma E|u_s|^p(\sigma - s)^{-2\alpha} ds d\sigma \\
&= \frac{T^{1-2\alpha}}{1-2\alpha} \int_0^T (\int_s^T (\sigma - s)^{-2\alpha} d\sigma) E|u_s|^p ds \\
&\le c_4 \int_0^T E|u_s|^p ds,
\end{aligned}$$

for some constant c_4. Similarly,

$$I_6 \le c_5 \int_0^T E|u_s|^r ds,$$

$$I_7 \le c_6 \int_{\{s \ge \theta\}} |E(D_s u_\theta)|^q ds d\theta,$$

and

$$I_8 \le c_7 \int_{\{r \vee s \ge \theta\}} E|D_r D_s u_\theta|^q dr ds d\theta,$$

for some constants c_5, c_6 and c_7. The proof is now complete. ∎

As a corollary, taking $q = 2$ we have the following result:

Theorem 3.2 *Let $p \in (2,4)$, $r = \frac{2p}{4-p}$. Let $u = \{u_s, s \in [0,T]\}$ be a stochastic process in the space $\mathbb{L}^F$ such that $\int_0^T E|u_s|^r ds < \infty$. Then $\int_0^t u_s dW_s$ is in L^p for all $t \in [0,T]$ and*

$$E(\sup_{0 \leq t \leq T} |\int_0^t u_s dW_s|^p) \leq K_p \{ \int_0^T E|u_s|^r ds + \int_{\{s \geq \theta\}} E|D_s u_\theta|^2 dsd\theta$$

$$+ \int_{\{r \vee s \geq \theta\}} E|D_r D_s u_\theta|^2 drdsd\theta \}, \tag{3.3}$$

where K_p is a constant depending only on p and T.

Remark Theorem 3.2 implies the continuity of the Skorohod process $\int_0^t u_s W_s$ assuming that $u \in \mathbb{L}^F$ and $\int_0^T E|u_s|^r ds < \infty$ for some $r > 2$. This result was proved in [AN] using Kolmogorov continuity criterion and the technique developed in [HN].

References

[AN] E. Alòs and D. Nualart, An extension of Itô's formula for anticipating processes. Preprint.

[dPT] G. da Prato and J. Zabcyck, *Stochastic equations in infinite dimensions*, Encyclopedia of Mathematics and its Applications **44**. Cambridge University Press, 1992.

[HN] Y. Hu and D. Nualart, Continuity of some anticipating integral processes. Preprint.

[Ma] P. Malliavin, Stochastic calculus of variations and hypoelliptic operators. In: *Proc. Inter. Symp. on Stoch. Diff. Equations, Kyoto 1976*, Wiley (1978), 195–263.

[Nu] D. Nualart, *The Malliavin Calculus and Related Topics*, Springer, 1995.

[NP] D. Nualart and E. Pardoux, Stochastic calculus with anticipating integrands, *Probab. Theory Rel. Fields*, **78** (1988), 535–581.

[Sk] A. V. Skorohod, On a generalization of a stochastic integral. *Theory Probab. Appl.* , **20** (1975), 219–233.

[Za] M. Zakai, Some moment inequalities for stochastic integrals and for solutions of stochastic differential equations. *Israel J. Math.* , **5** (1967), 170-176.

Facultat de Matemàtiques
Universitat de Barcelona
Gran Via 585,
08007 Barcelona, Spain

Received October 16,1996

On the Kinematics of Stochastic Mechanics

Michele Pavon

1. Introduction

In a series of recent papers [P1]-[P3] we have shown that the complex-ification of the velocity and momentum processes permits to effectively develop a Lagrangian and Hamiltonian formalism in stochastic mechanics [N1], [G],[N2]. In this paper, the kinematics employed in [P1]-[P3] is more thoroughly analyzed, particularly from the probabilistic viewpoint. The outline of the paper is as follows. In Section 2, we introduce the appropri-ate kinematics for finite-energy diffusions developing on [N3]. In Section 3, we discuss in detail the properties of the quantum noise and derive a fun-damental change of variables formula. In the following section, we consider the Markov case. In Section 5, for the purpose of comparison, we give a strong form of the classical Hamilton principle. In Section 6, we present the quantum Hamilton principle.

2. Kinematics of finite-energy diffusion processes

In this section, we review some essential concepts on the kinematics of diffusions as presented in [N3]. We then proceed to develop these concepts to obtain the time-symmetric kinematics adopted in [P1, P2, P3]. The latter is here more thoroughly analyzed particularly from the probabilistic viewpoint.

Let $(\Omega, \mathcal{E}, \mathbf{P})$ be a complete probability space. A stochastic process $\{\xi(t); t_0 \leq t \leq t_1\}$ mapping $[t_0, t_1]$ into $L_n^2(\Omega, \mathcal{E}, \mathbf{P})$ is called a *finite-energy diffusion* with constant diffusion coefficient $I_n \sigma^2$ if the path $\xi(\omega)$ belongs a.s. to $C([t_0, t_1]; \mathbb{R}^n)$ (n-dimensional continuous functions) and

$$\xi(t) - \xi(s) = \int_s^t \beta(\tau)d\tau + \sigma w_+(s, t), \quad t_0 \leq s < t \leq t_1, \qquad (2.1)$$

where the *forward drift* $\beta(t)$ is at each time t a measurable function of the past $\{\xi(\tau); 0 \leq \tau \leq t\}$, and $w_+(\cdot, \cdot)$ is a standard, n-dimensional

Wiener difference process with the property that $w_+(s,t)$ is independent of $\{\xi(\tau); 0 \le \tau \le s\}$. Moreover, β must satisfy the finite-energy condition

$$E\left\{\int_{t_0}^{t_1} \beta(\tau) \cdot \beta(\tau) d\tau\right\} < \infty. \tag{2.2}$$

We recall the characterizing properties of the n-dimensional *Wiener difference process* $w_+(s,t)$, see [N1, Chapter 11] and [N3, Section 1]. It is a process such that $w_+(t,s) = -w_+(s,t)$, $w_+(s,u) + w_+(u,t) = w_+(s,t)$, and that $w_+(s,t)$ is Gaussian distributed with mean zero and variance $I_n|s - t|$. Moreover, (the components of) $w_+(s,t)$ and $w_+(u,v)$ are independent whenever $[s,t]$ and $[u,v]$ don't overlap. Of course, $w_+(t) := w_+(t_0,t)$ is a standard Wiener process such that $w_+(s,t) = w_+(t) - w_+(s)$.

In [F], Föllmer has shown that a finite-energy diffusion also admits a reverse-time differential. Namely, there exists a measurable function $\gamma(t)$ of the future $\{\xi(\tau); t \le \tau \le t_1\}$ called *backward drift*, and another Wiener difference process w_- such that

$$\xi(t) - \xi(s) = \int_s^t \gamma(\tau) d\tau + \sigma w_-(s,t), \quad t_0 \le s < t \le t_1. \tag{2.3}$$

Moreover, γ satisfies

$$E\left\{\int_{t_0}^{t_1} \gamma(\tau) \cdot \gamma(\tau) d\tau\right\} < \infty, \tag{2.4}$$

and $w_-(s,t)$ is independent of $\{\xi(\tau); t \le \tau \le t_1\}$. Let us agree that dt always indicates a strictly positive variable. For any function f defined on $[t_0, t_1]$, let

$$d_+f(t) = f(t + dt) - f(t)$$

be the *forward increment* at time t, and

$$d_-f(t) = f(t) - f(t - dt)$$

be the *backward increment* at time t. For a finite-energy diffusion, Föllmer has also shown in [F] that the forward and backward drifts may be obtained as Nelson's conditional derivatives, namely

$$\beta(t) = \lim_{dt \searrow 0} E\left\{\frac{d_+\xi(t)}{dt} | \xi(\tau), t_0 \le \tau \le t\right\},$$

and

$$\gamma(t) = \lim_{dt \searrow 0} E\left\{\frac{d_-\xi(t)}{dt} | \xi(\tau), t \le \tau \le t_1\right\},$$

the limits being taken in $L_n^2(\Omega, \mathcal{B}, P)$. It was finally shown in [F] that the one-time probability density $\rho(\cdot, t)$ of $\xi(t)$ (which exists for every $t > t_0$) is absolutely continuous on $\mathbb{R}^n$ and the following relation holds $\forall t > 0$

$$E\{\beta(t) - \gamma(t)|\xi(t)\} = \sigma^2 \nabla \log \rho(\xi(t), t). \tag{2.5}$$

Corresponding to (2.1) and (2.3) are two change of variables formulas. Let $f : \mathbb{R}^n \times [0, T] \to \mathbb{R}$ be twice continuously differentiable with respect to the spatial variable and once with respect to time. Then, if ξ is a finite-energy diffusion satisfying (2.1) and (2.3), we have

$$f(\xi(t), t) - f(\xi(s), s) = \int_s^t \left(\frac{\partial}{\partial \tau} + \beta(\tau) \cdot \nabla + \frac{\sigma^2}{2} \Delta \right) f(\xi(\tau), \tau) d\tau$$

$$+ \int_s^t \sigma \nabla f(\xi(\tau), \tau) \cdot d_+ w_+(\tau), \tag{2.6}$$

$$f(\xi(t), t) - f(\xi(s), s) = \int_s^t \left(\frac{\partial}{\partial \tau} + \gamma(\tau) \cdot \nabla - \frac{\sigma^2}{2} \Delta \right) f(\xi(\tau), \tau) d\tau$$

$$+ \int_s^t \sigma \nabla f(\xi(\tau), \tau) \cdot d_- w_-(\tau), \tag{2.7}$$

where $d_+ w_+(t) := w_+(t, t+dt)$ and $d_- w_-(t) = w_-(t-dt, t)$. The stochastic integrals appearing in (2.6) and (2.7) are a (forward) Ito integral and a backward Ito integral, respectively, see [N3] for the details. Let us introduce the *current drift* $v(t) := (\beta(t) + \gamma(t))/2$ and the *osmotic drift* $u(t) := (\beta(t) - \gamma(t))/2$. Notice that, when σ tends to zero, v tends to $\dot{\xi}$ and u tends to zero. The semi-sum and the semi-difference of (2.6) and (2.7) give two more formulas:

$$f(\xi(t), t) - f(\xi(s), s) = \int_s^t \left(\frac{\partial}{\partial \tau} + v(\tau) \cdot \nabla \right) f(\xi(\tau), \tau) d\tau$$

$$+ \frac{\sigma}{2} \left[\int_s^t \nabla f(\xi(\tau), \tau) \cdot d_+ w_+ + \int_s^t \nabla f(\xi(\tau), \tau) \cdot d_- w_- \right], \tag{2.8}$$

$$0 = \int_s^t \left(u(\tau) \cdot \nabla + \frac{\sigma^2}{2} \Delta \right) f(\xi(\tau), \tau) d\tau$$

$$+ \frac{\sigma}{2} \left[\int_s^t \nabla f(\xi(\tau), \tau) \cdot d_+ w_+ - \int_s^t \nabla f(\xi(\tau), \tau) \cdot d_- w_- \right]. \tag{2.9}$$

Specializing (2.8) and (2.9) to $f(x, t) = x$, we get

$$\xi(t) - \xi(s) = \int_s^t v(\tau) d\tau + \frac{\sigma}{2} [w_+(s, t) + w_-(s, t)], \tag{2.10}$$

$$0 = \int_s^t u(\tau) d\tau + \frac{\sigma}{2} [w_+(s, t) - w_-(s, t)] \tag{2.11}$$

Let us multiply (2.11) by $-i$, and add it to (2.10). We get

$$\xi(t) - \xi(s) = \int_s^t [v(\tau) - iu(\tau)]d\tau + \sigma \left[\frac{(1-i)}{2} w_+(s,t) + \frac{(1+i)}{2} w_-(s,t) \right].$$
(2.12)

We call $v_q := v - iu$ the *quantum drift*, and

$$w_q(s,t) := \frac{1-i}{2} w_+(s,t) + \frac{1+i}{2} w_-(s,t) \tag{2.13}$$

the *quantum noise*. Notice that $w_q(\cdot,\cdot)$ is a difference process taking values in $\mathbb{C}^n$. We can then rewrite (2.12) as

$$\xi(t) - \xi(s) = \int_s^t v_q(\tau)d\tau + \sigma w_q(s,t). \tag{2.14}$$

3. The quantum noise and a change of variables formula

At first sight, the decomposition of the *real-valued* increments of ξ into the sum of two *complex* quantities in (2.14) might look somewhat odd. Nevertheless, this representation enjoys several important properties.

1. When σ^2 tends to zero, $v_q = v - iu$ tends to $\dot{\xi}$.

2. The quantum drift $v_q(t)$ contains at each time t precisely the same information as the pair $(v(t), u(t))$, and the quantum noise $w_q(s,t)$ contains the same information as the pair of "driving noises" $(w_+(s,t), w_-(s,t))$.

3. The representation (2.14), differently from (2.1) and (2.3) enjoys an important symmetry with respect to time, see Section 4.

The representation (2.14) has proven to be crucial in order to develop a Lagrangian and Hamiltonian dynamics formalism in the context of Nelson's stochastic mechanics, see [P1]-[P3]. In particular, to develop the second form of Hamilton's principle in [P1], the key tool has been a change of variables formula related to representation (2.14). In order to recall such a formula, we need first to define stochastic integrals with respect to the quantum noise w_q. Let us denote by $d_b f(t) := \frac{1-i}{2} d_+ f(t) + \frac{1+i}{2} d_- f(t)$ the *bi-directional increment* of f at time t, and recall that $d_+ w_q(t) := w_q(t, t+dt)$ and $d_- w_q(t) := w_q(t-dt, t)$. Then, from (2.13) and (2.11), we get

$$d_+ w_q(t) = \frac{1+i}{\sigma} u(t)dt + d_+ w_+,$$

$$d_- w_q(t) = \frac{-1+i}{\sigma} u(t) dt + d_- w_-.$$

These in turn give immediately the important relation:

Lemma 3.1

$$d_b w_q(t) := \frac{1-i}{2} d_+ w_+(t) + \frac{1+i}{2} d_- w_-(t). \tag{3.1}$$

From the definition (2.13), we also have the following property of the quantum noise.

Proposition 3.2 *Consider the real and imaginary parts of $w_q(s,t)$ $\Re w_q(s,t)$ and $\Im w_q(s,t)$, respectively. Then*

$$E\{\Re w_q(s,t) \Im w_q(s,t)^T\} = 0. \tag{3.2}$$

Let $f(x,t)$ be a measurable, $\mathbb{C}^n$-valued function such that

$$\mathbf{P}\left\{\omega : \int_{t_0}^{t_1} f(\xi(t),t) \cdot \overline{f(\xi(t),t)} dt < \infty\right\} = 1.$$

We then define the stochastic integral

$$\int_s^t f(\xi(\tau),\tau) \cdot d_b w_q(\tau) :=$$

$$\frac{1-i}{2} \int_s^t f(\xi(\tau),\tau) \cdot d_+ w_+(\tau) + \frac{1+i}{2} \int_s^t f(\xi(\tau),\tau) \cdot d_- w_-(\tau) \tag{3.3}$$

Thus, integration with respect to the "bi-directional increments" $d_b w_q$ of w_q is defined through a linear combination with complex coefficients of a forward and a backward Ito integral. Let $f(x,t)$ be a complex-valued function with real and imaginary parts of class $C^{2,1}$. Then, multiplying (2.9) by $-i$, and then adding it to (2.8), we get the *change of variables formula*

$$f(\xi(t),t) - f(\xi(s),s) =$$

$$\int_s^t \left(\frac{\partial}{\partial \tau} + [(v(\tau) - iu(\tau)] \cdot \nabla - \frac{i\sigma^2}{2}\Delta\right) f(\xi(\tau),\tau) d\tau \tag{3.4}$$

$$+ \int_s^t \sigma \nabla f(\xi(\tau),\tau) \cdot d_b w_q(\tau). \tag{3.5}$$

This formula proved to be the essential tool to develop the second (hydrodynamic) form of Hamilton's principle in stochastic mechanics [P1].

4. Markovian finite-energy diffusions

Let $\{\xi(t); t_0 \leq t \leq t_1\}$ be a finite-energy diffusion process with diffusion coefficient $I_n \sigma^2$. Let $\beta(t)$ and $\gamma(t)$ be its forward and backward drifts, respectively. The process $\xi(\cdot)$ is called *Markovian* if there exist two measurable functions $b_+(\cdot, \cdot)$ and $b_-(\cdot, \cdot)$ such that $\beta(t) = b_+(\xi(t), t)$ a.s. and $\gamma(t) = b_-(\xi(t), t)$ a.s., for all t in $[t_0, t_1]$. The duality relation (2.5) now reads

$$b_+(\xi(t), t) - b_-(\xi(t), t) = \sigma^2 \nabla \log \rho(\xi(t), t). \tag{4.1}$$

This immediately gives the *osmotic equation*

$$u(x, t) = \frac{\sigma^2}{2} \nabla \log \rho(x, t), \tag{4.2}$$

where $u(x, t) := (b_+(x, t) - b_-(x, t))/2$. The probability density $\rho(\cdot, \cdot)$ of $\xi(t)$ satisfies (at least weakly) the *Fokker-Planck equation*

$$\frac{\partial \rho}{\partial t} + \nabla \cdot (b_+ \rho) = \frac{\sigma^2}{2} \Delta \rho.$$

The latter can also be rewritten, in view of (4.1), as the *equation of continuity* of hydrodynamics

$$\frac{\partial \rho}{\partial t} + \nabla \cdot (v\rho) = 0, \tag{4.3}$$

where $v(x, t) := (b_+(x, t) + b_-(x, t))/2$.

Consider the *time reversal transformation* [G, p.289] $v(x, t) \rightarrow v'(x', t') = -v(x, t)$, $u(x, t) \rightarrow u'(x', t') = u(x, t)$, and, consequently, $v_q(x, t) = (v - iu)(x, t) \rightarrow (v' - iu')(x', t') = -(v + iu)(x, t) = -\overline{v_q(x, t)}$. Moreover, $w_+(s, t) \rightarrow -w_-(s, t)$, and $w_-(s, t) \rightarrow -w_+(s, t)$. Thus, $w_q(s, t) \rightarrow -\overline{w_q(s, t)}$, where overbar indicates conjugation. Thus, under time reversal, (2.14) transforms into

$$\xi(t) - \xi(s) = \int_s^t \overline{v_q(\xi(\tau), \tau)} d\tau + \sigma \overline{w_q(s, t)}. \tag{4.4}$$

Hence, (2.14), differently from (2.1) and (2.2), enjoys the time reversal invariance property.

Let us now recall how (2.6) and (2.7) are established for the Markovian diffusion $\xi(\cdot)$ and a sufficiently smooth $f(\cdot, \cdot)$. Expand f in a Taylor polynomial of degree two, and substitute into $d_+ f(\xi(t), t)$, to get

$$d_+ f = \frac{\partial f}{\partial t} dt + \nabla f \cdot d_+ \xi + \frac{\sigma^2}{2} \sum_{i,j=1}^{n} \frac{\partial^2 f}{\partial x_i \partial x_j} d_+ w_+^i d_+ w_+^j + o(dt), \tag{4.5}$$

$$d_-f = \frac{\partial f}{\partial t}dt + \nabla f \cdot d_-\xi + \frac{\sigma^2}{2}\sum_{i,j=1}^n \frac{\partial^2 f}{\partial x_i \partial x_j} d_- w_-^i d_- w_-^j + o(dt), \quad (4.6)$$

where we have used the facts

$$d_+\xi_+^i d_+\xi_+^j = d_+ w_+^i d_+ w_+^j + o(dt), \quad d_-\xi_+^i d_-\xi_+^j = d_- w_-^i d_- w_-^j + o(dt).$$

Here $o(dt)$ refers to the L^2 norm. Using the properties of w_+ and w_-

$$E\{d_+w_+(t)d_+w_+(t)^T|\xi(t)\} = I_n dt, \quad (4.7)$$
$$E\{d_-w_-(t)d_-w_-(t)^T|\xi(t)\} = -I_n dt, \quad (4.8)$$

we get (2.6) and (2.7). We now wish to understand (3.5), and in particular the coefficient of the Laplacian, in a similar fashion, namely from the basic probabilistic properties of the driving noise w_q. Starting from (2.12), we get

$$d_b f = \frac{\partial f}{\partial \tau}dt + \nabla f \cdot d_b\xi + \frac{\sigma^2}{2}\sum_{i,j=1}^n \frac{\partial^2 f}{\partial x_i \partial x_j} d_b w_q^i d_b w_q^j + o(dt). \quad (4.9)$$

We now need the following crucial result.

Lemma 4.1

$$E\{d_b w_q(t) d_b w_q(t)^T|\xi(t)\} = -iI_n dt. \quad (4.10)$$

Proof. Let us recall (3.1)

$$d_b w_q(t) := \frac{1-i}{2}d_+w_+(t) + \frac{1+i}{2}d_-w_-(t).$$

We now claim that $E\{d_+w_+(t)d_-w_-(t)^T|\xi(t)\} = 0$. Indeed, in view of

$$\xi(t+dt) - \xi(t) = \int_t^{t+dt} b_+(\xi(\tau),\tau)d\tau + \sigma w_+(t, t+dt)$$

we see that $d_+w_+(t) = w_+(t, t+dt)$ is measurable with respect to the "future" at time t $\sigma\{\xi(\tau); t \le \tau \le t_1\}$. Similarly, using the reverse-time representation, one sees that $d_-w_-(t)$ is measurable with respect to the "past" $\sigma\{\xi(\tau); t_0 \le \tau \le t\}$. By the Markov property, past and future of $\xi(\cdot)$ are conditionally independent given the present $\xi(t)$. Thus $E\{d_+w_+(t)d_-w_-(t)^T|\xi(t)\} = 0$, and, consequently, of the four terms of

$E\{d_b w_q(t) d_b w_q(t)^T | \xi(t)\}$ only two are nonzero. By (4.7) and (4.8), we finally get

$$E\{d_b w_q(t) d_b w_q(t)^T | \xi(t)\} = \frac{(1-i)^2}{4} I_n dt - \frac{(1+i)^2}{4} I_n dt = -i I_n dt$$

∎

By (4.10), equation (4.9) turns into

$$d_b f = \left[\frac{\partial f}{\partial \tau} + v_q \cdot \nabla f - \frac{i\sigma^2}{2} \Delta f \right] dt + \nabla f \cdot d_b w_q + o(dt). \tag{4.11}$$

Now let $C_b^2(\mathbb{R}^n; \mathbb{C})$ denote the complex, twice continuously differentiable, functions with compact support in $\mathbb{R}^n$. For $f \in C_b^2(\mathbb{R}^n; \mathbb{C})$, in view of (4.11), we define the *bi-directional generator* L_b of ξ by

$$L_b f = v_q \cdot \nabla f - \frac{i\sigma^2}{2} \Delta f. \tag{4.12}$$

The properties of the bi-directional generator, its relevance for various problems in quantum physics, and connections to the existing literature such as [AMU] will be discusses in a forthcoming paper.

5. The classical Hamilton principle

In the next section, we review the hydrodynamic form of Hamilton's principle in stochastic mechanics as established in [P1] (see [P3] for the corresponding particle form). In order to make the comparison with classical mechanics more transparent, we recall first the strong form of the classical principle as established in [P1, Section II].

Consider a dynamical system with configuration space $\mathbb{R}^n$. Let $L(x, v) := \frac{1}{2} mv \cdot v - V(x)$ be the Lagrangian, where $V(\cdot) : \mathbb{R}^n \to \mathbb{R}$ is of class C^1. Let $\mathcal{X}_{x_1}$ denote the class of all C^1 paths $x : [t_0, t_1] \to \mathbb{R}^n$ such that $x(t_1) = x_1$. Let $\mathcal{V}$ denote the family of continuous functions $v : [t_0, t_1] \to \mathbb{R}^n$. Consider the following problem: Minimize over $(x, v) \in (\mathcal{X}_{x_1} \times \mathcal{V})$

$$\int_{t_0}^{t_1} L(x(t), v(t)) \, dt + S_0(x(t_0)), \tag{5.1}$$

subject to the constraint

$$\dot{x}(t) = v(t), \quad \forall t \in [t_0, t_1]. \tag{5.2}$$

Notice that we have allowed a much larger class of paths than usual [GM], namely also trajectories solving a functional differential equation of the form $\dot{x}(t) = v(\{x(\tau); t_0 \leq \tau \leq t_1\})(t)$, where $v : C([t_0, t_1]; \mathbb{R}^n) \to C([t_0, t_1]; \mathbb{R}^n)$.

Theorem 5.1 *Suppose that $S(x, t)$ of class C^1 solves on $[t_0, t_1]$ the initial value problem*

$$\frac{\partial S}{\partial t} + \frac{1}{2m} \nabla S \cdot \nabla S + V(x) = 0, \tag{5.3}$$

$$S(x, t_0) = S_0(x). \tag{5.4}$$

Then any $x \in \mathcal{X}_{x_1}$ satisfying

$$\dot{x}(t) = \frac{1}{m} \nabla S(x(t), t), \tag{5.5}$$

solves together with $\frac{1}{m} \nabla S(x(t), t)$ the above minimum problem.

The crucial steps in the proof are the following two observations:

1. If the pair $(x, v) \in (\mathcal{X}_{x_1} \times \mathcal{V})$ satisfies the dynamical constraint (5.2), then

$$\frac{dS}{dt}(x(t), t) = \frac{\partial S}{\partial t}(x(t), t) + v(t) \cdot \nabla S(x(t), t); \tag{5.6}$$

2. The function $S(\cdot, \cdot)$ satisfies

$$\min_{v \in \mathbb{R}^n} \{\frac{1}{2} mv \cdot v - V(x) - \frac{\partial S}{\partial t}(x, t) - v \cdot \nabla S(x, t)\} = 0. \tag{5.7}$$

Let us now discuss the following question: when does the above result yield a minimizing pair (x, v)? If a C^1 solution $S(x, t)$ of (5.3)-(5.4) exists, then there are also solutions x of the terminal value problem (5.5) plus $x(t_1) = x_1$, and therefore minimizing pairs. The difficulty lies of course with the initial value problem (5.3)-(5.4) that in general only has a solution on some interval $[t_0, \bar{t})$, $\bar{t} < t_1$.

Suppose now that the terminal position of the particle is uncertain, namely we only know that $x(t_1)$ is distributed according to the probability density $\rho_1(\cdot)$. We then choose as class of motions $\mathcal{X}_{\rho_1}$ the class of all $\mathbb{R}^n$-valued stochastic processes on $[t_0, t_1]$ with a.s. C^1 paths such that $x(t_1)$ is distributed according to ρ_1. Let $\mathcal{V}$ denote the class of all $\mathbb{R}^n$-valued stochastic processes on $[t_0, t_1]$ with a.s. continuous paths. We consider the problem of minimizing over $(x, v) \in (\mathcal{X}_{\rho_1} \times \mathcal{V})$

$$E\left\{\int_{t_0}^{t_1} L(x(t), v(t)) \, dt + S_0(x(t_0))\right\}, \tag{5.8}$$

subject to the constraint

$$\dot{x}(t) = v(t), \text{ a.s.,} \quad \forall t \in [t_0, t_1]. \tag{5.9}$$

The same argument as before gives that, if a sufficiently smooth solution $S(x,t)$ of (5.3)-(5.4) exists, any $x \in \mathcal{X}_{\rho_1}$ satisfying a.s. on $[t_0, t_1]$ $\dot{x}(t) = \frac{1}{m}\nabla S(x(t), t)$, solves the minimum problem. Notice that $S(x,t)$ and $\rho(x,t)$ satisfy the system of equations

$$\frac{\partial S}{\partial t} + \frac{1}{2m}\nabla S \cdot \nabla S + V(x) = 0, \tag{5.10}$$

$$\frac{\partial \rho}{\partial t} + \nabla \cdot (v\rho) = 0, \tag{5.11}$$

$$v(x,t) = \frac{1}{m}\nabla S(x,t). \tag{5.12}$$

This system describes the *Hamilton-Jacobi fluid* of classical mechanics, cf.[G, Section 1].

6. The quantum Hamilton principle

Let us now go to the fully stochastic case where uncertainty also comes from the presence of a background noise. Let $\mathcal{X}_{\rho_1}$ denote the family of all finite-energy, $\mathbb{R}^n$-valued diffusions on $[t_0, t_1]$ with diffusion coefficient $I_n \frac{\hbar}{m}$, and having marginal probability density ρ_1 at time t_1. Let $\mathcal{V}$ denote the family of finite-energy, $\mathbb{C}^n$ - valued stochastic processes on $[t_0, t_1]$. Let $L(x,v) := \frac{1}{2}mv \cdot v - V(x)$ be defined on $\mathbb{R}^n \times \mathbb{C}^n$. Also let S_0 be a complex-valued function on $\mathbb{R}^n$. Consider the problem of extremizing on $(x, v_q) \in (\mathcal{X}_{\rho_1} \times \mathcal{V})$

$$E\left\{ \int_{t_0}^{t_1} L(x(t), v_q(t)) \, dt + S_0(x(t_0)) \right\} \tag{6.1}$$

subject to the constraint that

$$x \text{ has quantum drift (velocity) } v_q. \tag{6.2}$$

In [P1, Section VIII] the following result was established.

Theorem 6.1 *Suppose that $S_q(x,t)$ of class $C^{2,1}$ solves on $[t_0, t_1]$ the initial value problem*

$$\frac{\partial S_q}{\partial t} + \frac{1}{2m}\nabla S_q \cdot \nabla S_q + V(x) - \frac{i\hbar}{2m}\Delta S_q = 0, \tag{6.3}$$

$$S(x, t_0) = S_0(x), \tag{6.4}$$

and satisfies the technical condition

$$E\left\{\int_{t_0}^{t_1} \nabla S_q(x(t),t) \cdot \overline{\nabla S_q(x(t),t)}\, dt\right\} < \infty, \quad \forall x \in \mathcal{X}_{\rho_1}. \tag{6.5}$$

Then any $x \in \mathcal{X}_{\rho_1}$ *having quantum drift* $\frac{1}{m}\nabla S(x(t),t)$ *solves the extremization problem.*

The crucial steps in the proof are the following two observations:

1. If the pair $(x, v_q) \in (\mathcal{X}_{\rho_1} \times \mathcal{V})$ satisfies the dynamical constraint (5.2), then

$$E\left\{d_b S(x(t),t)\right\} = \left[\frac{\partial S}{\partial t}(x(t),t) + v_q(t) \cdot \nabla S(x(t),t) - \frac{i\hbar}{2m}\Delta S_q(x(t),t)\right] dt$$

$$+ o(dt); \tag{6.6}$$

2. The function $S_q(\cdot,\cdot)$ satisfies

$$\underset{v \in \mathbb{C}^n}{\text{extremize}}\left\{\frac{1}{2}mv\cdot v - V(x) - \frac{\partial S_q}{\partial t}(x,t) - v\cdot\nabla S_q(x,t) + \frac{i\hbar}{2m}\Delta S_q(x,t)\right\} = 0. \tag{6.7}$$

As for the classical case, let us now pose the question: when does the above result yield a minimizing pair (x, v_q)? As before, there are two separate issues. i) When does the initial value problem (6.3)-(6.4) have a $C^{2,1}$ solution on $[t_0, t_1]$ satisfying (6.5) ? ii) If such a solution exists, when does there exists a finite-energy, necessarily Markovian diffusion x having quantum drift $\frac{1}{m}\nabla S(x(t),t)$ and marginal density ρ_1 at time t_1? Curiously, it is now issue ii) that gives more trouble. Indeed, existence of a solution for the complicated nonlinear, complex Cauchy problem (6.3)-(6.4) is dealt with as follows. Let $\{\psi(x,t); t_0 \le t \le t_1\}$ be the solution of the *Schrödinger equation*

$$\frac{\partial \psi}{\partial t} = \frac{i\hbar}{2m}\Delta\psi - \frac{i}{\hbar}V(x)\psi, \tag{6.8}$$

with initial condition $\psi_0(x) := \exp\frac{i}{\hbar}S_0(x)$. If $\psi(x,t)$ never vanishes on $\mathbb{R}^n \times [t_0, t_1]$, and satisfies the condition

$$E\left\{\int_{t_0}^{t_1} \nabla\log\psi(x(t),t) \cdot \overline{\nabla\log\psi(x(t),t)}\, dt\right\} < \infty, \quad \forall x \in \mathcal{X}_{\rho_1}, \tag{6.9}$$

then $S_q(x,t) := \frac{\hbar}{i}\log\psi(x,t)$ satisfies (6.3)-(6.4) and (6.5). If, moreover, $\psi_0(x)$ has L^2 norm 1, and the terminal density satisfies $\rho_1(x,t) = |\psi(x,t_1)|^2$, then there does exist a Markov diffusion having the required quantum drift, namely the *Nelson process* associated to $\{\psi(x,t); t_0 \leq t \leq t_1\}$, see [P1] for the details. The compatibility condition $\rho_1(x,t) = |\psi(x,t_1)|^2$ is indispensable to give a positive answer to point ii) since fixing the quantum drift we are fixing both the forward and the backward drift. The condition that $\psi_0 = \exp \frac{i}{\hbar} S_0$ has norm one, instead, is necessary because of another peculiar aspect of the stochastic case. Namely, the fact that the continuity equation is contained in the Hamilton-Jacobi equation (6.3). Indeed, Born's relation $\rho(x,t) = |\psi(x,t)|^2$ must hold. Consequently, (6.3) alone gives the evolution of the Hamilton-Jacobi fluid in stochastic mechanics, see [G] and [P2] for further discussion on this point. The construction of the Nelson process corresponding to $\psi(x,t)$ in the case where $\psi(x,t)$ can vanish requires considerable effort. It is discussed in [C], [BCZ, Chapter IV], [DP], and references therein.

7. Closing comments

The quantum Hamilton principle was shown in [P1] to be, loosely speaking, equivalent to two other variational principles of the min-max type. The first one, called *the saddle-point action principle*, contains as special cases both the Guerra-Morato variational principle [GM] and Schrödinger original variational derivation of the time-independent equation, see e.g. [YM, p.118]. The second one, called *the saddle-point entropy production principle*, concerns the production of configurational entropy. The Nelson process appears then as a *saddle point equilibrium solution* for both stochastic differential games.

Since [N2], there has been an ongoing debate on whether the class of Markov diffusion processes was general enough to describe the motion of a quantum particle. For instance, alternative forms of stochastic mechanics based on the larger class of *Bernstein* or *reciprocal* processes have been proposed in [Z] and [LK].

Let us first analyze the classical case. We wish to stress here the fact that in the *hydrodynamic form* of Hamilton's principle the minimizing path satisfies the *first-order* differential equation (5.5) although we allowed the very large class $\mathcal{X}_{x_1}$ of trial processes. Hence, what is usually presented as a kinematical assumption, namely looking for the minimizing trajectory in the class of those satisfying a first-order equation of the form $\dot{\xi}(t) = v(\xi(t), t)$, can be obtained as a *result* of the variational principle.

Consider now the stochastic case. We see that the extremizing process, namely the Nelson process, satisfies a first order stochastic differential equation although we allowed as trial processes general finite-energy diffusions (this class includes reciprocal diffusions). It is therefore a Markov process, and the Markovianess is a *result* of the variational principle.

It seems to us that our treatment of the hydrodynamic form of Hamilton's principle in stochastic mechanics is the natural evolution of the remarkable intuition of Madelung at the dawn of quantum mechanics [M] through the work of Nelson [N1], [N2], Guerra [G], Guerra-Morato [GM], and many others. In spite of the differences we outlined above, the analogy between the classical and the quantum Hamilton principle appear considerable. Indeed, stochastic mechanics appears in this respect basically as classical mechanics with a different class of motions and different kinematics.

References

[AMU]   S. Albeverio, L. Morato and S. Ugolini, Non symmetric diffusions and related Hamiltonians, to appear.

[BCZ]   Ph. Blanchard, Ph. Combe and W. Zheng, *Math. and Physical Aspects of Stochastic Mechanics* , Lect. Notes in Physics vol. 281, Springer-Verlag, New York, 1987.

[C]   E. Carlen, Conservative diffusions, *Comm. Math. Phys.* **94** (1984), 293–315.

[DP]   G. F. Dell'Antonio and A. Posilicano, *Comm. Math. Phys.* **141** (1991), 599.

[F]   H. Föllmer, in: *Stochastic Processes - Mathematics and Physics* , Lect. Notes in Math. 1158 (Springer-Verlag, New York,1986), p. 119.

[G]   F.Guerra, Structural aspects of stochastic mechanics and stochastic field theory, *Phys.Rep.* **77** (1981)263.

[GM]   F.Guerra and L.Morato, Quantization of dynamical systems and stochastic control theory, *Phys.Rev.D* **27** (1983) 1774.

[LK]   B.C.Levy and A.J.Krener, Kinematics and dynamics of reciprocal diffusions, *J.Math.Phys.* **34** (1993)1846-1875.

[M]   E. Madelung, Quantentheorie in hydrodynamischer Form, *Z. Physik* **40** (1926), 322–326.

[N1]   E. Nelson,*Dynamical Theories of Brownian Motion*, Princeton University Press, Princeton, 1967.

[N2]   E. Nelson, *Quantum Fluctuations*, Princeton University Press, Princeton, 1985.

[N3]   E. Nelson, Stochastic mechanics and random fields, in*École d'Ètè de Probabilitès de Saint-Flour XV-XVII*, edited by P. L. Hennequin, Lecture Notes in Mathematics, Springer-Verlag, New York, 1988, vol.1362, pp. 428-450.

[P1]   M. Pavon, Hamilton's principle in stochastic mechanics, J. Math. Phys. **36** (1995), 6774–6800.

[P2]   M. Pavon, A new formulation of stochastic mechanics, Phys. Lett. A **209** (1995) 143–149, Erratum **211**(1996), 383.

[P3]   M. Pavon, Lagrangian dynamics for classical, Brownian and quantum mechanical particles, *J. Math. Phys.* **37** (1996), 3375–3388.

[YM]   W. Yourgrau and S. Mandelstam, *Variational Principles in Dynamics and Quantum Theory*, Pitman and Sons, London, 1960.

[Z]   J.C.Zambrini, *Phys. Rev.* **A33** (1986) 1532–1548.

Dipartimento di Elettronica e Informatica
Università di Padova
via Gradenigo 6/A, 35131 Padova
and LADSEB-CNR
Italy

Received October 1, 1996

Stochastic Equations in Formal Mappings

Igor Spectorsky [1]

1. Introduction

Let us consider a nonlinear stochastic equation

$$y(t) = y(0) + \int_0^t a(\tau)(y(\tau))d\tau + \int_0^t b(\tau)(y(\tau))dw(\tau), \qquad 0 \le t \le T \quad (1.1)$$

in Hilbert space Y, where:

- w is a Wiener process, associated with canonical triple $H_+ \subset H_0 \subset H_-$ with a Hilbert-Schmidt embedding (all Hilbert spaces are supposed to be real and separable);

- y is a non-anticipating process in Hilbert space Y;

- a and b are continuous mappings from $[0,T] \times Y$ into Y and $\mathcal{L}_2(Y)$ respectively.

To solve equation (1.1), we apply a method of power series. Some analyticity assumptions are given below.

Let function a and b be analytical with respect to $y \in Y$ for any t, and $a(t)(0) = 0$, $b(t)(0) = 0$. Therefore, there exist the following expansions in power series:

$$a(t,y) = \sum_{k \ge 1} a_k(t)(y, y, \ldots, y), \quad b(t,y) = \sum_{k \ge 1} b_k(t)(y, y, \ldots, y),$$

where $a_k(t)$ and $b_k(t)$ are k-linear continuous operators, acting from Y into Y and from Y into $\mathcal{L}_2(Y)$ respectively.

Suppose (1.1) has a unique solution $y(t) = S(t,0)(y) = S(t)(y)$, and the expansions $S(t)(y) = \sum_{k \ge 1} S_k(t)(y, y, \ldots, y)$ hold almost surely.

In this case $a(t) \circ S(t)(y)$ and $b(t) \circ S(t)(y)$ can be expanded in power series by y with expansion coefficients:

$$(a \circ S)_n = \sum_{k=1}^n \sum_{j_1+j_2+\ldots+j_k=n} a_k(S_{j_1}, \ldots, S_{j_k}),$$

[1]This work was supported, in part, by the International Soros Science Education Program (ISSEP) through grant N PSU051117.

$$(b \circ S)_n \;=\; \sum_{k=1}^{n} \sum_{j_1+j_2+\ldots+j_k=n} b_k(S_{j_1},\ldots,S_{j_k}).$$

Substituting power series for $a(y)$ and $b(y)$ in equation (1.1) and comparing development coefficients in both sides of the equation, one can easily obtain the following system of linear stochastic equations for n-linear continuous operators S_n, $n \geq 1$:

$$
\left\{
\begin{aligned}
S_1(t) \;&=\; id_Y + \int_0^t a_1(\tau)S_1(\tau)d\tau + \int_0^t b_1(\tau)(S_1(\tau),dw(\tau)),\\[4pt]
S_2(t) \;&=\; \int_0^t a_1(\tau)S_2(\tau)d\tau + \int_0^t b_1(\tau)(S_2(\tau),dw(\tau))+\\[2pt]
&\quad+\int_0^t a_2(\tau)(S_1(\tau),S_1(\tau))d\tau + \int_0^t b_2(\tau)(S_1(\tau),S_1(\tau))dw(\tau),\\[2pt]
S_3(t) \;&=\; \int_0^t a_1(\tau)S_3(\tau)d\tau + \int_0^t b_1(\tau)(S_3(\tau),dw(\tau))+\\[2pt]
&\quad+\int_0^t a_2(\tau)(S_1(\tau),S_2(\tau))d\tau + \int_0^t b_2(\tau)(S_1(\tau),S_1(\tau))dw(\tau)+\\[2pt]
&\quad+\int_0^t a_2(\tau)(S_2(\tau),S_1(\tau))d\tau + \int_0^t b_2(\tau)(S_2(\tau),S_1(\tau))dw(\tau)+\\[2pt]
&\quad+\int_0^t a_3(\tau)(S_1(\tau),S_1(\tau),S_1(\tau))d\tau+\\[2pt]
&\quad+\int_0^t b_3(\tau)(S_1(\tau),S_1(\tau),S_1(\tau))dw(\tau),\\[2pt]
&\quad\;\;\vdots\\[2pt]
S_n(t) \;&=\; \int_0^t \sum_{k=1}^{n} \sum_{j_1+j_2+\ldots+j_k=n} a_k(\tau)(S_{j_1}(\tau),S_{j_2}(\tau),\ldots,S_{j_k}(\tau))d\tau+\\[2pt]
&\quad+\int_0^t \sum_{k=1}^{n} \sum_{j_1+j_2+\ldots+j_k=n} b_k(\tau)(S_{j_1}(\tau),S_{j_2}(\tau),\ldots,S_{j_k}(\tau))dw(\tau),\\[2pt]
&\quad\;\;\vdots
\end{aligned}
\right.
$$

$$(1.2)$$

Note that system (1.2) remains valid if we don't worry about convergence of power series. In this case all power series are treated as formal mappings.

Definition 1.1 *A sequence* $a = (a_k)_{k\geq 1}$, *where* a_k *($k \geq 1$) are k-linear continuous mappings from Y into Z is called a formal mapping from Y into Z.*

Denote by $\mathcal{L}_\infty(Y,Z)$ *a space of formal mappings from Y into Z,* $\mathcal{L}_\infty(Y) = \mathcal{L}_\infty(Y,Y)$.

There exists a well-developed algebraic apparatus to deal with formal mappings. Composition and derivative operations for formal mappings as well as notation of identical formal mapping are introduced in [Da]. Here we consider stochastic equations in the space of formal mappings.

2. Recursion procedure

To implement the recursion algorithm, we rewrite (1.2) in more convenient form extracting 'linear' summands containing the terms $a_1 S_n$ and $b_1 S_n$.

$$
\begin{cases}
S_1(t) &= id_Y + \int\limits_0^t a_1(\tau)S_1(\tau)d\tau + \int\limits_0^t b_1(\tau)(S_1(\tau), dw(\tau)), \\
S_n(t) &= \int\limits_0^t a_1(\tau)S_n(\tau)d\tau + \int\limits_0^t b_1(\tau)(S_n(\tau), dw(\tau)) + \\
&+ \int\limits_0^t f_n(\tau)d\tau + \int\limits_0^t g_n(\tau)dw(\tau), \quad n \geq 2,
\end{cases} \tag{2.1}
$$

where
$$
f_n(t) = \sum_{k=2}^{n} \sum_{j_1+j_2+\ldots+j_k=n} a_k(t)(S_{j_1}(t), S_{j_2}(t), \ldots, S_{j_k}(t)),
$$
$$
g_n(t) = \sum_{k=2}^{n} \sum_{j_1+j_2+\ldots+j_k=n} b_k(t)(S_{j_1}(t), S_{j_2}(t), \ldots, S_{j_k}(t)).
$$

The first n equations of system (2.1) are closed with respect to S_k, $1 \leq k \leq n$ (for any n) because f_n and g_n contain S_k ($k \leq n-1$) only. So, it's possible to solve the system recursively: find S_1 from the first equation, then find S_2 from the second equation (using already found S_1), find S_3 from the third one (using S_1 S_2), etc.

It is important that the first equation of system (2.1) is linear homogeneous and all the rest (for $n \geq 2$) are linear nonhomogeneous ones.

If coefficients a_k and b_k ($k \geq 2$) are of Hilbert-Schmidt type then the following explicit formula for solution of linear nonhomogeneous stochastic equations can be used:

$$
S_n(t) = \int\limits_0^t S_1(t, \tau)f_n(\tau)d\tau + \int\limits_0^t S_1(t, \tau)g_n(\tau)\widehat{dw(\tau)} \tag{2.2}
$$

Here $\widehat{dw(\tau)}$ denotes the extended stochastic integral, introduced as adjoint to the stochastic derivative operator (see e.g. [DP]).

This formula also holds for the more general case when the 'drift' operator a_1 is nonbounded (see [Sp]).

Recall that f_n and g_n contain only a S_k with $k < n$. This fact allows us to use (2.2) recursively.

Due to formula (2.2), we can solve (2.1) using not only recursion procedure but also an explicit method connected with graph theory.

3. Graph method

Let's introduce the following notation

$$I_n(t) \circ (y_1 \otimes y_2 \ldots \otimes y_n) = \int_0^t S_1(t,\tau) a_n(\tau) \circ (y_1(\tau) \otimes y_2(\tau) \ldots \otimes y_n(\tau)) d\tau +$$

$$\int_0^t S_1(t,\tau) b_n(\tau) (y_1(\tau) \otimes y_2(\tau) \ldots \otimes y_n(\tau)) \widehat{dw(\tau)}, \qquad n \geq 2,$$

where y_k $(1 \leq k \leq n)$ are non-anticipating random processes in Y. Here we consider n-linear operators on Hilbert space Y as linear operators on a tensor power $Y^{\otimes n}$.

We put I_1 to be identical, i.e. $I_1(t)(y) = y(t)$.

We write expressions for the first several S_k $(k \geq 2)$ using (2.2). We suppose that S_1 is already found from the first linear homogeneous equation of system (2.1).

$$
\begin{aligned}
S_2(t) &= I_2(t) \circ S_1^{\otimes 2}; \\
S_3(t) &= I_3(t) \circ S_1^{\otimes 3} + I_2(t) \circ (I_2 \otimes I_1) \circ S_1^{\otimes 3} + I_2(t) \circ (I_1 \otimes I_2) \circ S_1^{\otimes 3}; \\
S_4(t) &= I_4(t) \circ S_1^{\otimes 4} + I_3(t) \circ (I_2 \otimes I_1 \otimes I_1) \circ S_1^{\otimes 4} \\
&+ I_3(t) \circ (I_1 \otimes I_2 \otimes I_1) \circ S_1^{\otimes 4} + I_3(t) \circ (I_1 \otimes I_1 \otimes I_2) \circ S_1^{\otimes 4} \\
&+ I_2(t) \circ (I_3 \otimes I_1) \circ S_1^{\otimes 4} + I_2(t) \circ (I_2 \otimes I_2) \circ S_1^{\otimes 4} \\
&+ I_2(t) \circ (I_1 \otimes I_3) \circ S_1^{\otimes 4} \\
&+ I_2(t) \circ (I_2 \otimes I_1) \circ (I_2 \otimes I_1 \otimes I_1) \circ S_1^{\otimes 4} \\
&+ I_2(t) \circ (I_2 \otimes I_1) \circ (I_1 \otimes I_2 \otimes I_1) \circ S_1^{\otimes 4} \\
&+ I_2(t) \circ (I_1 \otimes I_2) \circ (I_1 \otimes I_2 \otimes I_1) \circ S_1^{\otimes 4} \\
&+ I_2(t) \circ (I_1 \otimes I_2) \circ (I_1 \otimes I_1 \otimes I_2) \circ S_1^{\otimes 4};
\end{aligned}
$$

$\vdots$

For example, we rewrite S_2 without operators I_k.

$$S_2(t) = \int\limits_0^t S_1(t,\tau) a_2(\tau)(S_1(\tau), S_1(\tau)) d\tau + \int\limits_0^t S_1(t,\tau) b_2(\tau)(S_1(\tau), S_1(\tau)) \widehat{dw(\tau)}$$

Expressions for S_3 and S_4 in a form without I_k are very complicated and are not given here.

Note that summands $I_2(t) \circ (I_1 \otimes I_2) \circ (I_2 \otimes I_1 \otimes I_1) \circ S_1^{\otimes 4}$ and $I_2(t) \circ (I_2 \otimes I_1) \circ (I_1 \otimes I_1 \otimes I_2) \circ S_1^{\otimes 4}$ are not included in the expression for S_4. Writing the expression for S_n one can use the following mnemonic rule: 'Only I_1 or S_1 can be an argument of I_1'.

Each summand from the expressions for S_n can be associated with a graph of the certain type (a tree with n leaves). Moreover, there exists one-to-one correspondence of summands for S_n and graphs of a certain type (trees with n leaves taking into account the mnemonic rule).

Therefore, the problem of writing the expression for S_n turns out to be a problem to find all graphs of a certain type. The latter problem can be easily solved with the help of computer.

Here are associated trees for S_2 (Figure 3.1) and S_3 (Figure 3.2).

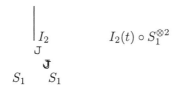

Figure 3.1: Associated tree for S_2

References

[Da]   Daletsky Yu. L., Algebra of compositions and nonlinear equations, *Kluwer Academic Publisher* (1992), 277–291.

[Sp]   Spectorsky I. Ya., Explicit formula for solution of linear nonhomo-geneous stochastic equation (in Russian) Deponed in State Science Technical Library of Ukraine 02-01-1996, 424 - UK 96

[DP]   DaletskiĭYu.L., Paramonova S.N., *Teor. Verojatnost. i Primenen.* 19, (1974), 845–849.

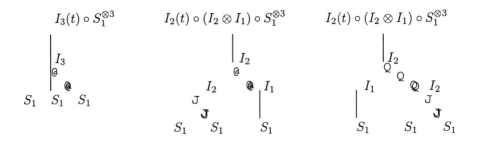

Figure 3.2: Associated trees for S_3

National Technical University of Ukraine
(Kiev Polytechnic Institute)
Current address: Department of Mathematical Methods of System Analysis
37, Prospect Pobedy
Kiev, Ukraine, 252056

Received September 16, 1996

On Fisher's Information Matrix of an ARMA Process

André Klein and Peter Spreij

Abstract

In this paper we study the Fisher information matrix for a stationary ARMA process with the aid of Sylvester's resultant matrix. Some properties are explained via realizations in state space form of the derivates of the white noise process with respect to the parameters.

1. Introduction

The Cramér-Rao bound is of considerable importance for evaluating the performance of (stationary) autoregressive moving average (ARMA) models, where the focus is on the error covariance matrix of the estimated parameters. See Cramér [Cr] and Rao [Ra]. For computing the Cramér-Rao bound and the asymptotic distribution of a Wald test statistic (Klein [Kl]) the inverse of Fisher's information matrix is needed. The latter is singular in the presence of common roots of the AR and the MA polynomial and vice versa. This fact is considered to be well-known in time series analysis, see [Po1] or [McL] for a proof along different lines than the ones we follow below.

In this paper we give an elementary proof of this equivalence by linking Fisher's information matrix to Sylvester's resultant matrix and an interpretation in terms of a state space realization.

In the literature the resultant matrix has been used in various studies in the fields of time series and systems theory. For instance, in [AS] this matrix shows up in a convergence analysis of maximum likelihood estimators of the ARMA parameters (more precisely in the study of the convergence of the criterion function), in Barnett [Ba1] a relationship between Sylvester's resultant matrix and the companion matrix of a polynomial is given. Kalman [Ka] has investigated the concept of observability and controllability in function of Sylvester's resultant matrix. Similar results can be found in Barnett [Ba2] which contains discussions on these topics and a number of further references.

Fisher's information matrix is studied in [Ro] for problems of local and global identifiability in a static context, whereas identifiability problems for parameterizations of linear (stochastic) systems are studied by Glover and Willems in [GW].

Furthermore in Söderström & Stoica [So] (pages 162 ff.) a discussion on overparameterization in terms of the transfer function of a system can be found.

We now introduce some notation. Consider the following two scalar polynomials in the variable z.

$$
\begin{align}
A(z) &= z^p + a_1 z^{p-1} + \ldots + a_p \tag{1.1}\\
C(z) &= z^q + c_1 z^{q-1} + \ldots + c_q \tag{1.2}
\end{align}
$$

By A^* and C^* we denote the reciprocal polynomials, so $A^*(z) = z^p A(z^{-1})$ and $C^*(z) = z^q C(z^{-1})$, and also write $a^T = [a_1, \ldots, a_p]$ and $c^T = [c_1, \ldots, c_q]$.

The Sylvester resultant matrix of A and C is defined as the $(p+q) \times (p+q)$ matrix

$$
S(a,c) =
\left.\begin{array}{l} q \left\{ \right. \\ \\ p \left\{ \right. \end{array}\right.
\begin{bmatrix}
1 & a_1 & \cdots & \cdots & a_p & & 0 \\
 & \ddots & \ddots & & & \ddots & \\
0 & & 1 & a_1 & \cdots & \cdots & a_p \\
\hdashline
1 & c_1 & \cdots & \cdots & c_q & & 0 \\
 & \ddots & \ddots & & & \ddots & \\
0 & & 1 & c_1 & \cdots & \cdots & c_q
\end{bmatrix}
\tag{1.3}
$$

In the presence of common roots of A and C the matrix $S(a,c)$ becomes singular. Moreover it is known (see e.g. [VW, page 106]) that

$$
\det S(a,c) = \prod_{i=1}^{p} \prod_{j=1}^{q} (\gamma_j - \alpha_i) \tag{1.4}
$$

where the α_i and the γ_j are the roots of A and C respectively.

Remark. The origin of Sylvester's matrix lies in the following problem, see [VW]. Find monic polynomials $K(z) = z^p + \sum_{i=1}^{p} k_i z^{p-i}$ and $L(z) = z^q + \sum_{i=1}^{q} l_i z^{q-i}$ such that $A(z)L(z) + C(z)K(z) = 0$. Writing the coefficients of K and L in column vectors k and l, one can cast this problem as solving a set of linear equations in k and l with $S(a,c)^T$ as coefficient matrix. Clearly, the solution of the problem is then given by the affine subspace $\begin{bmatrix} -a \\ c \end{bmatrix} + \ker S(a,c)^T$. Notice that for any solution K and L of the problem the rational function K/L coincides with A/C in the points where both are defined.

Remark. Let J_{p+q} the matrix in $\mathbf{R}^{(p+q)\times(p+q)}$ with ij-entries $\delta_{i,j+1}$ (a shifted identity matrix), $\alpha^T = (1, a_1, \ldots, a_p, \ldots, 0) \in \mathbf{R}^{p+q}$ and $\gamma^T = (1, c_1, \ldots, c_q, \ldots, 0) \in \mathbf{R}^{p+q}$. Then up to a permutation of its columns, the matrix $S(a, c)^T$ consists of the first $p + q$ columns of the controllability matrix of the pair $(J_{p+q}, (\alpha, \gamma))$,

2. A key result

First we specify Fisher's information matrix of an ARMA(p,q) process. Let A and C be the same monic polynomials as in the previous section. Consider then the stationary ARMA process y that satisfies

$$A^*(L)y = C^*(L)\varepsilon \tag{2.1}$$

with L the lag operator and ε a white noise sequence. We make the assumption (to give the expressions that we use below the correct meaning) that both A and C have zeros only inside the unit circle.

Let $\theta = (a_1, \ldots, a_p, c_1, \ldots, c_q)$ and denote by $\varepsilon_t^{\theta_i}$ the derivative of ε_t with respect to θ_i. Then we have

$$\varepsilon_t^{a_j} = \frac{1}{a(z)}\varepsilon_{t-j}$$

$$\varepsilon_t^{c_l} = -\frac{1}{c(z)}\varepsilon_{t-l}$$

With ε_t^θ the column vector with elements $\varepsilon_t^{\theta_i}$ the Fisher information matrix $F(\theta)$ is equal to $E\varepsilon_t^\theta \varepsilon_t^{\theta T}$. As can be found in for instance Klein & Mélard [KM] $F(\theta)$ then has the following block decomposition

$$F(\theta) = \begin{bmatrix} F_{aa} & F_{ac} \\ F_{ac}^T & F_{cc} \end{bmatrix} \tag{2.2}$$

where the matrices appearing here have the following elements

$$F_{aa}^{jk} = \frac{1}{2\pi i}\oint_{|z|=1}\frac{z^{j-k+p-1}}{A(z)A^*(z)}dz, (j, k = 1, \ldots, p)$$

$$F_{ac}^{jk} = \frac{-1}{2\pi i}\oint_{|z|=1}\frac{z^{j-k+q-1}}{C(z)A^*(z)}dz, (j = 1, \ldots, p, k = 1, \ldots, q)$$

$$F_{cc}^{jk} = \frac{1}{2\pi i}\oint_{|z|=1}\frac{z^{j-k+q-1}}{C(z)C^*(z)}dz, (j, k = 1, \ldots, q)$$

The key result of this paper is the easy to prove lemma 2.1 below. First we have to introduce some auxiliary notation. Write for each positive integer k

$$u_k(z) = [1, z, \ldots, z^{k-1}]^T$$
$$u_k^*(z) = [z^{k-1}, \ldots, 1]^T = z^{k-1}u_k(z^{-1})$$

and let

$$K(z) = A(z)A^*(z)C(z)C^*(z).$$

Define moreover

$$P(\theta) = \frac{1}{2\pi i} \oint_{|z|=1} \frac{u_{p+q}(z)u_{p+q}^*(z)^T}{K(z)} dz \qquad (2.3)$$

Notice that we can alternatively write

$$P(\theta) = \frac{1}{2\pi i} \oint_{|z|=1} \frac{u_{p+q}^*(z)u_{p+q}^*(\frac{1}{z})^T}{A(z)A(\frac{1}{z})C(z)C(\frac{1}{z})} \frac{dz}{z} \qquad (2.4)$$

Lemma 2.1 *The following factorization holds.*

$$F(\theta) = S(c, -a)P(\theta)S(c, -a)^T \qquad (2.5)$$

Proof. A simple computation shows that we can write $F(\theta)$ in matrix form as

$$F(\theta) = \frac{1}{2\pi i} \oint_{|z|=1} \frac{1}{K(z)} \begin{bmatrix} C^*(z)u_p(z) \\ -A^*(z)u_q(z) \end{bmatrix} \begin{bmatrix} C(z)u_p^*(z)^T & -A(z)u_q^*(z)^T \end{bmatrix} dz \qquad (2.6)$$

It also straightforward to verify that the following identities hold.

$$S(c, -a)u_{p+q}(z) = \begin{bmatrix} C^*(z)u_p(z) \\ -A^*(z)u_q(z) \end{bmatrix} \qquad (2.7)$$

$$S(c, -a)u_{p+q}^*(z) = \begin{bmatrix} C(z)u_p^*(z) \\ -A(z)u_q^*(z) \end{bmatrix} \qquad (2.8)$$

Hence equation (2.5) follows now immediately from equations (2.3), (2.6), (2.7) and (2.8). ∎

Remark. It follows that with probability one ε_t^θ belongs to the image space of $S(c, -a)$.

Corollary 2.2 *The Fisher information matrix of an ARMA(p,q) process with polynomials $A^*(z)$ and $C^*(z)$ of order p, q respectively becomes singular iff the polynomials A and C have at least one common root.*

Proof. Clearly the matrix $F(\theta)$ becomes singular if A and C have at least one common root in view of equation (1.4) and lemma 2.1. In order to prove the converse, we only have to prove that $P(\theta)$ is strictly positive definite, (again) because of (1.4) and (2.5). This can be shown via a straight forward computation (see also the next section for an alternative consideration):

Rewrite $P(\theta)$ as

$$P(\theta) = \frac{1}{2\pi i} \oint_{|z|=1} \frac{u_{p+q}(z)u_{p+q}(z^{-1})^T}{A(z)A(z^{-1})C(z)C(z^{-1})} z^{-1} dz$$

Take now $z = e^{i\phi}$, then we get

$$P(\theta) = \frac{1}{2\pi} \int_0^{2\pi} \frac{u_{p+q}(e^{i\phi})u_{p+q}(e^{-i\phi})^T}{A(e^{i\phi})A(e^{-i\phi})C(e^{i\phi})C(e^{-i\phi})} d\phi$$

which in turn can be rewritten as

$$P(\theta) = \frac{1}{2\pi} \int_0^{2\pi} \frac{u_{p+q}(e^{i\phi})}{A(e^{i\phi})C(e^{i\phi})} \overline{\left(\frac{u_{p+q}(e^{i\phi})}{A(e^{i\phi})C(e^{i\phi})}\right)}^T d\phi$$

Let now $x \in \mathbf{R}^{p+q}$ such that $x^T P(\theta) = 0$. Then it follows that $x^T \frac{u_{p+q}(e^{i\phi})}{A(e^{i\phi})C(e^{i\phi})} = 0$ for almost all ϕ. But this is clearly only possible if $x = 0$. So $P(\theta) > 0$. ■

Remark. In view of lemma 2.1, the nonsingularity of $P(\theta)$ and the first remark in the introduction we observe that for any $\begin{bmatrix} k \\ l \end{bmatrix}$ in the affine subspace $\begin{bmatrix} a \\ c \end{bmatrix} + \ker F(\theta)$ we find that the rational function $L(z)/K(z)$, where K and L depend on k and l as before, equals the transfer function $C(z)/A(z)$. This fact has also been noticed in [Po1], although the link with Sylvester's matrix is absent. A fairly explicit characterization of $\ker F(\theta)$ is given in [KS].

As a side remark we notice that the matrix $P(\theta)$ can be calculated by means of Cauchy's integral formula in the presence of common roots as follows. Let δ be a common root of A and C that appears as a zero of AC of order $l \geq 2$. Then

$$P(\theta) = \frac{1}{2\pi i} \oint_{|z|=1} \frac{f(z)}{(z-\delta)^l} dz$$

with

$$f(z) = \frac{u_{p+q}(z)u_{p+q}(z^{-1})^T}{A(z)A(z^{-1})C(z)C(z^{-1})} \frac{(z-\delta)^l}{z},$$

which is analytic in a disk of radius ρ around δ for sufficiently small ρ. Cauchy's theorem states that $P(\theta)$ is the sum of residuals, of which in particular the residual in δ can be computed as

$$f^{(l-1)}(\delta) = \frac{(l-1)!}{2\pi i} \oint_{|z-\delta|=\rho} \frac{f(z)}{(z-\delta)^l} dz. \tag{2.9}$$

It then follows that the more common roots A and C have, the less residuals are needed for the computation of $P(\theta)$.

As a corollary to lemma 2.1 we mention the following. Consider an AR process of order m, with AR polynomial $\tilde{A}^*(z)$ of order m. According to equation (2.5) and the fact that the Sylvester matrix is now the m-dimensional unit matrix, the Fisher information matrix $\tilde{F}$ becomes in this case

$$\tilde{F} = \frac{1}{2\pi i} \oint_{|z|=1} \frac{u_m(z)u_m^*(z)^T}{\tilde{A}(z)\tilde{A}^*(z)} dz \tag{2.10}$$

Take now in particular $\tilde{A}^*(z) = A^*(z)C^*(z)$ (so $m = p+q$), then it follows again from equation (2.10) and the fact that now $\tilde{A}(z)\tilde{A}^*(z) = K(z)$, that one has $P(\theta) = \tilde{F}$ and hence equation (2.5) reads

$$F(\theta) = S(c, -a)\tilde{F}S(c, -a)^T \tag{2.11}$$

So equation (2.11) gives a relationship between the Fisher information matrix of an ARMA(p,q) process and that of an appropriate AR($p+q$) process. See the next section for an explanation in state space terms of this phenomenon.

3. Computations in state space

We start with a realization of the ARMA process in state space form. Let $n = \max\{p, q\}$, $\alpha_n, \gamma_n \in \mathbf{R}^n$, $\alpha_n = [a_1, \ldots, a_n]^T$ and $\gamma_n = [c_1, \ldots, c_n]^T$, with zero entries for possibly previously undefined a_k or c_k (which happens only if $p \neq q$). We choose the following controllable realization.

$$X_{t+1} = AX_t + e\varepsilon_t \tag{3.1}$$
$$y_t = (\gamma_n - \alpha_n)^T X_t + \varepsilon_t \tag{3.2}$$

Here e is the first basis vector of the Euclidean space $\mathbf{R}^n$. A is then the matrix $J - e\alpha_n^T$, with J the shifted identity matrix, $J_{ij} = 1$ if $i = j+1$ and zero else. Below we also use the matrix F given by $F = J - e\gamma_n^T$.

Later on we will also use the notation e for the first basis vector in Euclidean spaces of possibly different dimension. Similarly, we denote by I the identity matrix of the appropriate size and 0 stands for the zero vector or matrix of appropriate dimensions. Occasionally these matrices and vectors will have a subscript, when it is necessary to indicate the sizes. Furthermore we will also use non-square 'identity' matrices, like I_{pn} which is the matrix in $\mathbb{R}^{p \times n}$ having ij-element δ_{ij}.

As before we denote by superscript partial derivatives with respect to a parameter. Let $Z_t = \text{vec}\,(X_t, X_t^{a_1}, \ldots, X_t^{a_p}, X_t^{c_1}, \ldots, X_t^{c_q})$. Then we can represent ε^θ after elementary computations as the output of the following system

$$Z_{t+1} = \mathcal{A}Z_t + e\varepsilon_t \tag{3.3}$$

$$\varepsilon_t^\theta = \mathcal{C}Z_t \tag{3.4}$$

where

$$\mathcal{A} = \begin{bmatrix} A & 0 & 0 \\ 0 & I_p \otimes F & 0 \\ -I_{qn} \otimes e_n & 0 & I_q \otimes F \end{bmatrix}$$

and

$$\mathcal{C} = \begin{bmatrix} I_{pn} & -I_p \otimes (\gamma_n - \alpha_n)^T & 0 \\ -I_{qn} & 0 & -I_q \otimes (\gamma_n - \alpha_n)^T \end{bmatrix}$$

If one computes the controllability matrix of this system, it is immediately seen that it contains a (middle) row of zero matrices. Therefore we replace it by the following system, using the same notation for the state variable and the coefficients.

$$Z_{t+1} = \mathcal{A}Z_t + e\varepsilon_t \tag{3.5}$$

$$\varepsilon_t^\theta = \mathcal{C}Z_t \tag{3.6}$$

where now

$$\mathcal{A} = \begin{bmatrix} A & 0 \\ -I_{qn} \otimes e_n & I_q \otimes F \end{bmatrix}$$

and

$$\mathcal{C} = \begin{bmatrix} I_{pn} & 0 \\ -I_{qn} & -I_q \otimes (\gamma_n - \alpha_n)^T \end{bmatrix}$$

Also this system is not controllable and we reduce it to one of lower order that is controllable by the same procedure that is used to decompose the state space a given system as a direct sum of the controllable subspace and its complement and the corresponding partitioning of the system matrices.

This is a well known procedure that we therefore only briefly sketch, leaving some computational details aside. We will work with the assumption that $c_q \neq 0$, although propositions 3.3 and 3.4 below and their consequences can also be proved if we drop this assumption by a little more complicated analysis.

First we compute the controllable subspace which is the linear span of the set $\{(I - Az)^{-1}e : z \in (-\delta, \delta)\}$, where δ is sufficiently small. Then $(I - Az)^{-1}e$ after a computation turns out to be equal to $(((I - Az)^{-1}e)^T, (-I_{qn}(I - Az)^{-1}e \otimes (I - Fz)^{-1}e)^T)^T$. Then the controllability matrix is computed by evaluating the derivatives of all orders of this function at $z = 0$. It is easily shown that the first $n + q$ columns of the controllability matrix span the controllable subspace and form a basis in the generic case where $c_q \neq 0$. In this case we form a non-singular matrix S consisting of these vectors augmented with a set of arbitrary independent vectors. Use S as a state space transformation to get a system described by the matrices $S^{-1}AS$, $S^{-1}e = e$ and CS. Then we first restrict the system to the controllable subspace and then by also restricting it further if $p < q$ in an appropriate way to a smaller subspace we get a new system (we use Z again to denote the state variable) that is given by

$$Z_{t+1} = \bar{A}Z_t + e\varepsilon_t \tag{3.7}$$
$$\varepsilon_t^\theta = \bar{C}Z_t \tag{3.8}$$

Here $\bar{A}$ is a companion matrix $\bar{A} = J - \hat{g}[0, \ldots, 0, 1]$, where $\hat{g}$ is the vector $\hat{g} = [g_{p+q}, \ldots, g_1]^T$ and the entries g_i are given by $z^{p+q} + \sum_{i=1}^{p+q} g_i z^{p+q-i} = A(z)C(z)$.

Finally we transform this system with the aid of the matrix $T \in \mathbb{R}^{(p+q)\times(p+q)}$ defined by its entries $T_{ij} = g_{j-i}$, with the convention that $g_0 = 1$ and $g_k = 0$ if $k < 0$. We then arrive at

Proposition 3.3 *The process ε^θ can be realized by the following stable and controllable system*

$$Z_{t+1} = \hat{A}Z_t + e\varepsilon_t \tag{3.9}$$
$$\varepsilon_t^\theta = \hat{C}Z_t, \tag{3.10}$$

where $\hat{A} = T^{-1}\bar{A}T = J - eg^T$ with $g^T = [g_1, \ldots, g_{p+q}]$, and $\hat{C} = S(c, -a)$. This system is observable iff the polynomials A and C have no common zeros.

Proof. The procedure outlined above has already shown the validity of equations (3.9) and (3.10). Controllability is obvious. If one computes the observability matrix one immediately sees that it has full rank iff the Sylvester matrix is invertible, so iff the A and C polynomials have no common factors. Furthermore the matrix $\hat{A}$ is stable, because its characteristic polynomial is $A(z)C(z)$ and we had assumed that all zeros of both A and C lie in the open unit disk. We skip the computations leading to $\hat{C} = S(c, -a)$.

∎

The result of the cascade of state space transformations leading to (3.9) and (3.10) can be summarized as follows. Start with the system described by equations (3.5) and (3.6). Apply a transformation with a nonsingular matrix $\mathcal{M} \in \mathbb{R}^{(n+n^2) \times (n+n^2)}$, that is such that its upperleft block of size $(p+q) \times (p+q)$ is given by

$$\begin{bmatrix} S_p(c) \\ 0_{q \times 1} \ I_q \ 0_{q \times (q-1)} \end{bmatrix}$$

This matrix is nonsingular under the condition that $c_q \neq 0$. Then it is a straightforward calculation to show that $\mathcal{M}^{-1} A \mathcal{M} = \hat{A}$ which has upperleft block of size $(p+q) \times (p+q)$ equal to $\hat{A}$.

Next we turn to the Lyapunov equation

$$P = \hat{A} P \hat{A}^T + ee^T, \tag{3.11}$$

because we see from proposition 3.3 that $E \varepsilon_t^\theta \varepsilon_t^{\theta T} = S(c, -a) P S(c, -a)^T$, with P the unique strictly positive solution of this Lyapunov equation, which exists since the pair $(\hat{A}, e)$ is controllable and $\hat{A}$ is stable. This solution is given (see [LR]) by

$$P = \frac{1}{2\pi i} \oint (z - \hat{A})^{-1} ee^T (\frac{1}{z} - \hat{A})^{-1} \frac{dz}{z}. \tag{3.12}$$

Because of the companion form of the matrix $\hat{A}$ we have $(z - \hat{A})^{-1} e = \frac{1}{A(z)C(z)} u_{p+q}^*(z)$. So we get from (2.4) that P is nothing else but the matrix P_θ from equation (2.3).

Notice also that the state process in (3.9) is equal to the ε^θ for an AR-process with AR-polynomial equal to $A^*(z)C^*(z)$, which gives an alternative explanation of (2.11).

The realization of the ε^θ process of proposition 3.3 is in an alternative way explained if we work with transfer functions. Consider first the system (3.5) and (3.6), and its transfer function ϕ. We had already computed the transfer function $(I - Az)^{-1}e$ from ε to Z as $(((I - Az)^{-1}e)^T, (-I_{qn}(I -$

$Az)^{-1}e \otimes (I - Fz)^{-1}e)^T)^T$. Premultiplying it with C gives

$$\phi(z) = \begin{bmatrix} \frac{u_p(z)}{A^*(z)} \\ \frac{-u_q(z)}{C^*(z)} \end{bmatrix} \tag{3.13}$$

On the other hand the transfer function $\hat{C}(I - \hat{A}z)^{-1}e$ of the system of proposition 3.3 is computed as $S(c, -a)\frac{u_{p+q}(z)}{A^*(z)C^*(z)}$. Using again the relations (2.7) and (2.8) we get the $\phi(z)$ of (3.13) back.

We got the system in proposition 3.3 starting from the controllable realization of the ARMA process y. Alternatively we could have started from its observable realization. Following a similar procedure one then arrives at another system that realizes the ε^θ process as its state process. Specifically, we have

Proposition 3.4 *The process ε^θ is the state process of the stable system given by*

$$\varepsilon^\theta_{t+1} = \tilde{A}\varepsilon^\theta_t + B\varepsilon_t, \tag{3.14}$$

where $\tilde{A} = \begin{bmatrix} A_0 & 0 \\ 0 & F_0 \end{bmatrix}$, $A_0 = J - e[a_1, \ldots, a_p] \in \mathbf{R}^{p \times p}$, $F_0 = J - e[c_1, \ldots, c_q] \in \mathbf{R}^{q \times q}$ and $B = \begin{bmatrix} e_p \\ -e_q \end{bmatrix}$. This system is controllable iff A and C have no common zeros. Moreover we have the relations $\tilde{A} = S(c, -a)\hat{A}$ with $\hat{A}$ as in proposition 3.3 and $B = S(c, -a)e$.

It follows from proposition 3.4 that $F(\theta)$ is also the solution to the Lyapunov equation

$$F = \tilde{A}F\tilde{A}^T + BB^T. \tag{3.15}$$

Again stability of the matrix $\tilde{A}$, which obviously has characteristic polynomial $A(z)C(z)$, ensures that this equation has a nonnegative definite solution, which is strictly positive definite if the pair $(\tilde{A}, B)$ is controllable. Let $R(\tilde{A}, B)$ be the controllability matrix of this pair and $R(\hat{A}, e)$ the nonsingular controllability matrix of the pair $(\hat{A}, e)$. Then we also have the relation $R(\tilde{A}, B) = S(c, -a)R(\hat{A}, e)$. Hence the pair $R(\tilde{A}, B)$ is controllable iff $S(c, -a)$ is nonsingular iff A and C have no common zeros, which is –of course– in perfect agreement with the previous results. Finally, we again recognize the factorization $F(\theta) = S(c, -a)P(\theta)S(c, -a)^T$ of lemma 2.1, because of the relations mentioned in proposition 3.4 and the facts that $P(\theta)$ solves equation (3.11) and $F(\theta)$ solves equation (3.15).

4. Final remarks

In a sense we considered in this paper an identifiability problem for ARMA processes with the emphasis on the role of Sylvester's resultant matrix related to Fisher's information matrix. Some additional considerations in terms of state space realization were given. The factorization given in lemma 2.1 can be extended to a more complicated one for ARMAX processes. This is discussed in [KS], where algebraic properties of Sylvester's matrix are stressed. Again non-singularity of Fisher's information matrix follows from the absence of common zeros of the three polynomials involved, although this result can alternatively be proved via a direct extension of the analysis in [Po1], communicated to us in [Po2].

References

[AS]   K.J. Åström and T. Söderström, Uniqueness of the Maximum Likelihood Estimates of the Parameters of an ARMA model, *IEEE Transactions on Automatic Control,* **AC-19** (1974), 769-773.

[Ba1]   S. Barnett, A new formulation of the the theorems of Hurwitz, Routh and Sturm, *J. Inst. Maths. Applics,* **8** (1971), 240 - 250.

[Ba2]   S. Barnett, Matrices, polynomials and linear time invariant systems, *IEEE Trans. Automat. Control,* **AC-18** (1973),1 - 10.

[Cr]   H. Cramér, *Mathematical methods of statistics,* Princeton NJ; Princeton University Press, 1951.

[GW]   K. Glover and J.C. Willems, Parametrizations of linear dynamical systems: canonical forms and identifiability, *IEEE Transactions on Automatic Control,* **AC-19** (1974), 640-646.

[Ka]   R.E. Kalman, Mathematical description of linear dynamical systems, *SIAM J. Contr.* **1** (1963), 152 - 192.

[KM]   A. Klein and G. Mélard, Fisher's information matrix for seasonal ARMA models, *J. Time Series,* **11** (1990), 231 - 237.

[Kl]   A. Klein, Hypothesis Testing of Common Roots, *Statistics and Probability Letters,* **20** (1994), 163-167.

[KS]   A. Klein and P.J.C. Spreij, On Fisher's Information Matrix of an ARMAX Process and Sylvester's Resultant Matrices, *Linear Algebra Appl.,* **237/238** (1996), 579-590.

[LR]   P. Lancaster and L. Rodman, *Algebraic Riccati Equations*, Oxford Science Publications, (1995).

[McL]   A.I. McLeod, A note on ARMA model redundancy, *J. of Time Series Analysis*, **14/2** (1993), 207-208.

[Po1]   B.M. Pötscher, The behaviour of the Lagrangian multiplier test in testing the orders of an ARMA model, *Metrika*, **32** (1985), 129-150.

[Po2]   B.M. Pötscher, *private communication.* (1996).

[Ra]   C.R. Rao (1965), *Linear statistical inference and its applications*, Wiley & Sons New York, 1965.

[Ro]   T.J. Rothenberg, Identification in parametric models, *Econometrica*, **39** (1971), 577-591.

[So]   T. Söderström and P. Stoica (1989), *System identification*, Prentice hall.

[VW]   B.L. van der Waerden (1966), *Algebra I*, Springer.

André Klein
Department of Actuarial Sciences,
Econometrics and Quantitative Methods
University of Amsterdam
Roetersstraat 11
1018 WB Amsterdam

Peter Spreij
Department of Econometrics
Vrije Universiteit
De Boelelaan 1105
1081 HV Amsterdam

Received September 28, 1996

Statistical Analysis of
Nonlinear and NonGaussian Time Series

T. Subba Rao

Abstract

We review recently developed methods for the analysis of station-ary nonlinear time series. We define unit root nonlinear (Bilinear) models. We also discuss higher order spectra for nonstationary process, and briefly consider the properties of time dependent Bilinear models. The estimation of time dependent parameters using wavelet expansions is also considered.

1. Stationary Time Series, Linear Models, Higher Order Spectra

Let $\{x_t,\ t = 0, \pm1, \pm2, ...\}$ be a discrete parameter time series satisfying the following conditions.

1. $E(x_t) = \mu$
2. $var(x_t) = \sigma_x^2 < \infty$.
3. $cov(x_t, x_{t+s}) = R(s),\ s = 0, \pm1, \pm2$.

We say the time series $\{x_t\}$ is second order stationary if the above conditions are satisfied. If we assume further, $\Sigma |R(s)| < \infty$, then we can define the Fourier transform.

$$f(w) = \frac{1}{2\pi} \sum_{-\infty}^{\infty} R(s)e^{-isw}. \, |w| \le \pi$$

and $\hspace{8cm}$ (1.1)

$$R(o) = \sigma_x^2 = \int_{-\pi}^{\pi} f(w)dw$$

If the time series is Gaussian, then the first two moments completely char-acterize its structure. Even without this assumption we can represent x_t (with $E(x_t) = 0$) in the form (under certain conditions on $f(w)$).2

$$x_t = \sum_{0}^{\infty} g_u e_{t-u} \ , \ \Sigma g_u^2 < \infty,$$ (1.2)

where (e_t) is a sequence of uncorrelated random errors with mean zero and variance σ_e^2. This follows from Wold's decomposition theorem and the condition $\int \ln f(w)dw > -\infty$. Though the above linear representation is very attractive, it is not very useful for statistical modeling purposes because the random variables $\{e_t\}$ are only mutually uncorrelated. We prefer that $\{e_t\}$ to be independent.

There is a significant difference between zero covariance and independence, as the former is a measure of only linear dependence. The only way to check whether the observations are uncorrelated or independent is through higher order spectra. We can define the higher order spectra as follows:

Consider the n dimensional vector $\underline{x}' = (x_{t_1}, x_{t_2}, ..., x_{t_n})$, and $M_x(t) = E(e^{\underline{t}'\underline{x}})$ be its n dimensional moment generating function. Let $\Psi_x(\underline{t}) = \ln M_x(t)$. The coefficient of $(t_1 t_2 ... t_n)$ in $\Psi_x(t)$ is known as the nth cumulant and is denoted by $cum\ (x_{t_1}, x_{t_2}, ..., x_{t_n})$. If $\{x_t\}$ is stationary up to the nth order moments, then

$$cum(x_{t_1}, x_{t_2}, ..., x_{t_n}) = c(t_n - t_1, t_n - t_2, ..., t_n - t_{n-1}) \qquad (1.3)$$

(see Brillinger and Rosenblatt, (1957) [BR], Subba Rao and Gabr (1984) [SG], Subba Rao (1996) [Su3]). If we assume the nth order cumulant $c(\tau_1, \tau_2, ..., \tau_{n-1})$ satisfies the condition $\Sigma |c(\tau_1, \tau_2, ..., \tau_{n-1})| < \infty$, then we can define the nth order cumulant spectra as

$$f_n(w_1, w_2, ..., w_{n-1}) = \frac{1}{(2\pi)^{n-1}} \Sigma...\Sigma c\,(\tau_1, \tau_2, ..., \tau_{n-1})\,e^{-i(w_1 t_1 + ... + w_{n-1}\tau_{n-1})}$$

$$(1.4)$$

Though we can define and estimate cumulant spectra of any order, for many purposes, it is sufficient to restrict up to the fourth order. If $n = 2$, we get the usual second order spectra and when $n = 3$, we get the bispectra defined by

$$f_3(w_1, w_2) = \frac{1}{(2\pi)^2} \Sigma\Sigma c(\tau_1, \tau_2)e^{-i(w_1\tau, + w_2\tau_2)}).$$

If the process $\{x_t\}$ is Gaussian, then $f_n(w_1, w_2, ..., w_{n-1}) = 0$ for all $n \geq 3$. This was the basis of Subba Rao and Gabr (1980) [SG] test for Gaussianity. If $\{x_t\}$ is a i.i.d sequence, then $f_n(w_1, w_2, ..., w_n) = \frac{k_n(x)}{(2\pi)^{n-1}}$, is independent of the frequencies. One can construct a statistical test based on the higher order spectra for testing independence.

Now let us suppose, x_t admits the representation

$$x_t = \sum_{o}^{\infty} g_u e_{t-u}, \qquad (1.5)$$

where $\{e_t\}$ is a sequence of independent, identically distributed random variables with $E(e_t) = 0$, $E(e_t^2) = k_2(e) = \sigma_e^2$ and $k_n(e)$ denoting the nth order cumulant. Then we can easily show

$$c(\tau_1, \tau_2, ..., \tau_{n-1}) = k_n(e) \sum_k g_k g_{k+\tau_1} \cdots, g_{k+\tau_{n-1}} \qquad (1.6)$$

$$f_n(w_1, w_2, ..., w_{n-1}) = \frac{k_n(e)}{(2\pi)^{n-1}} h(w_1) h(w_2) ... h(-w_1, w_2, ..., w_{n-1})$$

where $h(w) = \Sigma g_k e^{-ikw}$. A statistical test for linearity of time series can be constructed on the basis if the ratio $\frac{|f_n(w_1, w_2, ..., w_{n-1})|^2}{f(w_1) f(w_2) ... f(w_{n-1})}$ which is a constant of the process is linear obtained is of the form (1.5), but not Gaussian (see Subba Rao and Gabr, (1980) [SG] for further details).

Now consider the AR(p) model,

$$x_t + a_1 x_{t-1} ... + a_p x_{t-p} = e_t,$$

where $\{e_t\}$ is a sequence of independent, identically distributed random errors with mean zero and variance σ_e^2. It is known that the process is stationary if the roots of the polynomial $\varphi(z) = z^p + a_1 z^{p-1} + ... + a_p$ is strictly less than one. This nth order cumulant spectra of the above process can be written in the form (1.6) where

$$h(w) = (1 + a_1 e^{-iw} + a_2 e^{-2iw} + ... + a_p e^{-ipw})^{-1}.$$

Consider the nonlinear process

$$x_t = e_t + \beta e_{t-1} e_{t-2}, \qquad (1.7)$$

where $\{e_t\}$ is a sequence of independent random variables. it can be shown

$$cov(x_t, x_{t+s}) = 0 \text{ for } s \neq 0,$$

$$cum(x_t, x_{t+s_1}, x_{t+s_2}) \neq 0 \text{ for some } s_1 \text{ and } s_2.$$

In other words the time $\{x_t\}$ generated by (1.7) is like a white noise process. The only thing that distinguishes this process from a white noise process is the third order moments.

All the above process are stationary. Consider a process $\{y_t\}$ which is non stationary, but its first differences are stationary. In other words,

$$y_t - y_{t-1} = x_t$$

where $\{x_t\}$ is a white noise process, but with a nonlinear structure of the form (1.7). We have

$$y_t - y_{t-1} = e_t + \beta e_{t-1} e_{t-2}.$$

For this process $E(y_t) = 0$, $var(y_t) = t\sigma_e^2(1 + \beta^2\sigma_e^2)$. and hence as $t \to \infty$, $var(y_t) \to \infty$. This is an example of a nonlinear, nonstationary process similar to the unit root models. We now consider stationary nonlinear models.

2. Bilinear models.

Recently several nonlinear models have been proposed to describe non-linearity in time series. One such model is bilinear model for which many analytical properties can be obtained. (see Subba Rao, (1977) [Su1], (1981) [Su2], Subba Rao and Gabr, (1984) [SG]). A time series $\{x_t\}$ is said to be a bilinear process if x_t satisfies the nonlinear difference equation.

$$x_t + \sum_{i=1}^{p} a_i x_{t-i} = e_t + \sum_{j=1}^{q} b_j e_{t-j} + \sum_{i=1}^{P}\sum_{j=1}^{Q} c_{ij} x_{t-i} e_{t-j}, \qquad (2.1)$$

where $\{\dot{e}_t\}$ is a i.i.d sequence. The above model is usually denoted by $BL(p, q, p, Q)$. The properties of the model $BL(p, 0, p, 1)$ have been investigated by Subba Rao (1981) [Su2] and estimation of the parameters of these models have been investigated by Subba Rao and Gabr (1984) [SG]. In several papers, Terdik (1985) [Te] and Terdik and Subba Rao (1989) [TS] have looked into Ito Wiener expansions of these models. The properties of continuous bilinear models have been investigated by Terdik (1989) [Te].

Consider $BL(1, 0, 1, 1)$,

$$x_t - ax_{t-1} = bx_{t-1}e_{t-1} + e_1 \qquad (2.2)$$

It can be shown,

$$\mu \;\; = \;\; E(x_t) = \frac{b\sigma_e^2}{1 - a}$$

$$\mu_2' \;\; = \;\; E(x_t^2) = \frac{\sigma e^2(1 + 2b_{11}\sigma_e^2 + 4ab\mu)}{1 - a^2b^2\sigma_e^2}$$

and

$$\mu(1) \;\; = \;\; E(x_t x_{t+1}) = a\mu_2' + 2b\sigma_e^2\mu$$

$$\mu(s) \;\; = \;\; -a\mu(s - 1) + b\sigma_e^2\mu; \;\; s \geq 2$$

which implies

$$R(s) = -aR(s - 1); \;\; s \geq 2X_t$$

where $R(s) = cov(x_t, x_{t+s})$. In other words the process behaves like an ARMA (1,1).

Now consider a time series $\{y_t\}$ which is nonstationary, but can be made stationary after first differencing.

$$y_t - y_{t-1} = x_t \qquad (2.3)$$

where x_t satisfies the bilinear model $BL(1, 0, 1, 1)$,

$$x_t - ax_{t-1} = bx_{t-1}e_{t-1} + e_t \qquad (2.4)$$

Substituting for x_t from (2.3) in (2.4) we get,

$$y_t - (1+a)y_{t-1} + ay_{t-2} = by_{t-1}e_{t-1} - by_{t-2}e_{t-1} + e_t \qquad (2.5)$$

and it is $BL(2, 0, 2, 1)$. However, one of the roots of the polynomial $Z^2 - (1+a)z + a$ is equal to one, even though $|a| < 1$. We can define models of the form (2.5) as unit root bilinear models. The process is non-stationary. There is a huge literature on unit root linear models (Dickey and Fuller, (1979) [DF]), and one can investigate the properties and the usefulness of these type of models in economics. These models can be generalized to higher order Bilinear models where the AR part has roots on the unit circle. it will be interesting to study the statistical properties of these models. These will be considered in future publications. So far we considered stationary processes and nonstationary processes which are stationary after differencing. Though these models will be used in economics, but in communication engineering one deals with non stationarity of a different type. By this we mean the statistical properties, such as mean, variance, and other moments change with time, but they are finite, for each time t. In particular we consider process which are oscillatory, as defined by Priestley (1963) [Pr], and these will be considered below.

3. Nonstationary processes and higher order spectra

There are several definitions of power spectrum in the case of non stationary process (see Priestley, 1988 [Pr2] for a review). All these functions are aimed at defining a function of time and frequency which is positive. The time dependent spectrum $f_t(w)$ is used to study the local behaviour of the process. The Wigner-Ville spectrum is used widely in the engineering literature. We follow here the definition of Priestley (1963) [Pr]. Let $\{x_t\}$ be a zero mean, third order nonstationary process admitting the oscillatory process representations.

$$x_t = \int_{-\pi}^{\pi} e^{itw} A_t(w) dz(w) \tag{3.1}$$

where $\{dz(w)\}$ is an orthogonal process, and for each w, $A_t(w)$ is a deterministic function of w. We assume further that $A_t(w)$ does not oscillate too fast i.e. $A_t(w) = \int e^{it\theta} dk_w(\theta)$, where $\mid dk_w(\theta) \mid$ has an absolute maximum at $\theta = 0$. Priestley (1965) [Pr] defines the evolutionary spectrum as:

$$f_t(w) dw = |A_t(w)|^2 . d\mu(w) \tag{3.2}$$

where $d\mu(w) = E |dz(w)|^2$. Priestley (1965) [Pr] gives a method of estimation of $f_t(w)$. Priestley and Gabr (1993) [PG] defined the time dependent bispectrum as follows:

$$f_t(w_1, w_2) dw_1 dw_2 = A_t(w_1) A_t(w_2) A_t(-w_1 - w_2) d\mu(w_1, w_2) \tag{3.3}$$

where $d\mu(w_1, w_2) = E(dz(w_1) dz(w_2) dz(w_3))$ where $w_1 + w_2 + w_3 = 0 (\mathrm{mod}.2\pi)$. The normalized bispectral density function is defined as

$$h_t(w_1, w_2) = \frac{|f_t(w_1, w_2)|^2}{[f_t(w_1) f_t(w_2) f_t(-w_1 - w_2)]} \tag{3.4}$$

To estimate $f_t(w)$ and $f_t(w_1, w_2)$, we proceed as follows: (see Priestley and Gabr, (1993) [PG]).

Let

$$U(t, w_o) = \Sigma g_u X_{t-u} e^{-iw_o(t-u)} \tag{3.5}$$

where $\Gamma(\theta) = \Sigma g_u e^{iu\theta}$ is assumed to be highly concentrated at $\theta = 0$. Further, we will assume $\Sigma g_u^2 = \int |\Gamma(\theta)|^2 d\theta = 1$. Then the evolutionary spectral estimate is defined as:

$$\hat{f}_t(w) = \Sigma w_u |U(t-u, w)|^2 \tag{3.6}$$

We have

$$E[|U(t-u, w)|^2 \simeq f_t(w) \tag{3.7}$$

and hence

$$E[\hat{f}_t(w)] \simeq \bar{f}_t(w).$$

where $\bar{f}_t(w)$ is a smoothed form of the evolutionary spectrum of $f_t(w)$. Now consider the triple product

$$V(t, w_1, w_2) = U(t, w_1) U(t, w_2) U(t, -w_1 - w_2)$$

and

$$E[V(t, w_1, w_2)] \simeq f_t(w_1, w_2) \tag{3.8}$$

To evaluate the evolutionary spectrum and the bispectrum, we can use the relations (3.7) and (3.8) in situations where we cannot explicitly calculate $A_t(w)$ (see Subba Rao, 1995 [Su3]). let us now consider some examples to illustrate the methods of evaluation of $f_t(w)$ and $f_t(w_1, w_2)$ from the models.

Let $x_t = e_t + b_1(t)e_{t-1}$, where $\{e_t\}$ is a zero mean second order stationary process with spectral representation $e_t = \int e^{itw} dZ_e(w)$, where $E(dz_e(w)) = 0$, $E|dz_e(w)|^2 = \frac{\sigma_e^2}{2\pi}.dw$. Then we have, for the process x_t, $A_t(w) = (1 + b_1(t)e^{iw})$ and hence $f_t(w) = \frac{\sigma_e^2}{2\pi}|A_t(w)|^2$, where $A_t(w) = (1 + b_1(t)e^{itw})$ and similarly, the evolutionary bispectrum is given by, $f_t(w_1 \ w_2) = \frac{\mu_3}{(2\pi)^2}A_t(w_1)A_t \ (w_2)A_t(-w_1 - w_2), \mu_3 = E(e_t^3)$. If e_t is Gaussian, (x_t is Gaussian), $f_t(w_1, w_2) = 0$ for all w_1 and w_2. If $b_1(t) = b_1$ for all t, $B_t(w) = B(w)$. Hence $f_t(w_1, w_2)$ is independent of t. In other words the constancy of $f_t(w_1, w_2)$ implies the time series is stationary, but nonlinear If the time series $\{x_t\}$ is Gaussian, the second order stationarity can be tested using evolutionary spectrum $f_t(w)$. (see Priestley and Subba Rao, (1969) [PS]).

Consider the time dependent AR model,

$$
\begin{aligned}
x_t &= a_1(t)x_{t-1} + e_t, \ | \ a_1(t) \ |< 1 \\
&= \sum_0^\infty b_t(u)e_{t-u} \tag{3.9}
\end{aligned}
$$

where

$$b_t(u) = a_1(t)a_1(t-1)...a_1(t-u)$$

From (3.9), we have

$$x_t = \int e^{itw} A_t(w) dz_e(w) \tag{3.10}$$

where $A_t(w) = \Sigma b_t(u)e^{iuw}$ and hence $f_t(w) = \frac{\sigma_e^2}{2\pi}|\Sigma b_t(u)e^{iuw}|^2$. If $\{a_1(t)\}$ are changing slowly, $b_t(u) \sim a_1^u(t)$ and hence $f_t(w) \simeq \frac{\sigma_e^2}{2\pi}|1 - a_1(t)e^{itw}|^{-2}$. It is not always easy to obtain a solution in the form (3.9) for higher order time dependent AR models. In general, if we have the time dependent linear process representation

$$x_t = \sum_o^\infty g_t(u)e_{t-u} \tag{3.11}$$

where $\{e_t\}$ are i.i.d random variables with $E(e_t) = 0$, $E(e_t^2) = \sigma_e^2$, $E(e_t^3) = \mu_3$ then we can show $f_t(w) = \frac{\sigma_e^2}{2\pi}|A_t(w)|^2$, $f_t(w_1, w_2) = \frac{\mu_3}{(2\pi)^2}A_t(w_1)A_t(w_2)A_t(-w_1-w_2)$, where $A_t(w) = \sum_o^\infty g_t(u)e^{-iuw}$. If $\mu_3 = 0$, then $f_t(w_1, w_2) = 0$ for all w_1 and w_2. If x_t is linear (but non Gaussian), then we have $\frac{|f_t(w_1,w_2)|^2}{f_t(w_1)f_t(w_2)f_t(-w_1-w_2)} = $ does not depend on t, w_1, w_2.

In principle, one can construct statistical tests for linearity (and Gaussianity) of nonstationary time series using the evolutionary bispectrum as in Subba Rao and Gabr (1980) [SG]).

4. Nonstationary and nonlinear time series - Volterra series

Our object in this section is to show through examples that evolutionary higher order spectra play a significant role in the analysis of nonstationary, nonlinear time series.

Consider the nonstationary process, $x_t = c_t e_t$, where $\{e_t\}$ is a i.i.d sequence with $E(e_t) = 0$, $E(e_t^2) = \sigma_e^2$, $E(e_t^3) = \mu_3$. Then we can show $f_t(w) = c_t^2 \cdot \frac{\sigma_e^2}{2\pi}$, independent of the frequency w, but depends on t. We call such a process nonstationary white noise process. Consider the nonlinear nonstationary process $x_t = e_t + \beta_t e_{t-1} e_{t-2}$, where $\{e_t\}$ is defined earlier. We can show $E(x_t) = 0$, $E(x_t x_{t+s}) = 0$ for all s $\neq 0$. Hence $f_t(w) \simeq \sigma_e^2(1 + \beta_t^2 \sigma_e^2)$, independent of w. In other words the process behaves like a nonstationary white noise process, even though the process is nonlinear. However, if we calculate the evolutionary bispectral density function, we can show that $f_t(w_1, w_2)$ is a function of w_1 and w_2. This shows clearly that higher order spectra are useful in detecting the nonlinear, nonstationary time series.

Consider the time dependent bilinear model $TBL(1, 0, 1, 1)$.

$$x_t = a(t)x_{t-1} + d_1(t)x_{t-1}e_{t-1} + e_t \qquad (4.1)$$

where $[e_t\}$ is an i.i.d sequence. For convenience we assume e_t is $N(0, \sigma_e^2)$. We follow the notation introduced by Subba Rao (1981) for the stationary bilinear models. We want to seek a solution of (4.1). Consider the perturbed model

$$x_t = a(t)x_{t-1} + \lambda d_1(t)x_{t-1}e_{t-1} + \lambda e_t \qquad (4.2)$$

and λ is a perturbation parameter, and the solution we seek is in the form $x_t = \sum_{j=1}^\infty \lambda^j x_j(t)$. Substitute for x_t in (4.2) and equate the coefficients of

$\lambda^j (j = 1, 2, ...)$ both sides; we obtain a sequence of successive equations of the form

$$
\begin{aligned}
x_1(t) &= a(t)x_2(t-1) + e_t \\
&= \Sigma b_t(u)e_{t-u}
\end{aligned}
\tag{4.3}
$$

$$
x_2(t) = a(t)x_2(t-1) + d_1(t)x_1(t-1)e_{t-1}
\tag{4.4}
$$

and for all j,

$$
x_j(t) = a(t)x_j(t-1) + d_1(t)x_{j-1}(t-1)e_{t-1}
\tag{4.5}
$$

From (4.4), we have

$$
x_2(t) = \sum_u^{\infty} b_t(u)\eta_1(t-u-1),
$$

where $\eta_1(t-1) = d_1(t)x_1(t-1)e_{t-1}$. We can show

$$
x_2(t) = \sum_u \sum_v b_t(u)b_{t-u-1}(v)d_1(t-u)e_{t-u-1}e_{t-u-1-v}.
$$

Similarly we can solve $x_j(t)$ in terms of $x_{j-1}(t)$ etc. This suggests that we can write the general solution of the above equation (4.2) in the Volterra form

$$
x_t = \Sigma b_t(u)e_{t-u} + \Sigma\Sigma b_t(u,v)e_{t-u}e_{t-v} + \dots
$$

$$
+ \Sigma \dots \Sigma b_t(u_1, u_2 \dots, u_k)e_{t-u_1}e_{t-u_2} \dots e_{t-u_k} + \dots
\tag{4.6}
$$

where $\{b_t(u)\}$, $\{b_t(u,v)\}, \dots$ are time dependent, Volterra kernels. The problem of interest is to identify these kernels given the cumulants of $\{x_t\}$. If e_t is Gaussian, we can write alternatively,

$$
\begin{aligned}
x_t &= \Sigma g_t(u)H^{(1)}(\dot{e}_{t-u}) + \Sigma\Sigma g_t(u,v)H^{(2)}(e_{t-u}, e_{t-v}) \\
&\quad + \dots + \Sigma \dots \Sigma g_t(u_1, u_2 \dots, u_r)H^{(r)}(e_{t-u_1}, e_{t-u_2}, \dots e_{t-u_r}) \\
&\quad + \dots
\end{aligned}
\tag{4.7}
$$

which is the Wiener expansion, with time dependent Wiener kernels.

Suppose we have a nonlinear, nonstationary dynamical system satisfying the equation

$$x_t = \Sigma g_t(u) H^{(1)}(\eta(t-u) + \Sigma\Sigma g_t(u,v) H^{(2)}(\eta(t-u), \eta(t-v))$$
$$+ .. \tag{4.8}$$

where the input $\{\eta(t)\}$ and the output $\{x_t\}$ and both are observed.

The time dependent kernels (or equivalently the transfer functions) can be obtained from the higher order cross covariances between the input $\{\eta_t\}$ and the output $\{x_t\}$. These problems will be pursued in a future publication.

5. Estimation of the time dependent parameter

The problem of estimation of the time dependent parameters of linear models have been first considered by Subba Rao (1970) [Su1], Hussain and Subba Rao (1976) [HS]. In recent years, the problems have received enormous attention from the engineers and mathematical statisticians as seen by many papers (for Grenier (1983) [Gr], Dahlhaus (1995) [Da] etc.. We briefly describe the approach proposed by Subba Rao (1970) [Su1]. Consider the model

$$x_t = a(t)x_{t-1} + e_t \; ; \; t = 1, 2, .., N$$

Subba Rao (1970) [Su1] has approximated $a(t) = a_1 + a_2 t + a_3 \frac{t^2}{2}$, and then considered the estimation of (a_1, a_2, a_3) by the method of weighted least squares.

$$Q = \sum_{-m}^{m} w_h(x_{t-h} + a_1(t-h)x_{t-h-1})^2 \tag{5.1}$$

where the weights $w_m = w_{-m}$ and m is chosen such that over the domain of minimization $\{a(t)\}$ is slowly changing. The advantage of the above procedure is that one can obtain Durbin-Levinson type of recursive algorithms for the time dependent parameters, similar to the stationary case, and these results were reported in the PhD thesis by Hussain (1973) [Hu]. Subba Rao (1970) [Su1] has defined the weighted likelihood function.

$$L_w = \{f(e_{t-h})\}^{w_h} \{f(e_{t-h+1})\}^{w_{h-1}} \ldots \{f(e_{t+h})\}^{w_h} \tag{5.2}$$
$$= \{f(x_{t-h}/x_{t-h+1})\}^{w_h} \{f(x_{t-h+1}/x_{t-h+2})\}^{w_{h-1}} \ldots \{f(x_{t+h}/x_{t+h-1})\}^{w_h}$$

where $f(e_t)$ is the probability density function of e_t, and $f(x_t/x_{t-1})$ is the conditional density of x_t given x_{t-1}. The maximization of the above function is same as the minimization of the weighted least squares criterion Q.

In the above method of estimation, the idea is to expand $a(t)$ in terms of the basis elements. $\{1, t, \frac{t^3}{2}\}$. In fact, one can expand $a(t)$ in terms of any orthonormal basis $\{f_j(t)\}$. For example, one can write $a(t) = \sum_{j=0}^{m} a_j f_j(t)$, where $\{f_o(t), f_1(t) \ldots\}$ form an orthonormal basis. The usual functions chosen are Legendre functions, prolate spheroidal sequences (Grenier, (1983) [Gr]) etc.

In recent years, wavelet methods have been widely used as an alternative to Fourier method. Wavelets are another form of orthogonal basis which have the property of dilation and translation (see Subba Rao and Indukumar, (1996) [SI]). In view of these they can form a useful class for the expansion of the time dependent parameters. We illustrate this with a simple example. Let us write

$$a_1^\ell(t) = \sum_{-m}^{m} b_{\ell,n} \varphi_{\ell,n}(t) \tag{5.3}$$

where $\{\varphi_{\ell,n}(t)\}$ are scale transforms, $\varphi_{\ell,n}(t) = 2^{\ell/2}\varphi(2^\ell t - n)$. $(n = 0, \pm 1, \pm 2, \ldots)$. As $\ell \to \infty$, $a_1^{(\ell)}(t) \to a_1(t)$. From orthogonality of $\varphi_{\ell,n}(t)$, we have $b_{\ell,m} = \int a_1^\ell(t)\varphi_{\ell,m}(t)dt$. We can now minimize $Q = \sum_{t=1}^{N}[x_t - (\Sigma b_{\ell,n}\varphi_{\ell,n}(t))x_{t-1}]^2$ with respect to $b_{\ell,n}$, $(n = -m, -(m-1), \ldots, m$. Differentiating Q with respect to $b_{\ell,k}$ and equating to zero leads to the normal equations.

$$\Sigma \varphi_{\ell,n}(t)x_t x_{t-1} = \Sigma b_{\ell,n}[\sum_{t} \varphi_{\ell,n}(t)\varphi_{\ell,k}(t)x_{t-1}^2]; \quad k = -M, -(M-1), \ldots, M) \tag{5.4}$$

Let

$$a_{\ell,k} = \sum_{t} \varphi_{\ell,k}(t)x_t x_{t-1}$$

$$d_{k,n} = \sum_{t} \varphi_{\ell,n}(t)\varphi_{\ell,k}(t)x_{t-1}^2$$

Therefore, we can write (5.4) as

$$\underline{a}_{\ell,k} = \underline{x}_k' \underline{\theta}_\ell, \quad (k = 1, 2, \ldots, M^2) \tag{5.5}$$

where

$$x_k' = [d_{k,-M}, d_{k,-(M-1)} \ldots, d_{k,M}]$$
$$\underline{\theta}_\ell' = [b_{\ell,-M}, b_{\ell,-(M-1)} \ldots, bM].$$

We can now write (5.5) as

$$\underline{A\theta}_\ell = \underline{a}_\ell,$$

or
$$\underline{\theta}_\ell = \underline{A}^{-1}\underline{a}_\ell.$$

From the above we obtain

$$a_1^{(\ell)}(t) = (b_{\ell,-M}, \ldots, b_{\ell,M}) \begin{pmatrix} \varphi_{\ell-M}(t) \\ \varphi_{\ell,M}(t) \end{pmatrix}$$
$$= \underline{\varphi}_\ell' \underline{\Psi}_\ell(t) = \underline{a}_\ell' \underline{A}_\ell^1 \cdot \underline{\Psi}_\ell(t).$$

where $\underline{\Psi}_\ell'(t) = (\varphi_{\ell-M}(t), \ldots, \varphi_{\ell,M}(t)$.

If $a_1^{(\ell)}(t) = a_1$ for all t, it suggests that the process is stationary. These estimates can be used to detect change in time series. The method can be extended to nonlinear models, linear ARMA models. These will be considered in future publications.

References

[BR]   D. R. Brillinger, and M. Rosenblatt, *Asymptotic theory of k-th order spectrum in spectral analysis of time series*, ed. by B. Harris. John Wiley, 1957, 153-188.

[Da]   R. Dahlhaus, Fitting of time series models to nonstationary time series. Tech Report. University of Heidelberg, 1995.

[DF]   D. Dickey, and W. A. Fuller, Distribution of the estimates for autoregressive time series with a unit root, *Jour. Amer. Statist. Assoc.*, **74** (1979), 427-431.

[Gr]   Y. Grenier, Time dependent ARMA modeling of nonstationary signals, *IEEE ASSP*, **31** (1983), 899 - 911.

[Hu]   M. Y. Hussain, *Goodness of fit tests in nonstationary time series models*, unpublished PhD thesis submitted to UMIST 1973.

[HS]   M. Y. Hussain, and T. Subba Rao, The estimation of autoregressive, moving average and mixed autoregressive - moving average systems with time dependent parameters of nonstationary time series, *Int. Journ. Control.*, **23/5** (1976), 647-656.

[Pr]   M. B. Priestley, Evolutionary spectra and nonstationary process, *J. Roy. Stats. for Soc. B Series*, **27** (1965), 204-237.

[PS]   M. B. Priestley, and T. Subba Rao, A test for stationarity of time series. *J. Roy. Stats. Soc. B Series*, **31** (1969), 140-149.

[Pr1]   M. B. Priestley, *Spectral analysis and time series*, Academic Press, 1981.

[Pr2]   M. B. Priestley, *Nonlinear and nonstationary time series analysis* Academic Press, 1988.

[PG]   M. B. Priestley, and M. M. Gabr, *Bispectral analysis of nonstationary processes*, Multivariate Analysis: Future directions. ed. by C.R. Rao. North Holland, 1993.

[SS]   T. Subba Rao,and M. Eduarda A. da Silva, Identification of bilinear time series model BL (p,o, p,1) *Statistica Sinica*, **2** (1992), 465-478.

[Su1]   T. Subba Rao, The fitting of nonstationary time series models with time dependent parameters, *J. Roy Stats. Soc. B. series*, **32** (1970), 312-322.

[Su2]   T. Subba Rao, On the theory of bilinear models, *J. Roy. Stats. Soc. B series*, **43** (1981), 244-255.

[SG]   T. Subba Rao, and M. M. Gabr, *An introduction to bispectral analysis and bilinear time series models*, Lecture notes in statistics, Springer-Verlag, Berlin, 1984.

[Su3]   T. Subba Rao, Analysis of nonstationary and nonlinear signal. *Applied stochastics and optimisation International Congress on Industrial and Applied Mathematics*, Hamburg, 1-8 July.

[SI]   T. Subba Rao, and K. C. Indukunmar, Spectral and Wavelet methods for the analysis of nonlinear and nonstationary time series, *J. Franklin Inst.*, **333(b)** (1996), 425-452.

[Te]   Gy. Terdik, Transfer functions and conditions for stationarity of bilinear models with Gaussian residuals, *Proc. Roy. Soc. London. A series*, **400** (1985), 351-330.

[TS]   Gy. Terdik, and T. Subba Rao, On Wiener-Ito representation and the best linear predictors for bilinear time series, *J. Appl. Prob.*, **26** (1989), 274-286.

[Te2]   Gy. Terdik, *Stationary Solutions for bilinear systems with constant coefficients*. Progress in Probability. **W/18**. Birkhäuser, Boston.

Department of Mathematics
UMIST
Manchester M60 1QD
United Kingdom

Received October 1, 1996

Bilinear Stochastic Systems
with Long Range Dependence
in Continuous Time [1]

E. Iglói and Gy. Terdik

1. Introduction

Application of a long range dependent model includes several fields of science and economics as geophysics, hydrology, turbulence, weather and so on. Recently it has been successfully used for modeling network traffic data (see [WTLW]). The basic stochastic process of this kind is the fractional Brownian motion defined in [MvN]. The fractional Brownian motion is given as a particular fractional operator on the standard Brownian motion. The linear or Gaussian parametric models of long range dependent phenomena are both the linear stochastic differential equations with fractional Brownian motion input and the fractional operator on the solution of a linear stochastic differential equation (see [C]). Actually these two types of processes are equivalent. Because most of the observations are not Gaussian there is a need nonlinear modeling of long range dependence. One possibility is to get rid of Gaussianity is the bilinear model started by Subba Rao [SR] in the discrete time case. The easy way to get a long range non-Gaussian process is to apply the fractional operator to the solution of the bilinear SDE. It is more painful to consider a bilinear SDE with fractional Brownian motion input.

In this paper we start with the bilinear SDE with white noise input and list the basic ideas leading to the stationary solution given in both time domain and chaotic frequency domain forms. The fractional integral operator on this stationary solution is applied and its basic properties are pointed out. In section three the bilinear SDE with fractional Brownian motion input is considered. The problem of stochastic integration by the fractional Brownian motion is solved and the stationary solution of the SDE is explicitly given in the case when the coefficient of the bilinear term is pure imaginary.

[1]Supported by OTKA grant #T19501

2. White noise input

2.1. Time domain. Consider the SDE

$$dy_t = (\mu + \alpha y_t)\, dt + (\beta + \gamma y_t)\, dw_t, \tag{2.1}$$

where w_t is Brownian motion (Bm), i.e. standard Wiener process with variance σ^2. The σ^2 is considered as 1; otherwise one can use the transformation y_t/σ, μ/σ, β/σ, $\gamma\sigma$. The equation (2.1) is a linear differential equation, nevertheless it is called bilinear one in system theory to differentiate between the situations when γ is zero and nonzero, i.e. when the solution is Gaussian and nonGaussian. The solution of equation (2.1) is well known (see [K], pp. 111.),

$$y_t = e^{\left(\alpha - \frac{\gamma^2}{2}\right)t + \gamma w_t}\left(y_0 + (\mu - \beta\gamma)\int_0^t e^{-\left(\alpha - \frac{\gamma^2}{2}\right)s - \gamma w_s}\, ds + \beta \int_0^t e^{-\left(\alpha - \frac{\gamma^2}{2}\right)s - \gamma w_s}\, dw_s\right)$$

When $\gamma = 0$, $\alpha < 0$ and $\beta \neq 0$, that provides the stationary Gaussian Ornstein–Uhlenbeck process.

We are interested in the stationary nonGaussian solution of (2.1). Therefore it is necessary to assume that $\mu^2 + \beta^2 > 0$, $\gamma \neq 0$ and not only $\alpha < 0$ but $2\alpha + \gamma^2 < 0$ as well. The starting value y_0 is also well defined and then the stationary physically realizable solution of (2.1) is

$$y_t = (\mu - \beta\gamma)\int_{-\infty}^t e^{(\alpha - \frac{\gamma^2}{2})(t-s) + \gamma(w_t - w_s)}\, ds + \beta \int_{-\infty}^t e^{(\alpha - \frac{\gamma^2}{2})(t-s) + \gamma(w_t - w_s)}\, dw_s. \tag{2.2}$$

Note that $\mu\gamma \neq \alpha\beta$ must hold, otherwise (2.1) has only the degenerated solution $y_t = -\frac{\beta}{\gamma} = -\frac{\mu}{\alpha}$.

From now on we shall assume that $\beta = 0$ and $\mu \neq 0$, otherwise the following transformation is applied

$$\tilde{y}_t = \frac{\gamma\mu}{\gamma\mu - \alpha\beta} y_t + \frac{\beta}{\gamma}.$$

The expectation of y_t can be calculated from (2.1) as

$$\mathsf{E}\, y_t = -\frac{\mu}{\alpha}.$$

One can also easily get the covariance function of y_t directly from (2.1) and the variance from (2.2).

$$R(t) = R(0)e^{\alpha|t|} = \frac{-\mu^2\gamma^2}{\alpha^2(2\alpha + \gamma^2)}e^{\alpha|t|}, \quad t \in \mathbb{R} \tag{2.3}$$

2.2. Frequency domain. Suppose that there exists a stationary physically realizable solution of equation (2.1) which is subordinated to the input process w_t. The frequency domain representation theorem says (see [D]), that all such solutions can be put into the form

$$y_t = \sum_{k=0}^{\infty} \int_{R^k} \exp(it\Sigma\omega_{(k)}) f_k(\omega_{(k)}) W(d\omega_{(k)}),\qquad (2.4)$$

where $\omega_{(k)} = (\omega_1, \omega_2, ...\omega_k)$, $\Sigma\omega_{(k)} = \sum_{j=1}^{k}\omega_j$ and $W(d\omega_{(k)})$ is the k dimensional multiple Wiener-Ito measure according to the Wiener process w_t. The representation (2.4) is unique up to the permutation of the variables of the transfer functions f_k. Now it follows from the diagram formula (see [M]), that the following recursion is valid for the transfer functions f_k (see [T]).

$$f_0 = -\frac{\mu}{\alpha},\quad f_k(\omega_{(k)}) = \frac{\gamma f_{k-1}(\omega_{(k-1)})}{i\Sigma\omega_{(k)} - \alpha},\quad k \geq 1.\qquad (2.5)$$

Only to make it easier to understand the later formula (3.5), we remark that one can check, using a little algebra, that the symmetrized version $\tilde{f}$ of these transfer functions can be written in the form

$$\tilde{f}_k(\omega_{(k)}) = \mu\frac{\gamma^k}{k!}\int_0^{\infty} e^{\alpha u}\prod_1^k \frac{1 - e^{-iu\omega_j}}{i\omega_j} du.\qquad (2.6)$$

The spectrum of y_t can be calculated from the covariance function (2.3), i.e.

$$\varphi(\omega) = \int_R e^{-it\omega} R(t)dt = \frac{2\mu^2\gamma^2}{\alpha(2\alpha + \gamma^2)}\frac{1}{\omega^2 + \alpha^2},\quad \omega \in \mathbb{R}.\qquad (2.7)$$

It should be noted that there is no difference between the spectrum of an Ornstein Uhlenbeck process and (2.7); therefore it is necessary to consider higher order spectra for bilinear processes.

2.3. Fractional integral of the bilinear process. There are two ways to combine the fractional integration and the bilinear process. One is the fractional integration with a bilinear process input and the other one is the bilinear process with fractional Brownian motion input. The scheme of these constructions is the following.

Bm $\longrightarrow$ bilinear process $\longrightarrow$ fractionally integrated bilinear process

Bm $\longrightarrow$ fractional Bm $\longrightarrow$ bilinear process with fractional Bm input

The second construction is rather complicated so we are going to sketch it in section 3. In this subsection the first one, which is simpler, will be explained.

For this aim, let us consider a stationary process x_t with spectral representation

$$x_t = \int_{\mathbb{R}} e^{it\omega} f^{(x)}(\omega) V(d\omega),$$

where $V(\omega)$ is a (not necessarily Gaussian) complex valued process with orthogonal increments. Let us define for $0 < h < 1$ the h-th order fractional integral process of x_t, in a manner analogous to the fractional integration of deterministic functions in the Riemann-Liouville sense (see [R]),

$$y_t \doteq (I^{(h)} x)_t \doteq \frac{1}{\Gamma(h)} \int_{-\infty}^{t} (t-s)^{h-1} x_s ds, \tag{2.8}$$

where $\int_{-\infty}^{t} = \text{l.i.m.}_{N \to \infty} \int_{-N}^{t}$ and $\int_{-N}^{t} (t-s)^{h-1} x_s ds$ is an L_2–integral.

Using the formula $\int_{0}^{\infty} u^{h-1} e^{-iu\omega} du = \Gamma(h)(i\omega)^{-h}$ (see [Z]) we get the spectral representation

$$y_t = \int_{\mathbb{R}} e^{it\omega} f^{(x)}(\omega)(i\omega)^{-h} V(d\omega).$$

Thus, if

$$\int_{\mathbb{R}} \left| f^{(x)}(\omega) \right|^2 |\omega|^{-2h} d\omega < \infty, \tag{2.9}$$

then the definition (2.8) of y_t is correct and y_t is a stationary process.

Now, let x_t be the bilinear process (2.2) after subtraction of its expectation. Since condition (2.9) holds for $0 < h < \frac{1}{2}$ because of (2.7), we have defined for $0 < h < \frac{1}{2}$ the h-th order fractionally integrated bilinear process. It is a stationary process the spectrum of which is

$$\varphi^{(y)}(\omega) = \frac{2(\mu\gamma)^2}{\alpha(2\alpha + \gamma^2)} \frac{|\omega|^{-2h}}{\omega^2 + \alpha^2}, \quad \omega \neq 0,$$

and this behaves as $|\omega|^{-2h}$ near zero, i.e. y_t is a long range dependent process.

3. Bilinear process with fractional Brownian motion input

3.1. Fractional Brownian motion and stochastic integration with respect to it. The definition of the fractional Brownian motion with parameter $h \in \left(-\frac{1}{2}, \frac{1}{2}\right)$ due to Mandelbrot and Van Ness is the following.

$$w_t^{(h)} = \frac{1}{\Gamma(1+h)} \left\{ \int_{-\infty}^{0} \left[(t-s)^h - (-s)^h \right] dw_s + \int_{0}^{t} (t-s)^h dw_s \right\},$$

$t \in \mathbb{R}$. It is the h-th fractional integral process of the Brownian motion, adjusted to zero at zero (see [MvN]). Clearly, $w_t^{(0)} = w_t$. The most interesting properties of $w_t^{(h)}$ which are related to our subject, are the following (see [MvN]). a.) $w_t^{(h)}$ has stationary increment processes, b.) $w_t^{(h)}$ is almost surely continuous and L_2–continuous, c.) $\mathsf{E}\, w_t^{(h)} = 0$, $\mathsf{E}\, (w_{t_1}^{(h)} w_{t_2}^{(h)}) = \frac{1}{2}(|t_1|^{2h+1} + |t_2|^{2h+1} - |t_1 - t_2|^{2h+1})$.

The connection of w_t and $w_t^{(h)}$ can be formulated in the frequency domain too. Namely,

$$w_t = \int_{\mathbb{R}} \frac{e^{it\omega} - 1}{i\omega} W(d\omega), \qquad w_t^{(h)} = \int_{\mathbb{R}} \frac{e^{it\omega} - 1}{i\omega} (i\omega)^{-h} W(d\omega).$$

Unfortunately, like the non–fractional Brownian motion w_t, in any $t \in \mathbb{R}$, $w_t^{(h)}$ is almost surely not differentiable. And what is more, $w_t^{(h)}$ is not semimartingale, and this fact makes it necessary to build a completely new theory of stochastic integration with respect to $w_t^{(h)}$. There is not enough space in this paper to explain it in detail therefore it will be published somewhere else soon. We remark only that the main tools of this integration theory are based on the chaotic representation

$$y_t = \sum_{k=0}^{\infty} \int_{\mathbb{R}^k} e^{it\Sigma\omega_{(k)}} f_k(\omega_{(k)}) \prod_{1}^{k} (i\omega_j)^{-h} W(d\omega_{(k)}) \tag{3.1}$$

of the stationary nonanticipative functionals y_t of $w_t^{(h)}$ on the one hand and the diagram formula on the other hand.

3.2. The bilinear differential equation with Gaussian fractional noise input. The bilinear SDE with Gaussian fractional noise input is

$$dy_t = (\alpha y_t + \mu)dt + \gamma y_t dw_t^{(h)}, \tag{3.2}$$

interpreting the stochastic integral in accordance in the above mentioned sense.

If $0 \neq \gamma \in \mathbb{R}$, then we suspect that (3.2) has no stationary solution. Now if $0 > \alpha \in \mathbb{R}$ and γ is imaginary, that is, if we replace the original γ with $i\gamma$, where $\gamma \in \mathbb{R}$, then we have found the unique solution. Namely, the SDE

$$dy_t = (\alpha y_t + \mu) \, dt + i\gamma y_t \, dw_t^{(h)} \tag{3.3}$$

has the solution

$$y_t = \mu \int_{-\infty}^{t} e^{\alpha(t-s) + i\gamma(w_t^{(h)} - w_s^{(h)})} \, ds, \tag{3.4}$$

which is unique in the linear space of stationary nonanticipative processes, for which the stochastic integration with respect to $w_t^{(h)}$, can be defined. In (3.4), the integral is an L_2–integral similarly to (2.2).

The chaotic representation of the solution (3.4) is (3.1) with transfer functions

$$f_k(\omega_{(k)}) = \mu \frac{(i\gamma)^k}{k!} \int_0^\infty e^{\alpha u - \frac{\gamma^2}{2} u^{2h+1}} \prod_1^k \frac{1 - e^{-iu\omega_j}}{i\omega_j} \, du \prod_1^k (i\omega_j)^{-h}. \tag{3.5}$$

Remark that the replacement $\gamma \to i\gamma$ can be made in the non–fractional case as well; we need only substitute γ with $i\gamma$ everywhere. In the light of this, let us notice that there is an essential difference between (3.4) and (2.2) and also between (3.5) and (2.6). Remember that we are dealing with the simplified equation, i.e. $\beta = 0$. For example in (2.2), there is a $\frac{\gamma^2}{2}(t-s)$ term in the exponent, not so as in (3.4). The cause of this fact is that in the non–fractional case, the Ito–formula holds, while in the fractional case the ordinary deterministic differentiation rules. The difference arises from the formula

$$2 \int_{\mathbb{R}} \frac{1 - e^{-it\omega}}{i\omega} |\omega|^{-2h} d\omega = \frac{d}{dt} t^{2h+1}, \quad t > 0, \quad 0 < h < \frac{1}{2}, \tag{3.6}$$

which can be interpreted as

$$2\mathsf{E} \int_0^T w_t^{(h)} dw_t^{(h)} = \int_0^T \frac{d}{dt} t^{2h+1} dt = \int_0^T \frac{d}{dt} \mathsf{E} \left(w_t^{(h)} \right)^2 dt.$$

For $h = 0$, the integral in (3.6) is not meaningful and $2\mathsf{E} \, w_t dw_t = 0 \neq dt = \frac{d}{dt} \mathsf{E} \, w_t^2 dt$. This is the reason for the break in the form of the solution when $h \to 0$.

Arising from this,

$$\left(w_T^{(h)}\right)^2 = \int\limits_0^T d\left(w_t^{(h)}\right)^2 = \int\limits_0^T 2w_t^{(h)} dw_t^{(h)}$$

holds in the sense of the L_2–limit of the Riemann–Stieltjes approximating sums. Thus, $w_t^{(h)}$ behaves as if it had a zero quadratic variation process. This also explains why the deterministic differentiation rule does hold.

Let us return to the solution (3.4). Denote by $y_t^{(1)}$ and $y_t^{(2)}$ the real and the imaginary part of y_t. The expectation and the covariance function of y_t can be calculated either by (3.4) or by (3.5). Let us define the function

$$K(u) = e^{\alpha u - \frac{\gamma^2}{2} u^{2h+1}}, \quad u \geq 0.$$

With this notation,

$$\mathsf{E}\, y_t = \mu \int\limits_0^\infty K(u) du$$

One gets the most simple formulae for the covariances when $\mu \in \mathbb{R}$. In this case, for example, the variances of the component processes are

$$\mathsf{Var}\, y_t^{(1)} = \mu^2 \int\limits_0^\infty \int\limits_0^\infty \left[\mathrm{ch}\left(\gamma^2 G(u_1, u_2)\right) - 1\right] K(u_1) K(u_2) du_1 du_2$$

$$\mathsf{Var}\, y_t^{(2)} = \mu^2 \int\limits_0^\infty \int\limits_0^\infty \mathrm{sh}\left(\gamma^2 G(u_1, u_2)\right) K(u_1) K(u_2) du_1 du_2$$

where $G(u_1, u_2) = \frac{1}{2}(u_1^{2h+1} + u_2^{2h+1} - |u_1 - u_2|^{2h+1})$. The formulae for the covariances are of the same type but a bit longer so we omit them. Just remark that the cross–covariance function, that is the covariance between $y_0^{(1)}$ and $y_t^{(2)}$, is zero. This is an immediate consequence of (3.5).

If $\mu \in \mathbb{R}$, then the spectra of $y_t^{(1)}$ and of $y_t^{(2)}$ are the following.

$$\varphi^{(1)}(\omega) = \mu^2 \int\limits_0^\infty \int\limits_0^\infty \left[\mathrm{ch}_\otimes\left(\gamma^2 A^{(u_1, u_2)}\right)(\omega) - 1\right] K(u_1) K(u_2) du_1 du_2$$

$$\varphi^{(2)}(\omega) = \mu^2 \int\limits_0^\infty \int\limits_0^\infty \mathrm{sh}_\otimes\left(\gamma^2 A^{(u_1, u_2)}\right)(\omega) K(u_1) K(u_2) du_1 du_2$$

where

$$A^{(u_1, u_2)}(\omega) = -\frac{1 - e^{-iu_1\omega}}{i\omega}\overline{\left(\frac{1 - e^{-iu_2\omega}}{i\omega}\right)} |\omega|^{-2h},$$

and ch $_\otimes(g)(\omega) = \sum\limits_{k=0}^{\infty} \frac{1}{(2k)!} g^{\otimes(2k)}(\omega)$ and sh $_\otimes(g)(\omega) = \sum\limits_{k=0}^{\infty} \frac{1}{(2k+1)!} g^{\otimes(2k+1)}(\omega)$
are the convolution cosine hyperbolic and sine hyperbolic of a function $g(\omega)$, respectively.

References

[C]   F. Comte, Simulation and estimation of long memory continuous time models, *J. Time Ser. Anal.* **17/1** (1996).

[D]   R. L. Dobrushin, Gaussian and their subordinated generalized fields, *Ann. of Prob.* **7** (1979).

[K]   G. Kallianpur, *Stochastic Filtering Theory*, Applications of Mathematics. **13**, ed. A. V. Balakrishnan, Springer–Verlag, 1980.

[M]   P. Major, *Multiple Wiener–Ito integrals*, Lecture Notes in Math. **849.**, Springer–Verlag, 1981.

[MvN] B. B. Mandelbrot and J. W. van Ness, Fractional Brownian motions, fractional noises and applications, *SIAM Rev.* **10** (1968).

[R]   B. Ross, A brief history and exposition of the fundamental theory of fractional calculus, in: *Fractional calculus and its applications*, Lecture notes in mathematics, **457.** Springer–Verlag, 1975.

[SR]   T. Subba Rao, On the theory of bilinear time series models, *J. of Royal Stat. Soc., Ser B,* **43** 244-255

[T]   Gy. Terdik, Stationary solutions for bilinear systems with constant coefficients, in:*Seminar on Stochastic processes, 1989. Progress in Probability,* editors E. Cinlar, K.L. Chung, R.K. Getoor Birkhäuser, Boston, 1990. 196-206.

[WTLW] W. Willinger, M. S. Taqqu, W. E. Leland and D. W. Wilson, Self–similarity in high–speed packet traffic: analysis and modeling of ethernet traffic measurements, *Statistical Science* **10/1** (1995).

[Z]   A. Zygmund, *Trigonometric Series*, Vol I, II. Cambridge, Univ. Press, 1968.

E. Igóli and Gy. Terdik
Lajos Kossuth University of Debrecen
Center for Informatics and Computing
H–4010 Debrecen, Pf. 58,
Hungary

Received September 30, 1996

On Support Theorems for Stochastic Nonlinear Partial Differential Equations

Krystyna Twardowska

1. Introduction.

The aim of this paper is to present some versions of the Stroock and Varadhan support theorem [SV] in infinite dimensions. First, a theorem on the support for some stochastic nonlinear partial differential equations is examined. We consider a model similar to that in [Pa] and [Tw1]. Second, the support theorem for the stochastic Navier-Stokes equations is given. We consider a model similar to that in [Tw2]. In the proofs of our support theorems we generalize the method of Mackevičius [Ma] and Gyöngy [Gy] to our infinite-dimensional models. In this aim we also prove some modified versions of the approximation theorems of Wong-Zakai type investigated in [Tw1] and [Tw2], respectively.

2. Definitions and notation.

Let $(\Omega, F, (F_t)_{t \in [0,T]}, P)$ be a filtered probability space on which an increasing and right-continuous family $(F_t)_{t \in [0,T]}$ of sub-σ-algebras of F is defined such that F_o contains all P-null sets in F. Let $L(X, Y)$ denote the vector space of continuous linear operators from X to Y, where X and Y are arbitrary Banach spaces (we put $L(X) = L(X, X)$); $L^p(\Omega, X)$, $\infty \geq p \geq 1$, denotes the space of equivalence classes of random variables with values in X which are p-integrable (essentially bounded for $p = \infty$) with respect to the measure P. Moreover, $L^2(X, Y)$ is the Hilbert space of Hilbert-Schmidt operators with the norm $\| \cdot \|_{HS}$, where X and Y are arbitrary separable Hilbert spaces.

Let H and K be real separable Hilbert spaces with scalar products $(\cdot, \cdot)_H$, $(\cdot, \cdot)_K$ and with orthonormal bases $\{l_n\}_{n=1}^{\infty}$, $\{k_n\}_{n=1}^{\infty}$, respectively. We consider a K-valued Wiener process $(w(t))_{t \in [0,T]}$, adapted to the family F_t, with nuclear covariance operator $J \in L(K)$. It is known that there are real-valued independent Wiener processes $\{w^j(t)\}_{j=1}^{\infty}$ on [0,T] such that $w(t) = \sum_{j=1}^{\infty} w^j(t)k_j$ almost everywhere in $(t, \omega) \in [0, T] \times \Omega$, where $\{k_j\}_{j=1}^{\infty}$ is an orthonormal basis of eigenvectors of J corresponding to eigenvalues $\{\lambda_j\}_{j=1}^{\infty}$, $\sum_{j=1}^{\infty} \lambda_j \langle \infty$, $E[\Delta w^i \Delta w^j] = (t - s)\lambda_i \delta_{ij}$ for $\Delta w^j = w^j(t) - w^j(s)$ and $s \langle t$ (δ_{ij} is the Kronecker delta). We put $w^{(m)}(t) = \sum_{j=1}^{m} w^j(t)k_j = \sum_{j=1}^{m}(w(t), k_j)_K k_j$.

Define the n-th F_t-adapted approximation of the processes $(w(t))_{t\in[0,T]}$ and $(w^m(t))_{t\in[0,T]}$, respectively, by

$$w_{(n)}(t) = \sum_{j=1}^{\infty} w_{(n)}^j(t)k_j \qquad (w_{(n)}^{(m)}(t) = \sum_{j=1}^{m} w_{(n)}^j(t)k_j), \tag{2.1}$$

with $w_{(n)}^j(t) = w^j * p_{1/n}(t) = \int_0^{1/n} w^j(t-s)p_{1/n}(s)ds$ ($*$ denotes the convolution) for $j = 1, 2, ...$, $p_{1/n}(s) = (1/n)^{-1}p((1/n)^{-1}s)$, $s \in [0, 1/n]$, where p is a nonnegative C^{∞}-function with support in $[0,1]$ and such that $\int_0^1 p(s)ds = 1$ (we put $w^j(s) = 0$ for $s\langle 0)$.

For any $z \in C([0,T], R)$, $n\rangle0$ and $p_{1/n}$ introduced above we define

$$f^{1/n}(z) = z * p_{1/n}, \quad g^{1/n}(z) = [(-(1/n)^{-1}) \vee (z \wedge (1/n)^{-1})] * p_{1/n}.$$

Let $\varphi \in C^1([0,T], K)$ be a deterministic function and $w_{1/n}^j$ be a solution of the following Volterra equation:

$$w_{1/n}^j(t) = w^j(t) - (\varphi(t), k_j)_K + g^{1/n}(w_{1/n}^j)(t). \tag{2.2}$$

By a standard Picard argument for any $j \geq 1$ there exists a unique solution $w_{1/n}^j \in C((0,T), R)$ of equation (2.2). Define

$$\xi_j^{1/n}(t) = \frac{d}{dt}\{(\varphi(t), k_j)_K - g^{1/n}(w_{1/n}^j)(t)\}, \tag{2.3}$$

$$\rho_j^{1/n} = \exp\{\int_0^T \xi_j^{1/n}(t)dw^j(t) - \frac{1}{2}\int_0^T \xi_j^{1/n}(t)^2dt\}. \tag{2.4}$$

We observe that $\xi_j^{1/n}$ are bounded functions for any $n\rangle0$ and $j \geq 1$. Then, by the Girsanov theorem, the following process

$$w_n(t) = \sum_{j=n}^{\infty} w^j(t)k_j + \sum_{j=1}^{n-1} w_{1/n}^j(t)k_j = w(t) - \sum_{j=1}^{n-1} w^j(t)k_j + \sum_{j=1}^{n-1} w_{1/n}^j(t)k_j \tag{2.5}$$

is a Wiener process on K with covariance operator J under the probability $P^n(dw) = P(dw)\rho_n(w)$, where $\rho_n = \prod_{j=1}^{n} \rho_j^{1/n}$. Indeed, from (2.2) and (2.3) we have $w_n(t) = w(t) - \sum_{j=1}^{n-1}(\varphi(t), k_j)_K - g^{1/n}(w_{1/n}^j)(t))$
$= w(t) - \sum_{j=1}^{n-1}\int_0^t \xi_j^{1/n}(s)ds$.

3. Support theorem for nonlinear SPDE.

We consider a normal triple $V \subset H = H^* \subset V^*$, where V is a real separable Banach space and H is a Hilbert space. The pairing between V and V^* is denoted by $\langle \cdot, \cdot \rangle$.

We consider the following stochastic differential equation

$$du(t) + A(t, u(t))dt + B(t, u(t))dw(t) = f(t, w(t))dt, u(0) = u_0, \qquad (3.1)$$

where $(u(t))_{t \in [0,T]}$ is an H-valued stochastic process and
(A1) $u_0 \in L^2(\Omega, F_0, P; H)$.

Moreover, we consider equations

$$du_{(n)}(t) + A(t, u_{(n)}(t))dt + B(t, u_{(n)}(t))dw_{(n)}(t) = f(t, w_{(n)}(t))dt, u_{(n)}(0) = u_0, \qquad (3.2)$$

$$d\hat{u}(t) + A(t, \hat{u}(t))dt + B(t, \hat{u}(t))dw(t) + \frac{1}{2}Idt = f(t, w(t))dt, \hat{u}(0) = u_0, \quad (3.3)$$

with $I = \widetilde{tr}(JDB(t, \hat{u}(t))B(t, \hat{u}(t)))$ to be described later in this section.

Let $P_{\hat{u}}$ be the probability law of the solution $\hat{u} = (\hat{u}(t))_{t \in [0,T]}$ to (3.3) and

$$S_1 = \text{supp } P_{\hat{u}} \text{ in } G, \qquad (3.4)$$

$$S_2 = cl\{u^\varphi = u^\varphi(u_0, \varphi), \ \varphi \in \mathbf{H}, \ u_0 \in H\} = cl\{\widetilde{S_2}\} \text{ (closure in } G), \qquad (3.5)$$

where $G = C([0, T], H)$, $\mathbf{H}$ is an arbitrary subset of H-valued absolutely continuous functions vanishing at $t = 0$ and which contains every C^∞ function $y : [0, T] \to H$ with $y(0) = 0$; $u^\varphi = (u^\varphi(t))_{t \in [0,T]}$ is the solution of the nonrandom equation corresponding to (3.1) with $dw(t)$ replaced by $\dot{\varphi}(t)dt$.

We assume ([Tw1], [Tw3]) that the family of operators $A(t, \cdot) : V \to V^*$ defined for almost every $t \in (0, T)$ and for some $p\rangle 1$ has the following properties:
(A2) growth restriction: $\exists \beta \rangle 0, \forall u \in V : \|A(t, u)\|_{V^*} \le \beta \|u\|_V^{p-1}$,
(A3) hemicontinuity: the mapping $R \ni \theta \to \langle A(t, u + \theta v), w \rangle \in R$ is continuous for all $u, v, w \in V$,
(A4) measurability with respect to t.

The family of operators $B(t, \cdot) : V \to L^2(K, H)$ defined for almost every $t \in (0, T)$ satisfies:
(A5) boundedness: $\exists \widetilde{L}\rangle 0, \forall u \in V : \|B(t, u)\|_{L^2(K,H)}^2 \le \widetilde{L}$,
(A6) the operator $B(t, \cdot) \in C_b^1$, the derivative is bounded in the Hilbert-Schmidt topology and is globally Lipschitz; the boundedness of $DB(t, \cdot)$ is meant on V in the sense of the norm in H: $\exists \widetilde{\widetilde{L}}\rangle 0, \forall u \in V, h \in H: \|DB(t, u)h\|_{HS} \le \widetilde{\widetilde{L}}\|h\|_H$,
(A7) measurability with respect to t.

Moreover, we assume
(A8) coercivity: $\exists \alpha \rangle 0, \lambda$ and $\nu, \forall u \in V : 2\langle A(t, u), u \rangle + \lambda \|u\|_H^2 + \nu \ge \alpha \|u\|_V^2 +$

$\|B(t,u)\|_{HS}^2$,

(A9) monotonicity: $\forall u, v \in V : 2\langle A(t,u) - A(t,v), u-v \rangle + \lambda\|u-v\|_H^2 \geq \|B(t,u) - B(t,v)\|_{HS}^2$,

(A10) $f \in L^{p'}((0,T) \times K; V^*)$, where $\frac{1}{p} + \frac{1}{p'} = 1$, and f is nonanticipating,

(A11) the operator $A(t, \cdot)$ and the function f are of class C_b^1, the operator $B(t, \cdot)$ is of class C_b^2 (in their standard operator norms) and the second derivative is globally Lipschitz.

Now we observe that the Fréchet derivative $DB(t, h_1)$ is in $L(V, L(K,H))$ for $h_1 \in V$ and almost every t. We consider the composition $DB(t, h_1) \circ B(t, h_1) \in L(K, L(K,H))$. Let $\Psi \in L(K, L(K,H))$ and define $B_{\widetilde{h_1}}(h, h') := (\Psi(h)(h'), \widetilde{h_1})_H \in R$ for $h, h' \in K$. By the Riesz theorem, for every $\widetilde{h_1} \in H$ there exists a unique operator $\widetilde{\Psi}(\widetilde{h_1}) \in L(K)$ such that for all $h, h' \in K$, we get $(\widetilde{\Psi}(\widetilde{h_1})(h), h')_K = (\Psi(h)(h'), \widetilde{h_1})_H$. Now, the covariance operator J has the finite trace and therefore the mapping $\widetilde{\xi} : H \ni \widetilde{h_1} \to tr(J\widetilde{\Psi}(\widetilde{h_1})) \in R$ is a linear bounded functional on H. Therefore, using the Riesz theorem we find a unique $\widetilde{\widetilde{h_1}} \in H$ such that $\widetilde{\xi}(\widetilde{h_1}) = (\widetilde{\widetilde{h_1}}, \widetilde{h_1})_H$. Denote $\widetilde{\widetilde{h_1}} = \widetilde{tr}(J\Psi)$. We observe that $(\widetilde{\widetilde{h_1}}, \widetilde{h_1})_H$ is the trace of the operator $J\widetilde{\Psi}(\widetilde{h_1}) \in L(K)$ but $\widetilde{tr}(J\Psi)$ is merely a symbol for $\widetilde{\widetilde{h_1}}$, i.e., for the correction term in equation (3.3).

The well measurable solution $u(t) \in L^p((0,T) \times \Omega, V) \cap L^2(\Omega, C((0,T), H))$ to (3.1) is understood in the following sense for $y \in Y \subset H$ (Y is an everywhere dense, in the strong topology, subset of H): $(y, u(t,\omega))_H = (y, u_0(\omega))_H - \int_0^t \langle A(s, u(s,\omega)), y \rangle ds - (y, \int_0^t B(s, u(s,\omega))dw(s))_H + \langle f(t,\omega), y \rangle$ (see [Tw1], [Tw3]). The uniqueness of the solutions is understood in the sense of trajectories. It is easy to verify that equation (3.3) containing the correction term satisfies assumptions (A8) and (A9). So we can use the Krylov and Rozovskii [KR], and Pardoux [Pa] (Theorem 3.1, p.105) results to obtain the existence and uniqueness of solution to (3.3) under assumptions (A1)-(A10). The existence and uniqueness of the solution $u_{(n)} \in L^p((0,T), V)$ of equation (3.2) such that $\dot{u}_{(n)} \in L^{p'}((0,T), V^*)$ is ensured ([Tw1], [Tw3]).

Let us denote by $V_m = H_m = V_m^*$ the vector space spanned by the vectors $l_1, ..., l_m$ and let $P_m \in L(H, H_m)$ be the orthogonal projection. We assume that $l_m \in H$ for every $m \in N$. Otherwise the equalities $V_m = H_m = V_m^*$ would not be satisfied. We extend P_m to an operator $V^* \to V_m^*$ by $\widetilde{P_m}u = \sum_{j=1}^m \langle u, l_j \rangle l_j$ for $u \in V^*$. We denote by K_m the vector space spanned by the vectors $k_1, ..., k_m$. Let $\Pi_m \in L(K; K_m)$ be the orthogonal projection. Now, we define the families of operators $A^m(t, \cdot) : V^* \to V_m^*$ by $A^m(t,u) := \widetilde{P_m}A(t,u)$ for $u \in V_m$ and $B^m(t, \cdot) : H_m \to L^2(K_m, H_m)$ by $B^m(t,u) := P_m B(t,u)$ for $u \in H_m$. Let $w^m(t)$ be the Wiener process with values in K_m defined by $w^m(t) = \Pi_m w(t)$. Moreover, we put $f^m = P_m f \in L^{p'}(\Omega \times K_m, V_m^*)$, $u_0^m = P_m u_0 \in L^2(\Omega, H_m)$.

Now, we consider the following stochastic differential equation in $\mathbb{R}^m$ for the process $v^m(t) = (v_1^m(t), ..., v_m^m(t)) \in H_m$:

$$dv_i^m(t) + A_i^m(t, v^m(t))dt + B_i^m(t, v^m(t))dw^m(t) + \frac{1}{2}I^m dt$$
$$= f_i^m(t, v^m(t))dt, \ i = 1, ..., m, \tag{3.6}$$

where $I^m = \sum_{j=1}^m \sum_{l=1}^m \lambda_j \frac{\partial B_{ij}^m(t, v^m(t))}{\partial v_l^m} B_{lj}^m(t, v^m(t))$. For every $n \in N$, we also consider (the deterministic on every $(t_{i-1}^n, t_i^n]$ for almost every $\omega \in \Omega$) approximation equation

$$dv_{(n)}^m(t) + A^m(t, v_{(n)}^m(t))dt + B^m(t, v_{(n)}^m(t))dw_{(n)}^m(t) = f(t, w_{(n)}^m(t))dt,$$
$$v_{(n)}^m(0) = u_0^m. \tag{3.7}$$

Proposition 3.1 *Let $\hat{u}(t)$ and $u_{(n)}(t)$ be the solutions of equations (3.3) and (3.2), respectively, under assumptions (A1)-(A10). Take approximations $w_{(n)}(t)$ of the Wiener process $w(t)$ given by (2.2). Then for every $\varepsilon \rangle 0$ we have $\lim_{n \to \infty} P[\sup_{0 \le t \le T} \|u_{(n)}(t) - \hat{u}(t)\|_H > \epsilon] = 0$.*

Proof. We may repeat the proof of Theorem 5.1 in [Tw1] and deduce that for each $t \in [0, T]$, $0\langle T\langle \infty$, $\lim_{n \to \infty} E[\||u_{(n)}(t) - \hat{u}(t)\|_H^2] = 0$. Put $(u_{(n)}(t) - \hat{u}(t), h)$ $= (u_{(n)}(t) - v_{(n)}^m(t), h) + (v_{(n)}^m(t) - v^m(t), h) + (v^m(t) - \hat{u}(t), h)$, $h \in H$. Now we modify Lemma 4.3 from [Tw1] (it shows that for every $t \in [0, T]$, $0\langle T\langle \infty$, $\lim_{n \to \infty} E[\|v^m(t) - \hat{u}(t)\|_H^2] = 0$). The functions $v^m(t)$, $\hat{u}(t)$ have continuous modifications which satisfy the estimates, $E[\sup_{0 \le t \le T} \|v^m(t)\|_H^2] \le C$, $E[\sup_{0 \le t \le T} \|\hat{u}(t)\|_H^2] \le C$, $C\rangle 0$ (see Theorems 4.1 and 2.3 in [KR], respectively). From this we obtain $E[\sup_{0 \le t \le T} \|v^m(t) - \hat{u}(t)\|_H^2] \le C$. Next we observe that the Wiener process satisfies the α-Hölder condition with $\alpha \langle \frac{1}{2}$. Indeed, the asymptotics of the modulus of the continuity of the Wiener path for $h\rangle 0$, $h \to 0$, is $(2h \ln \frac{1}{h})^{1/2}$, independently of $\omega \in \Omega$. Now we compute from (3.6) and from the boundedness assumptions on A^m, B^m, DB^m and f^m that $\|v^m(t_1) - v^m(t_2)\|_{L^2} \le C\Phi(t_1 - t_2)$, where $\lim_{s \to 0} \Phi(s) = 0$. Indeed,

$$\|v^m(t_1) - v^m(t_2)\|_{L^2} \le$$
$$\| \int_{t_1}^{t_2} A_i^m(s, v^m(s))ds\|_{L^2} + \| \int_{t_1}^{t_2} B_i^m(s, v^m(s))dw^{(m)}(s)\|_{L^2} +$$
$$+\| \int_{t_1}^{t_2} \sum_{j=1}^m \sum_{l=1}^m \lambda_j \frac{\partial B_{ij}^m(s, v^m(s))}{\partial v_l^m} B_{lj}^m(s, v^m(s))ds\|_{L^2} +$$
$$+\| \int_{t_1}^{t_2} f_i^m(s, w^{(m)}(s))ds\|_{L^2} \le C \mid t_1 - t_2 \mid +C(t_1 - t_2)^{1/2} + C.$$

So this family of processes satisfies the Arzeli theorem. Therefore, for every given $\varepsilon\rangle 0$ we have: (*) $\lim_{m \to \infty} P[\sup_{0 \le t \le T} \|v^m(t) - \hat{u}(t)\|_H > \epsilon] = 0$. Similarly, we modify Lemma 4.4 from [Tw1] (it shows that for each $t \in [0, T]$,

$0\langle T\langle\infty$, $\lim_{m\to\infty} E[\|v^m_{(n(m))}(t) - u_{(n(m))}(t)\|^2_H] = 0$, where $\{n(m)\}_{m=1,2,...}$ is an arbitrary sequence) to obtain, for each $t \in [0,T]$, $0\langle T\langle\infty$, and given $\varepsilon\rangle 0$: (**) $\lim_{n\to\infty} P[\sup_{0\leq t\leq T} \|v^m_{(n(m))}(t) - u_{(n(m))}(t)\|_H > \epsilon] = 0$ independently of n. Indeed, we repeat the above estimations and we also observe that $\|\int_{t_1}^{t_2} B^m_i(s, v^m(s))dw^m_{(n(m))}(s)\|_{L^2} \leq C(t_1 - t_2)^{1/2}$ because the modulus of continuity of all approximations $w^{(m)}_{(n)}(t)$ of $w^{(m)}(t)$ is smaller than the modulus of continuity of $w^{(m)}(t)$.

Now, we modify the result in [IW], Chapter VI, Section 7, Theorem 7.2, that is, we understand the convergence in the following sense: (***) $\lim_{n\to\infty} E[\sup_{0\leq t\leq T} \|v^m_{(n(m))}(t) - v^m(t)\|^2_{R^m}] = 0$, using the approximation (2.1) of $w(t)$ instead of piecewise linear intrpolation and assuming that the Wiener process has different variances in different directions, that is, it satisfies $E[\Delta w^i \Delta w^j] = (t - s)\lambda_i\delta_{ij}$. From this it follows that the expression c_{ij} defined by (7.6) in [IW], Chapter VI, Section 7, is of the form: $c_{ij} = s_{ij} + \frac{1}{2}\delta_{ij}\lambda_i$, $s_{ij} = 0$, $i, j = 1, ..., r$. Therefore, the correction term I^m is now like in (3.6).

By (*)–(***) the proof is complete.   ∎

Lemma 3.2 *Let E^n be the expected value operator with respect to the probability P^n, $n \geq 1$. Then for each $t \in [0,T]$, $0\langle T\langle\infty$, $\lim_{n\to\infty} E^n[\|u_{(n)}(t) - \widehat{u}(t)\|^2_H] = 0$.*

Proof. The proof is the simple consequence of the approximation theorem of Wong-Zakai type in [Tw1], Theorem 5.1 but instead of the fixed probability P we examine the family $\{P^n\}$, $n \geq 1$.   ∎

Theorem 3.3 *Let $u_0 \in H$. Let the operators A, B and the function $f(t)$ satisfy assumptions (A2)-(A11) and let $\widehat{u}(t)$ and $u_{(n)}(t)$ be the solutions of equations (3.3) and (3.2), respectively. Then $S_1 = S_2$.*

Proof. In order to prove that $S_1 \subseteq S_2$ we deduce from Proposition 3.1 that $P_{u_{(n)}} \to P_{\widehat{u}}$ weakly, where $P_{u_{(n)}}$ is the distribution of $u_{(n)}$ in G and $P_{\widehat{u}}$ is the distribution of $\widehat{u}$ in G. It is obvious that if $w_{(n)}(t)$ is the approximation of $w(t)$ such that $w^j_{(n)}(t) = f^{1/n}(w^j)(t)$ so $u_{(n)} \in \widetilde{S}_2$. Therefore, since the set S_2 is closed and because of the equivalent condition of the weak convergence, we obtain that $1 = \limsup_{n\to\infty} P_{u_{(n)}}(S_2) \leq P_{\widehat{u}}(S_2)$. From this we get $P_{\widehat{u}}(S_2) = 1$. Therefore, from the definition of S_1 we conclude that $S_1 \subseteq S_2$.

To prove that $S_2 \subseteq S_1$ we fix a smooth function $\varphi = w_n$ given by (2.5), $\varphi(0) = 0$, like in definition of S_2. By Lemma 3.2, and repeating the argumentation from the proof of Proposition 3.1, for a fixed $\varepsilon\rangle 0$ and a certain $n \geq 1$ we get $P^n(\sup_{0\leq t\leq T} \|u_{(n)}(t) - \widehat{u}(t)\|_H\langle\varepsilon\rangle)0$. But $P^n \ll P$, so for any $\varepsilon\rangle 0$ we obtain $P(\sup_{0\leq t\leq T} \|u_{(n)}(t) - \widehat{u}(t)\|_H\langle\varepsilon\rangle)0$. It means that $u_{(n)} \in S_1$ for any $u_{(n)} \in S_2$.

Hence $S_2 \subseteq S_1$ and finally $S_1 = S_2$, which completes the proof. ∎

4. Support theorem for Navier-Stokes equations

We consider a real separable Banach spaces V and W which are continuously and densely embedded in the Hilbert space H; $W \subset V \subset H = H^* \subset V^* \subset W^*$. The pairing between V and V^* (as well as between W and W^*) is denoted by $\langle \cdot, \cdot \rangle$. Consider now the following stochastic Navier-Stokes equation

$$du - \nu \Delta u dt + (u \cdot \Delta)u dt + \nabla p dt + B(u)dw(t) = f(t, w(t)),$$

with $u = 0$ on $\Sigma = [0, T] \times \partial O$, $u(0) = u_0$ in O, div $u = 0$ in $[0, T] \times O$, where O is an open bounded set of R^2 with a regular boundary ∂O, $u = u(t, x)$ is the velocity vector field of a fluid and $p = p(t, x)$ is the pressure. We reduce this equation to the following abstract form ([Tw2] and [Tw3], p.347-350)

$$du(t) + A(t)u(t)dt + G(u(t))dt + B(t, u(t))dw(t) = f(t, w(t))dt, u(0) = u_0 \quad (4.1)$$

where $(u(t))_{t \in [0,T]}$ is an H-valued stochastic process satisfying $(A1)$. Moreover, we consider equation

$$d\hat{u}(t) + A(t)\hat{u}(t)dt + G(\hat{u}(t))dt + B(t, \hat{u}(t))dw(t) + \frac{1}{2}I dt = f(t, w(t))dt, \hat{u}(0) = u_0. \quad (4.2)$$

We assume ([Tw2], [Tw3]) that the family of operators $A(t) \in L(V, V^*)$ defined for almost every $t \in (0, T)$ has properties $(A4)$, $(A8)$ and

$(\tilde{A}1)$ ∃ $\beta\rangle 0$, $\forall u \in V : \|A(t, u)\|_{V^*} \leq \beta \|u\|_V$.

The bilinear continuous mapping $G : V \times V \to W^*$ satisfies:

$(\tilde{A}2)$ $\forall u \in V, v \in W : \langle G(u, v), v \rangle = 0$,

$(\tilde{A}3)$ boundedness: ∃$\tilde{C}\rangle 0$, $\forall u \in V : \|G(u, v)\|_{W^*} \leq \tilde{C}\|u\|_H^{1/2}\|v\|_H^{1/2}\|u\|_V^{1/2}\|v\|_V^{1/2}$.

The family of operators $B(t, \cdot) : V \to L^2(K, H)$ defined for almost every $t \in (0, T)$ satisfies assumptions $(A5)$-$(A7)$. Finally, we assume

$(\tilde{A}4)$ $f \in L^2((0, T) \times \Omega, V^*)$ and f is nonanticipating.

Put $G(u) = G(u, u)$ for every $u \in V$. Define the predictable solution $u(t) \in L^2((0, T) \times \Omega, V) \cap L^2(\Omega, L^\infty((0, T), H))$ to (4.1) similarly like to (3.1) (see [CC], [Tw3]). The existence and uniqueness of the solutions is ensured ([Tw2]).

We prove the following support theorem analogously as Theorem 3.3 using Theorem 5.1 in [Tw2]. We only refer to [FG] (Appendix 1)for the estimates $E[\sup_{0 \leq t \leq T} \|v^m(t)\|_H^2] \leq C$, $E[\sup_{0 \leq t \leq T} \|\hat{u}(t)\|_H^2] \leq C$, $C\rangle 0$, where $v^m(t)$ is defined like in (3.6) (see [Tw2]).

Theorem 4.1 Let $u_0 \in H$. Let the operators A, B, G and the function $f(t)$ satisfy assumptions $(\tilde{A}1)$-$(\tilde{A}4)$, $(A4)$-$(A8)$,$(A11)$ and $\hat{u}(t)$ be the solution of equation (4.2). Let S_1 and S_2 be defined as in (3.4) and (3.5), respectively. Then $S_1 = S_2$.

References

[CC]   M. Capiński and N. Cutland, Stochastic Navier-Stokes equations, *Acta Applicandae Math.* 25 (1991), 59-85.

[FG]   F. Flandoli and D. Gątarek, Martingale and stationary solutions for stochastic Navier-Stokes equations. *Preprints di Matematica No. 14. Scuola Normale Superiore Pisa*, 1994, 1-33.

[Gy]   I. Gyöngy, The stability of stochastic partial differential equations and applications. Theorems on supports, in: *Lecture Notes Math.* 1390, Springer, Berlin, 1989, 91-118.

[IW]   N.Ikeda and S.Watanabe, *Stochastic Differential Equations and Diffusion Processes*, North-Holland, Amsterdam, 1981.

[KR]   N.U. Krylov and B.L. Rozovskii, On stochastic evolution equations. *Itogi Nauki i Techniki. Teor. Veroyatnost.* Moscow, 14 (1979), 71-146.

[Ma]   V. Mackevičius, On the support of a solution of stochastic differential equations, *Liet. Mat. Rink.* 26(1) (1986), 91-98 (in Russian).

[Pa]   E.Pardoux, Equations aux dérivées partielles stochastiques non linéaires monotones. *Thèse. Univ. Paris Sud*, 1975.

[SV]   D.W. Stroock and S.R.S. Varadhan, On the support of diffusion processes with applications to the strong maximum principle, in: *Proc. 6th Berkeley Sympos. Math. Statist. Probab. III, Univ. California Press, Berkeley*, 1972, 333-359.

[Tw1] K. Twardowska, An approximation theorem of Wong-Zakai type for nonlinear stochastic partial differential equations, *Stochastic Anal. Appl.* 13(5), (1995), 601-626.

[Tw2] K. Twardowska, An approximation theorem of Wong-Zakai type for stochastic Navier-Stokes equations, *Rend. Sem. Math. Univ. Padova* 96 (1996), 15-36.

[Tw3] K. Twardowska, Wong-Zakai approximations for stochastic differential equations, *Acta Applicandae Math.* 43, (1996), 317-359.

Institute of Mathematics,
Warsaw University of Technology,
Poland

Received September 30, 1996

Excitation and Performance in Continuous-time Stochastic Adaptive LQ-control[1]

Zs. Vágó

1. Introduction

The purpose of the paper is to present a new methodology for the analysis of the interaction of identification and control. Earlier works on this interaction were presented in [AW, A2, G3, Ke]. Our approach is based on techniques of stochastic complexity (cf. [GV1]).

This work is a follow-up of [GSVS], where we considered a minimum variance adaptive controller in discrete time. The novelty of the present paper is that we extend our result to continuous time systems, for which an LQ controller is applied. Also, we determine an optimal level of excitation.

Consider a linear stochastic control system described by the stochastic differential equation:

$$dy_t = a^* y_t dt + b^* u_t dt + \sigma_w dw_t, \qquad (1.1)$$

where (y_t) is the output process, (u_t) is the input process and (w_t) is a standard Wiener process. We assume, that the system is stable, i.e. $a^* < 0$. The initial state $y(0)$ is assumed to be Gaussian with mean 0. The classical LQ problem is to determine an admissible input signal (u_t), such that the following expected loss is minimal:

$$\frac{1}{T} \mathrm{E} \int_0^T (y_t^2 + \lambda u_t^2) dt, \qquad \lambda > 0.$$

It is well-known (cf. e.g. [A1]), that (u_t) is found by a static linear feedback $u_t = -k_t y_t$, where k_t converges to some limiting value k^* as $T \to \infty$, which can be computed by first finding the positive solution of the algebraic Riccati-equation

$$0 = 2a^* s + 1 - \frac{s^2 b^{*2}}{\lambda},$$

[1]This research was supported by the Hungarian Science Foundation (OTKA) T 020984.

and then setting $k^* = b^* s^*/\lambda$. In the scalar case we can get a second order equation directly for k^*:

$$\lambda k^2 - 2\lambda k a^*/b^* - 1 = 0. \tag{1.2}$$

The solution of (1.2) is denoted by $k^* = k^*(a^*, b^*)$. It is known, that the closed loop system is stable, i.e. $a^* - b^* k^*(a^*, b^*) < 0$ (cf. [FR]).

Now we consider the problem of LQ control of an unknown linear stochastic control system, i.e. we assume that a^* and b^* are unknown. A standard approach would then be to estimate a^* and b^*, get the estimated values say $\hat{a}_t$ and $\hat{b}_t$ and then define the controller by

$$u_t = -\hat{k}_t y_t$$

where $\hat{k}_t = k^*(\hat{a}_t, \hat{b}_t)$. This is called a certainty equivalence design principle. If we replace $\hat{k}_t$ by a fixed k such that $a^* - b^* k < 0$, then the parameters a^* and b^* are not identifiable because of the linear independence of y and u, hence it is impossible to improve the apriori estimations of a^* and b^*. A standard method to overcome this difficulty is to inject a known external signal, called the dither which will however cause the performance degrade. The trade-off between identifiability and performance degradation is controlled by the variance of the dither. The purpose of this paper is to analyze the effect of the dither and to determine the optimal level of excitation.

2. The frozen parameter closed loop process

Let the parameters of the true system be $\theta^* = (a^*, b^*)$ and take a tentative value $\theta = (a, b)$ with $a < 0$. Consider the closed loop system defined by the control law

$$u_t = -k(\theta)y_t + \sigma_v v_t \qquad \sigma_v \neq 0 \tag{2.1}$$

where $k(\theta) = k^*(a, b)$ is the admissible solution of the Riccati-equation $\lambda k^2 - 2\lambda ka/b - 1 = 0$ (cf. (1.2)) and v_t is an external excitation.

In this paper (v_t) is a process generated by a stable linear filter

$$dv_t = \frac{\mu}{2} v_t dt + |\mu|^{1/2} d\overline{w}_t,$$

where $(\overline{w}_t)$ is a standard Wiener process, independent of (w_t). Then $\mathrm{E}v_t^2 = 1$, independently of $\mu < 0$. This parameter can be chosen arbitrarily.

Then the closed-loop system is described by the following equations, in which $^-$ indicates that θ is "frozen":

$$d\bar{y}_t(\theta) = a^* \bar{y}_t(\theta)dt + b^* \bar{u}_t(\theta)dt + \sigma_w dw_t \tag{2.2}$$
$$\bar{u}_t(\theta) = -k(\theta)\bar{y}_t(\theta) + \sigma_v v_t. \tag{2.3}$$

Thus we get

$$d\bar{y}_t(\theta) = (a^* - b^*k(\theta))\bar{y}_t(\theta)dt + b^*\sigma_v v_t dt + \sigma_w dw_t.$$

We shall use notations $\alpha(\theta) = a^* - b^*k(\theta)$ and $\alpha^* = a^* - b^*k(\theta^*)$. We assume, that the closed loop system is stable, i.e. $\alpha^* < 0$.

3. Off-line fixed gain estimation of the parameters

In this section we define an off-line estimation of the system parameters. Consider any test value $\theta = (a, b)$ such that $a < 0$ and $a^* - b^*k(\theta) < 0$. The set of all such parameters is denoted by D. Let $D_0 \subset D$ be a compact domain such that $\theta^* \in$ int D_0. For any $\theta \in D_0$ invert the system (1.1) to reconstruct dw_t. We define $d\bar{\varepsilon}_t(\theta)$ by (cf. [GV2])

$$\sigma_w d\bar{\varepsilon}_t(\theta) = d\bar{y}_t(\theta) - a\bar{y}_t(\theta)dt - b\bar{u}_t(\theta)dt. \tag{3.1}$$

Let $\gamma > 0$ be a forgetting rate and define the cost function by

$$V_T^\gamma(\theta) = \int_0^T e^{-\gamma(T-t)}\gamma\frac{1}{2}\frac{1}{dt}(d\varepsilon_t^2(\theta) - dw_t^2),$$

where past information is weighted down by an exponential filter. It is easy to see, that $EV_T^\gamma(\theta)$ is minimized for $\theta = \theta^*$. Then the fixed gain estimator $\widehat{\theta}_T^\gamma$ of θ^* is defined as the minimizing value of the above cost function. More exactly $\widehat{\theta}_T^\gamma$ is the solution of the equation

$$\frac{\partial}{\partial\theta}V_T^\gamma(\theta) = 0,$$

if the solution is unique in D_0, otherwise we define $\widehat{\theta}_T^\gamma$ arbitrarily subject to the condition that $\widehat{\theta}_T^\gamma \in D_0$ a.s. and $\widehat{\theta}_T^\gamma$ must be a random variable.

The off-line estimator for the present problem is a purely mathematical construction, but its analysis can be useful for the understanding of adaptive control methods. Namely, it is hoped ([GGM]), that certain recursive estimation methods, that are called quasi-Newton, yield an estimator process which is very close to the off-line estimator process defined above. This connection has been completely worked out for discrete-time and decreasing gain estimators in [G1]. In analogy with [GV2] the following strong approximation theorem can be proved:

Theorem 3.1 *The estimation error can be written as*

$$\widehat{\theta}_T^\gamma - \theta^* = -(R^*)^{-1}\int_0^T e^{-\gamma(T-t)}\gamma\dot{\bar{\varepsilon}}_{\theta t}(\theta^*,\theta^*)dw_t + O_M(\gamma) + O_M(e^{-\gamma't}).$$

where

$$R^* = \mathrm{E}\dot{\bar{\varepsilon}}_{\theta t}(\theta^*)\dot{\bar{\varepsilon}}_{\theta t}^T(\theta^*).$$

Here the subscript θ denotes derivation with respect to θ.

In the above theorem we used the following convention: if (ξ_t) is a stochastic process, and (c_t) is a positive function then we write $\xi_t = O_M(c_t)$ if $\xi_t/c_t = O_M(1)$, which in turn means that the process (ξ_t/c_t) is M-bounded, i.e. for all $1 \le q < \infty$

$$M_q(\xi) = \sup_{t \ge 0} \mathrm{E}^{1/q}|\xi_t|^q < \infty.$$

Using the above theorem the covariance matrix of the estimator can be easily computed as follows; namely we have with some $\varepsilon > 0$

$$\mathrm{Cov}\ (\hat{\theta}_t^\gamma - \theta^*) = \frac{\gamma}{2}(R^*)^{-1}(1 + O(\gamma^\varepsilon)).$$

4. The matrix R^*

In this section we derive a convenient expression for R^* in terms of the system parameters and in term of level of excitation.

It is easy to see, that the gradient $\bar{\varepsilon}_{\theta t}(\theta^*)$ is the same as the gradient of the open loop residual $d\bar{y}_t - a\bar{y}_t dt - b\bar{u}_t dt$ at $\theta = \theta^*$, when the dependence of $\bar{y}$ and $\bar{u}$ on θ is depressed, i.e. $\sigma_w d\bar{\varepsilon}_{\theta,t}(\theta^*) = (-y_t dt, \ -u_t dt)^T$ with $y_t = \bar{y}_t(\theta^*), u_t = \bar{u}_t(\theta^*)$. Thus

$$\sigma_w^2 R^* = \begin{pmatrix} \mathrm{E}(y_t^2) & \mathrm{E}(y_t u_t) \\ \\ \mathrm{E}(y_t u_t) & \mathrm{E}(u_t^2) \end{pmatrix}.$$

Using $u_t = -k^* y_t + \sigma_v v_t$, with $k^* = k^*(\theta^*)$ we get that $\mathrm{E}(y_t u_t) = -k^* \mathrm{E}(y_t^2) + \sigma_v \mathrm{E}(y_t v_t)$ and $\mathrm{E}(u_t^2) = k^{*2}\mathrm{E}(y_t^2) + \sigma_v^2 - 2k^*\sigma_v \mathrm{E}(y_t v_t)$. We determine $\mathrm{E}(y_t^2)$ and $\mathrm{E}(y_t v_t)$ by considering the two dimensional system

$$\begin{pmatrix} dy_t \\ dv_t \end{pmatrix} = \begin{pmatrix} \alpha^* & b^*\sigma_v \\ 0 & \frac{\mu}{2} \end{pmatrix} \begin{pmatrix} y_t\ dt \\ v_t\ dt \end{pmatrix} + \begin{pmatrix} \sigma_w & 0 \\ 0 & |\mu|^{1/2} \end{pmatrix} \begin{pmatrix} dw_t \\ d\bar{w}_t \end{pmatrix},$$

Solving the corresponding Lyapunov-equation and separating the effect of the system noise (w_t) and the effect of dither we write

$$\sigma_w^2 R^* = \sigma_w^2 R^{*w} + \sigma_v^2 R^{*v}.$$

where

$$R^{*w} = \frac{1}{-2\alpha^*} \begin{pmatrix} 1 \\ -k^* \end{pmatrix} (1, \ -k^*)$$

$$R^{*v} = \frac{2b^*}{\alpha^*(\mu + 2\alpha^*)} \begin{pmatrix} b^* & -a^* \\ -a^* & b^*/\lambda \end{pmatrix} + \begin{pmatrix} 0 & 0 \\ 0 & 1 \end{pmatrix}.$$

We can see, that the rank of R^{*w} is 1, thus R^* becomes singular when $\sigma_v = 0$. This expression for R^{*w} is analogous with the expression obtained in discrete time in [GSVS], where v was a discrete time white noise process.

As the parameter $\mu < 0$ can be chosen arbitrarily, we shall consider the case when $|\mu| \to \infty$. Then $\mathrm{E}(y_t v_t) = 0$ and

$$\det R^* = \frac{\sigma_v^2 \big(\mathrm{E}(y_t^2) - (\mathrm{E}(y_t v_t))^2 \big)}{\sigma_w^4} = \frac{\sigma_v^2}{-2\alpha}, \tag{4.1}$$

$$\mathrm{adj}\, R^* = \begin{pmatrix} k^* \\ 1 \end{pmatrix} (k^*, \ 1) + \frac{\sigma_v^2}{\sigma_w^2} \begin{pmatrix} 1 \\ 0 \end{pmatrix} (1, \ 0). \tag{4.2}$$

5. Parameter uncertainty and performance

In the first part of this section we compute the sensitivity of the performance criterion with respect to the controller. Let us now drop the dependence of k on θ and consider the closed loop system

$$d\bar{y}_t = a^* \bar{y}_t dt + b^* \bar{u}_t dt + \sigma_w dw_t,$$
$$\bar{u}_t = -k\bar{y}_t + \sigma_v v_t,$$

where k is arbitrary, satisfying $a^* - b^* k < 0$. Let us denote the corresponding input and output processes of the above closed loop system by $\bar{y}_t(k)$ and $\bar{u}_t(k)$, and define

$$\Phi(k) = \mathrm{E}(\bar{y}_t^2(k) + \lambda \bar{u}_t^2(k)),$$

assuming stationarity. Separating the effect of (w_t) and (v_t) we write

$$\Phi(k) = \Phi^w(k)\sigma_w^2 + \Phi^v(k)\sigma_v^2. \tag{5.1}$$

We get with $\alpha = a^* - b^* k$

$$\Phi^w(k) = \frac{1 + \lambda k^2}{-2\alpha} \tag{5.2}$$

$$\Phi^v(k) = (1 + \lambda k^2) \frac{2b^{*2}}{\alpha(\mu + 2\alpha)} + \frac{4k\lambda b^*}{\mu + 2\alpha} + \lambda. \tag{5.3}$$

Let $k^*(\theta)$ denote the solution of the Riccati-equation with coefficients computed from θ. To compute the effect of parameter uncertainty onto performance assume that the estimator $\widehat{\theta} = \widehat{\theta}_s^\gamma$ for some fixed s is used for control for $t \geq s$, and investigate the difference

$$\mathrm{E}\big(\overline{y}_t(\widehat{\theta}_t^\gamma)^2 + \lambda \overline{u}_t(\widehat{\theta}_t^\gamma)^2\big) - \mathrm{E}\big(\overline{y}_t(\theta^*)^2 + \lambda \overline{u}_t(\theta^*)^2\big).$$

This difference can be handled using a second-order Taylor-series expansion. A crucial quantity is the second order sensitivity matrix defined by

$$T^* = \frac{\partial^2}{\partial \theta^2} \Phi(k^*(\theta))|_{\theta=\theta^*}.$$

Combining the techniques of [GV1, G2] we come to the following conclusion:

Theorem 5.1 *Under the conditions of Theorem 3.1 we have with some* $\varepsilon > 0$

$$\mathrm{E}\big(\overline{y}_t(\widehat{\theta}_t^\gamma)^2 + \lambda \overline{u}_t(\widehat{\theta}_t^\gamma)^2\big) - \mathrm{E}\big(\overline{y}_t(\theta^*)^2 + \lambda \overline{u}_t(\theta^*)^2\big) = \frac{\gamma}{4} \operatorname{Tr} T^*(R^*)^{-1} \sigma_w^2 (1 + o(\gamma^\varepsilon)).$$

The expression $\operatorname{Tr} T^*(R^*)^{-1}$ will be called the *normalized error of adaptation*. To get an explicit expression for it let us now consider the computation of T^*. Since Φ depends on θ through the control parameter k, T^* can be written as

$$T^* = \frac{\partial}{\partial \theta}(\Phi_k \cdot k_\theta)\big|_{\theta=\theta^*} = \Phi_{kk}(k^*)k_\theta k_\theta^T\bigg|_{\theta=\theta^*} + \Phi_k(k^*)k_{\theta\theta}\bigg|_{\theta=\theta^*}, \qquad (5.4)$$

where $k^* = k^*(\theta^*)$ and subscripts k and kk denote first and second order derivatives with respect to k.

The derivatives of $\Phi(k)$ can be computed using formulas (5.2) and (5.3). We get $\Phi_k^w(k) = -R(k)/(2b^{*2}\alpha^2)$, where $R(k) = \lambda b^{*2}k^2 - 2\lambda a^* b^* k - b^{*2}$ is the left hand side of the Riccati-equation (1.2) multiplied by b^{*2}. Thus $\Phi_k^w(k^*) = 0$ and using this fact we easily get

$$\Phi_{kk}^w(k^*) = -\frac{R'(k^*)}{2b^*\alpha^{*2}} = -\frac{\lambda}{b^*\alpha^*}$$

The derivatives of $\Phi^v(k)$ can be computing more easily, if we write it as $\Phi^v(k) = \frac{2R(k)}{\alpha(\mu+2\alpha)} + \lambda$. Since $R'(k^*) = -2\lambda\alpha^* b^*$, $R''(k^*) = 2\lambda b^{*2}$ we get $\Phi_k^v(k^*) = -4\lambda b^*/(\mu+2\alpha)$ and $\Phi_{kk}^v(k^*) = 4\lambda b^{*2}(\alpha^*(\mu+2\alpha^*)) + O(|\mu|^{-2})$. Thus we get for $|\mu| \to \infty$

$$\Phi_k(k^*) = 0$$

$$\Phi_{kk}(k^*) = \frac{\lambda\sigma_w^2}{-\alpha^*}.$$

To get the gradient of k we have to differentiate the Riccati-equation (1.2). After some computations we get

$$k_\theta(\theta^*) = \frac{k^*}{-\alpha^*} \begin{pmatrix} 1 \\ -a^*/b^* \end{pmatrix}. \tag{5.5}$$

Finally, as $\Phi_k(k^*) = 0$, (5.4), we can write T^* as

$$T^* = \Phi_{kk}(k^*)k_\theta k_\theta^T \Big|_{\theta=\theta^*} = \frac{\lambda\sigma_w^2 k^{*2}}{-\alpha^{*3}} \begin{pmatrix} 1 \\ -a^*/b^* \end{pmatrix}(1 - a^*/b^*).$$

6. Optimal choice of the dither

Due to Theorem 5.1, if we use the fixed gain estimator of the parameter with forgetting factor γ, we get that the loss in performance compared to the use of the optimal feedback gain is approximately $\frac{\gamma}{4}\mathrm{Tr}\,(T^*(R^*)^{-1})\sigma_w^2$. Thus we can compute an optimal value for the variance of the dither by minimizing the overall cost given by

$$\sigma_v^2\Phi^v(k^*) + \frac{\gamma}{4}\sigma_w^2\mathrm{Tr}\,(T^*(R^*)^{-1}).$$

This can be written as the sum of a constant term and of the *cumulative loss* due to the dither, the latter being

$$\lambda\sigma_v^2 + \frac{\gamma}{4}\frac{2\lambda k^{*2}\sigma_w^4}{b^{*2}\sigma_v^2}.$$

The minimum of this expression is achieved at

$$\sigma_{v,\mathrm{opt}}^2 = \gamma^{1/2}\sigma_w^2\frac{k^*}{|b^*|},$$

the optimal value of the cumulative loss is

$$\lambda\gamma^{1/2}\sigma_w^2\frac{k^*}{|b^*|}.$$

This expression depends on the parameters of the true system, and the value of the corresponding optimal k^*. In practice we recommend to use estimated values of a^* and b^*.

References

[A1]   K.J. Åström, *Introduction to stochastic control theory*, Academic Press, 1970.

[A2]   K.J. Åström, Matching criteria for control and identification, In C. Praagman J.W. Niueuwenhuis and H.L. Trentelman, editors, *Proceedings of the 2nd European Control Conference, Groningen, Holland*, pages 248–251, 1993.

[AW]   K.J. Åström and B. Wittenmark, Problems of identification and control, *J. Math. Anal. Appl.*, **34** (1971), 90–113.

[FR]   W.H. Fleming and R.W. Rishel, *Deterministic and Stochastic Optimal Control*, Springer-Verlag, 1975.

[G1]   L. Gerencsér, Strong approximation results in estimation and adaptive control, In L. Gerencsér and P.E. Caines, editors, *Topics in Stochastic Systems: Modelling, Estimation and Adaptive Control*, pp. 268–299, Springer-Verlag Berlin, Heidelberg, 1991.

[G2]   L. Gerencsér, Predictive stochastic complexity associated with fixed gain estimators, In C. Praagman J.W. Niueuwenhuis and H.L. Trentelman, editors, *Proceedings of the 2nd European Control Conference, Groningen*, (1993), 1673–1677.

[G3]   M. Gevers, Identification for control, In Cs.Bányász, editor, *5-th IFAC Symposium on Adaptive Systems in Control and Signal Processing* (1995), 1–12.

[GGM]   L. Gerencsér, I. Gyöngy, and Gy. Michaletzky, Continuous-time recursive maximum-likelihood method. A new approach to Ljung's scheme, In L. Ljung and K.J. Åström, editors, *Proc. of the 9th Triennial World Congress of IFAC, Budapest*, pp. 75–77. Pergamon Press, Oxford, 1984.

[GV1]   L. Gerencsér and Zs. Vágó, Model selection in continuous time, In *Proceedings of the 30th CDC*, (1991), 959–962.

[GV2]   L. Gerencsér and Zs. Vágó, Fixed gain estimation in continuous time. *Acta Applicaende Mathematica*, **35** (1994), 153–164.

[GSVS] L. Gerencsér, J.H. van Schuppen, Zs. Vágó, and M. Sétáló, *Performance of a Minimum Variance Adaptive Controller with Noise Injection*, 44–48. 1995.

[Ke]   L. Keviczky, Combined identification and control: another way, In Cs.Bányász, editor, *5-th IFAC Symposium on Adaptive Systems in Control and Signal Processing*, (1995), 13–30.

Computer and Automation Institute of
the Hungarian Academy of Sciences
H-1111 Budapest, Kende 13-17, Hungary.

Received September 30 ,1996

Invariant Measures for Diffusion Processes
in Conuclear Spaces

Jie Xiong

1. Introduction

Let Φ be a nuclear space whose topology is given by an increasing sequence $\|\cdot\|_p$, $p \geq 0$, of Hilbertian norms. Let Φ_p be the completion of Φ with respect to the norm $\|\cdot\|_p$. Let Φ_{-p}, Φ' be the dual spaces of Φ_p, Φ respectively.

Let X be a Φ'-valued process governed by the following stochastic differential equation (SDE):

$$X_t = X_0 + \int_0^t A(s, X_s)ds + \int_0^t B(s, X_s)dW_s \qquad (1.1)$$

where $A : \mathbf{R}_+ \times \Phi' \to \Phi'$ and $B : \mathbf{R}_+ \times \Phi' \to L(\Phi', \Phi')$ are two measurable mappings, X_0 is a Φ'-valued random variable and W is a Φ'-valued Wiener process with covariance Q. For the definition of the Φ'-valued Wiener process, we refer the reader to the book of Kallianpur and Xiong [KX2].

Motivated from neurophysiological applications, the existence and uniqueness for the solution of the SDE (1.1) has been obtained by Kallianpur, Mitoma and Wolpert [KMW]. Based on diffusion approximation techniques, Kallianpur and Xiong [KX1] established the same results under weaker conditions.

In this paper, we first characterize the invariant measure of the Markov process X when the mappings A, B do not depend on the time variable t. Let $\bar{A}$ be the generator of X and let μ be a Borel probability measure on Φ' such that

$$\int_{\Phi'} (\bar{A}F)(v)\mu(dv) = 0 \qquad \forall F \in \mathrm{Dom}(\bar{A}). \qquad (1.2)$$

Then μ is an invariant measure of X. However, the domain of $\bar{A}$ is difficult to describe. On the other hand, it follows from Itô's formula that "smooth" cylinder functions F are in the domain of $\bar{A}$ and $\bar{A}F$ can be easily calculated. The aim of the present article is to show that if (1.2) holds for all "smooth" cylinder functions F, then μ is an invariant measure for the process.

Secondly, we are interested in $\mathcal{L}(X_t)$, the distribution of X_t. We shall prove that $\mathcal{L}(X_t)$ is the unique solution of a deterministic differential equation in measure space.

2. Preliminaries

In this section, we state some results concerning the solution of the SDE (1.1). We refer the reader to [KX2] for the proofs.

Let $\theta_p : \Phi_{-p} \to \Phi_p$ be the isometry defined in [KX2]. Let $H_Q \subset \Phi'$ be a Hilbert space such that W is an H_Q-cylinder Brownian motion. To establish the existence and uniqueness for the solution of (1.1), we make the following assumptions

(D): $\forall T > 0$, $\exists p_0 = p_0(T)$ such that $\forall p \geq p_0$, $\exists q = q(p) \geq p$ and $K_T \in L^1([0, T])$ with the following properties:

(D1) (Continuity) $\forall t \in [0, T]$, the maps $A(t, \cdot) : \Phi_{-p} \to \Phi_{-q}$ and $B(t, \cdot) : \Phi_{-p} \to L_{(2)}(H_Q, \Phi_{-p})$ are continuous.

(D2) (Coercivity) $\forall t \in [0, T]$ and $\phi \in \Phi$, we have

$$2A(t, \phi)[\theta_p \phi] \leq K_T(t) \left(1 + \|\phi\|_{-p}^2\right).$$

(D3) (Growth) $\forall t \in [0, T]$ and $v \in \Phi_{-p}$, we have

$$\|A(t, v)\|_{-q}^2 \leq K_T(t) \left(1 + \|v\|_{-p}^2\right)$$

and

$$\|B(t, v)\|_{L_{(2)}(H_Q, \Phi_{-p})}^2 \leq K_T(t) \left(1 + \|v\|_{-p}^2\right).$$

(D4) (Monotonicity) $\forall t \in [0, T]$, $v_1, v_2 \in \Phi_{-p}$, we have

$$2 \langle A(t, v_1) - A(t, v_2), v_1 - v_2 \rangle_{-q}$$
$$+ \|B(t, v_1) - B(t, v_2)\|_{L_{(2)}(H_Q, \Phi_{-q})}^2 \leq K_T(t) \|v_1 - v_2\|_{-q}^2.$$

(D5) (Initial) $\exists r_0$ such that $E\|X_0\|_{-r_0}^2 < \infty$.

When $\sup_{t \in [0, T]} K_T(t) < \infty$, $\forall T > 0$, the following theorem has been proved in [KX2]. The arguments employed there also apply to the present case.

Theorem 2.1 *Under Assumptions (D), the SDE (1.1) has a unique solution X. Further, $\forall T > 0$, $X|_{[0,T]} \in C([0, T], \Phi_{-p_1(T)})$ a.s. and*

$$E \sup_{0 \leq t \leq T} \|X_t\|_{-r_1(T)}^2 < \infty$$

where $r_1(T) = \max(p_0(T), r_0)$ and $p_1(T) \geq r_1(T)$ such that the canonical injection from $\Phi_{-r_1(T)}$ to $\Phi_{-p_1(T)}$ is Hilbert-Schmidt.

Now we consider the martingale problem corresponding to (1.1). Let

$$\mathcal{D}_0^2(\Phi') = \left\{ F : \Phi' \to \mathbf{R} \; \begin{array}{l} \exists h \in C_0^2(\mathbf{R}) \text{ and } \phi \in \Phi \text{ such} \\ \text{that } F(v) = h(v[\phi]), \; \forall v \in \Phi' \end{array} \right\}.$$

For $F \in \mathcal{D}_0^2(\Phi')$, consider a map $\mathcal{A}_t F : \Phi' \to \mathbf{R}$ defined by

$$\mathcal{A}_t F(v) \equiv A(t,v)[\phi] h'(v[\phi]) + \frac{1}{2} Q \left(B(t,v)'\phi, B(t,v)'\phi \right) h''(v[\phi]) \quad (2.1)$$

where $B(t,v)' : \Phi \to \Phi$ is the dual operator of $B(t,v)$. For a measurable Φ'-valued function y on $[0, \infty)$, let

$$M^F(y)_t \equiv F(y_t) - F(y_0) - \int_0^t \mathcal{A}_s F(y_s)ds. \quad (2.2)$$

We shall denote the probability measure $\mathcal{L}(X_0)$ on Φ_{-r_0} by ν_0 and make the Assumptions (D) throughout the rest of this article.

Definition 2.2 *A Φ'-valued progressively measurable process Y is a solution to the martingale problem for $(\mathcal{A}_t, \nu_0)$ if $\mathcal{L}(Y_0) = \nu_0$ and $\forall F \in \mathcal{D}_0^2(\Phi')$, $M^F(Y)$ is a martingale.*

If Y has continuous paths a.s., then Y is a solution to the $C([0, \infty), \Phi')$-martingale problem for $(\mathcal{A}_t, \nu_0)$. D-martingale problem can be defined similarly.

Theorem 2.3 *Let $T > 0$ and $p \geq p_1(T)$. Regard ν_0 as a probability measure on Φ_{-p}. Suppose that Y is a solution to the $C([0, T], \Phi_{-p})$-martingale problem for $(\mathcal{A}_t, \nu_0)$ such that*

$$E \sup_{0 \leq t \leq T} \|Y_t\|_{-p}^2 < \infty.$$

Then Y has a $\Phi_{-p_1(T)}$-continuous modification which is equal to $X|_{[0,T]}$ in distribution.

3. Martingale problem

In this section, we consider the martingale problem for $(\mathcal{A}_t, \nu_0)$ in detail. We shall replace the conditions of Theorem 2.3 by weaker conditions. For simplicity of notations, we assume that $T = 1$ and write $K_T(t)$, $p_1(T)$, $r_1(T)$ by $K(t)$, p_1, r_1 respectively.

Let $p \geq p_1$ and let Y be a Φ_{-p}-valued solution to the martingale problem for $(\mathcal{A}_t, \nu_0)$. Let

$$\tau_n = \inf \left\{ t \in [0,1] : \int_0^t (1 + K(s)) \|Y_s\|_{-p}^2 ds > n \right\}, \qquad (3.1)$$

and $\tau_n = \infty$ if the set on the right hand side of the above equation is empty.

Lemma 3.1 *If $\tau_n \to \infty$ a.s., then $\exists r \geq p$ such that*

$$E \sup_{0 \leq t \leq 1} \|Y_t\|_{-r}^2 < \infty. \qquad (3.2)$$

Sketch of the proof: Let $\{h_m\} \subset C_0^2(\mathbf{R})$ be an increasing sequence such that $h_m(x) \to x^2$, $h'_m(x) \to 2x$, $h''_m(x) \to 2$ and

$$\sup\{|x^{-1}h'_m(x)| + |h''_m(x)| : x \in \mathbf{R}, n \geq 1\} < \infty.$$

By the martingale property of $M^{F_m}(Y)$, we have

$$
\begin{aligned}
Eh_m(Y_{t \wedge \tau_n}[\phi]) &= Eh_m(Y_0[\phi]) + E \int_0^{t \wedge \tau_n} h'_m(Y_s[\phi]) A(s, Y_s)[\phi] ds \\
&\quad + E \int_0^{t \wedge \tau_n} \frac{1}{2} h''_m(Y_s[\phi]) Q \left(B(s, Y_s)'\phi, B(s, Y_s)'\phi \right) ds.
\end{aligned}
$$

Taking $m \to \infty$, we have

$$
\begin{aligned}
EY_{t \wedge \tau_n}[\phi]^2 &= EY_0[\phi]^2 + E \int_0^{t \wedge \tau_n} 2Y_s[\phi] A(s, Y_s)[\phi] ds \qquad (3.3) \\
&\quad + E \int_0^{t \wedge \tau_n} Q \left(B(s, Y_s)'\phi, B(s, Y_s)'\phi \right) ds.
\end{aligned}
$$

Let $q \geq p$ be given by Assumptions (D) and $r' \geq q$ be such that the canonical injection from Φ_{-q} to $\Phi_{-r'}$ is Hilbert-Schmidt. Taking $\phi = \phi_j^{r'}$ and adding on both side of (3.3), we have

$$E\|Y_{t \wedge \tau_n}\|_{-r'}^2 \leq E\|Y_0\|_{-r'}^2 + \int_0^t K(s)(1 + E\|Y_{s \wedge \tau_n}\|_{-r'}^2) ds. \qquad (3.4)$$

It follows from Gronwall's inequality and $\tau_n \to \infty$ that

$$\sup_{0 \leq t \leq 1} E\|Y_t\|_{-r'}^2 < \infty.$$

Making use of the martingale properties of $M^{F^2}(Y)$ and $M^F(Y)$, it follows from Itô's formula that

$$\langle M^F(Y) \rangle_t = \int_0^t Q \left(B(s, Y_s)'\phi, B(s, Y_s)'\phi \right) (h'(Y_s[\phi]))^2 ds. \qquad (3.5)$$

Applying Doob's inequality to $M^F(Y)$, we have

$$E \sup_{0 \leq t \leq 1} F(Y_t)^2 \leq 16Eh(Y_0[\phi])^2 + 16E \int_0^1 (h'(Y_s[\phi]))^2 (A(s, Y_s)[\phi])^2 ds$$

$$+4E \left(\int_0^1 |h''(Y_s[\phi])| Q \left(B(s, Y_s)'\phi, B(s, Y_s)'\phi \right) ds \right)^2$$

$$+64E \int_0^1 Q \left(B(s, Y_s)'\phi, B(s, Y_s)'\phi \right) (h'(Y_s[\phi]))^2 ds \quad (3.6)$$

It then follows from similar arguments as those leading to (3.4) that

$$E \sup_{0 \leq t \leq 1} \|Y_t\|_{-r}^2 \leq 16E\|Y_0\|_{-r}^2 + 80 \int_0^1 K(s)ds \sup_{0 \leq t \leq 1} (1 + E\|Y_s\|_{-r}^2) < \infty$$

where $r \geq r'$ such that the canonical injection from $\Phi_{-r'}$ to Φ_{-r} is Hilbert-Schmidt. ∎

Theorem 3.2 *If $\tau_n \to \infty$ a.s., then Y has a Φ_{-p_1}-continuous modification which is equal to $X|_{[0,1]}$ is distribution. As a consequence, the $D([0,1], \Phi_{-p})$-martingale problem for (A_t, ν_0) is well-posed.*

Sketch of the proof: Let $r'' \geq r$ be such that the canonical injection from Φ_{-r} to $\Phi_{-r''}$ is compact. Making use of (3.2), it follows from Theorem 3.6 in (Ethier and Kurtz [EK], p178) that Y has a $D([0,1], \Phi_{-r''})$-modification. Keeping the same notation, we now show that $Y \in C([0,1], \Phi_{-r''})$ a.s.. By (3.2) again, we only need to prove that $\forall \phi \in \Phi$, $Y.[\phi] \in C([0,1], \mathbf{R})$ a.s..

Let $\sigma_n = \inf \{t \in [0,1] : \|Y_t\|_{-r''} > n\}$. Since $Y \in D([0,1], \Phi_{-r''})$ a.s., $\{\sigma_n\}$ increases to ∞ a.s.. Note that $Y._{\wedge \sigma_n}[\phi]$ is an Itô process with covariance a and drift b given by

$$a(t) = Q \left(B(t, Y_t)'\phi, B(t, Y_t)'\phi \right) 1_{t < \sigma_n} \text{ and } b(t) = A(t, Y_t)[\phi] 1_{t < \sigma_n}.$$

It follows from Exercise 4.6.3 in (Stroock and Varadhan [SV], p92) that $Y._{\wedge \sigma_n}[\phi] \in C([0,1], \mathbf{R})$. Letting $n \to \infty$ we see that $Y.[\phi] \in C([0,1], \mathbf{R})$ a.s.

The conclusion of the theorem then follows from Theorem 2.3. ∎

4. Main results

Since $\mathcal{D}_0^2(\Phi')$ is not an algebra, we extend A_t to a larger domain $\mathcal{C}_0^2(\Phi')$ by

$$\mathcal{C}_0^2(\Phi') = \left\{ F : \Phi' \to \mathbf{R} \quad \begin{array}{l} \exists m \in \mathbf{N}, \ h \in C_0^2(\mathbf{R}^m) \text{ and } \phi_1, \cdots, \phi_m \in \Phi \\ \text{such that } F(v) = h(v[\phi_1], \cdots, v[\phi_m]), \ \forall v \in \Phi' \end{array} \right\}$$

and

$$A_t F(v) \equiv \sum_{i=1}^{m} A(t,v)[\phi_i] \partial_i h (v[\phi_1], \cdots, v[\phi_m]) \tag{4.1}$$

$$+ \frac{1}{2} \sum_{i,j=1}^{m} \partial_{ij} h (v[\phi_1], \cdots, v[\phi_m]) Q (B(t,v)'\phi_i, B(t,v)'\phi_j).$$

It is easy to show that X is still a solution to the martingale problem for (A_t, ν_0).

Let $\mathcal{P}^2(\Phi_{-p})$ be the collection of all Borel probability measures μ on Φ_{-p} with finite second moment $\gamma_p^2(\mu) = \int \|v\|_{-p}^2 \mu(dv)$. Then $\mathcal{P}^2(\Phi_{-p})$ is a metric space with metric d_p given by

$$d_p^2(\rho_1, \rho_2) = \inf \left\{ \begin{array}{l} \int_{\Phi_{-p}} \int_{\Phi_{-p}} \|x - y\|_{-p}^2 R(dxdy) : \\ R(dx, \Phi_{-p}) = \rho_1(dx), R(\Phi_{-p}, dy) = \rho_2(dy) \end{array} \right\},$$

$\forall \rho_1, \rho_2 \in \mathcal{P}^2(\Phi_{-p})$. Note that $\{\rho_n\} \subset \mathcal{P}^2(\Phi_{-p})$ is relatively compact if it is relatively compact in weak topology and $\{\gamma_p(\rho_n)\}$ is bounded. For this fact and other related properties for probability metrics, we refer the reader to Szulga [Sz], Tanaka [Ta] and Zolotarev [Zo].

Now we assume that A, B do not depend on t and denote A_t by A.

Theorem 4.1 *If* $\mu \in \mathcal{P}^2(\Phi_{-r_0})$ *and*

$$\int_{\Phi'} AF(v)\mu(dv) = 0 \qquad \forall F \in C_0^2(\Phi'), \tag{4.2}$$

then μ *is an invariant measure of* X.

Proof: It is easy to verify the conditions of Theorem 3.1 in Bhatt and Karandikar [BK] for A. Therefore, there exists a stationary solution Y to the martingale problem for (A, μ). Note that

$$E \int_0^1 (1 + K(s))\|Y_s\|_{-p_1}^2 ds = \int_0^1 (1 + K(s))ds \gamma_{p_1}^2(\mu) < \infty$$

and hence, $\tau_n \to \infty$ a.s.. Let X be the solution of (1.1) with initial measure μ. Then, by Theorem 3.2, X is a continuous modification of Y and hence, μ is an invariant measure of X. $\blacksquare$

Theorem 4.2 *Let* $p \geq p_1$ *and* $\nu \in C([0,1], \mathcal{P}^2(\Phi_{-p}))$. *If* $\forall t \in [0,1]$, $F \in C_0^2(\Phi')$

$$\langle \nu_t, F \rangle = \langle \nu_0, F \rangle + \int_0^t \langle \nu_s, AF \rangle \, ds, \tag{4.3}$$

then $\nu_t = \mathcal{L}(X_t)$, $\forall t \in [0,1]$.

Sketch of the proof: Extend ν to $t \in [0, \infty)$ by $\nu_{t+1} \equiv \mathcal{L}(X_t^1)$ where X_t^1 is the solution of (1.1) with initial distribution ν_1. Then $\nu \in C([0, \infty), \mathcal{P}^2(\Phi_{-p}))$ and (4.3) holds for $t \in [0, \infty)$. Further, making use of Itô's formula for X_t^1, we can show that $\gamma_p^2(\nu_t) \le K_1 e^{K_2 t}$, $\forall t \ge 0$, where K_1, K_2 are two constants.

For $\lambda > K_2$, $g_1 \in C_0^2(\Phi')$ and $g_2 : \{-1, 1\} \to \mathbf{R}$, let

$$\tilde{\mathcal{A}}(g_1 g_2)(v, n) = g_2(n)(\mathcal{A} - \lambda)g_1(v) + \lambda g_2(-n) \langle \nu_0, g_1 \rangle.$$

Let $\mu \in \mathcal{P}^2(\Phi_{-p} \times \{-1, 1\})$ be defined by

$$\mu = \lambda \int_0^\infty e^{-\lambda t} \nu_t dt \times \left(\frac{1}{2}\delta_1 + \frac{1}{2}\delta_{-1} \right). \tag{4.4}$$

Then $\int \tilde{\mathcal{A}}(g_1 g_2) d\mu = 0$ and hence, there exists a stationary solution (Y, N) to the martingale problem for $(\tilde{\mathcal{A}}, \mu)$. Similar to the proof of Theorem 3.2, we can show that (Y, N) has a $D([0, \infty), \Phi_{-r''} \times \{-1, 1\})$-modification. By Theorem 10.3 in ([EK], p257), we see that the $D([0, \infty), \Phi_{-r''} \times \{-1, 1\})$-martingale problem for $(\tilde{\mathcal{A}}, \mu)$ is well-posed.

Let $S_k = \eta_1 + \cdots + \eta_k$ where $\{\eta_k\}$ is an i.i.d. sequence with $Poisson(\lambda)$ as its common distribution. Let $\{X^k\}_{k \ge 1}$ be a sequence of independent copies of X. Let X^0 be the solution of (1.1) with X_0 and W replaced by a random variable X_0^0 and a Wiener process W' respectively such that X^0 is independent of $\{X^k\}_{k \ge 1}$ and $\mathcal{L}(X_0^0) = \mathcal{L}(Y_0)$. Define

$$\tilde{N}_t = (-1)^k N_0, \quad \tilde{Y}_t = X_{t-S_k}^k \quad \forall t \in [S_k, S_{k+1}),\ k \ge 0.$$

It is easy to show that $(\tilde{Y}, \tilde{N})$ is a strong Markov process and a solution to the $D([0, \infty), \Phi_{-r''} \times \{-1, 1\})$-martingale problem for $(\tilde{\mathcal{A}}, \mu)$. Therefore, by the same arguments as in Proposition 9.19 in ([EK],p252) that the stationary distribution is unique.

Let $\tilde{\nu}_t = \mathcal{L}(X_t)$. Then $\tilde{\nu}$ is a solution of (4.3). Let $\tilde{\mu}$ be defined by (4.4) with ν replaced by $\tilde{\nu}$. Then $\mu = \tilde{\mu}$ and hence,

$$\int_0^\infty e^{-\lambda t} \nu_t dt = \int_0^\infty e^{-\lambda t} \tilde{\nu}_t dt, \quad \forall \lambda > K_2.$$

Therefore, $\nu_t = \tilde{\nu}_t$, $\forall t \in [0, 1]$. ∎

Note that the $\mathbf{R}_+ \times \Phi'$-valued process (t, Y_t) is a solution to the martingale problem for $\left(\frac{\partial}{\partial t} + \mathcal{A}_t, \delta_0 \times \nu_0 \right)$. The following theorem can be proved by similar arguments as in the proof of Theorem 4.2. We refer the reader to Horowitz and Karandikar [HK] for a similar treatment where they have studied the locally compact case.

Theorem 4.3 *Let* $p \geq p_1$ *and* $\nu \in C\big([0,1], \mathcal{P}^2(\Phi_{-p})\big)$. *If* $\forall t \in [0,1]$, $F \in \mathcal{C}_0^2(\Phi')$

$$\langle \nu_t, F \rangle = \langle \nu_0, F \rangle + \int_0^t \langle \nu_s, \mathcal{A}_s F \rangle \, ds, \qquad (4.5)$$

then $\nu_t = \mathcal{L}(X_t)$, $\forall t \in [0,1]$.

References

[BK]   A. Bhatt and R. Karandikar, Invariant measures and evolution equations for Markov processes characterized via martingale problem, *Annals of Probability*, **21** (1993), 2246–2268.

[KMW] G. Kallianpur, I. Mitoma and R. L. Wolpert, Diffusion equation in duals of nuclear spaces, *Stochastics*, **29** (1990), 1–45.

[EK]   S. Ethier and T. Kurtz, *Markov Processes: Characterization and Convergence*, John Wiley and Sons, 1986.

[HK]   J. Horowitz and R.L. Karandikar, Martingale problems associated with the Boltzman equation, in *Seminar on Stochastic Processes*, ed. E. Cinlar, K.L. Chung and R.K. Getoor, Birkhäuser, Boston, 1990, 75–122,

[JS]   J. Jacod and A.N. Shiryaev, *Limit theorems for stochastic processes*, Springer-Verlag, Berlin, 1987.

[KX1]   G. Kallianpur and J. Xiong, Diffusion approximation of nuclear space-valued stochastic differential equations driven by Poisson random measures, *Annals of Applied Probability*, **5** (1995), 493–517.

[KX2]   G. Kallianpur and J. Xiong, *Stochastic Differential Equations in Infinite Dimensional Spaces*, IMS Lecture Notes-Monograph Series **26**, Institute of Mathematical Statistics, 1995.

[SV]   D.W. Stroock and S.R.S. Varadhan, *Multidimensional Diffusion Processes*, Springer-Verlag, 1979.

[Sz]   A. Szulga, On the Wasserstein metric, *Transactions of the 8th Prague Conference*, (1978), 267–273.

[Ta]   H. Tanaka, Probabilistic treatment of the Boltzmann equation of Maxwellian molecules, *Z. Wahrscheinlichkeitstheorie verw. Gebiete*, **66** (1978), 67–105.

[Zo]   V.M. Zolotarev, Probability metrics, *Theory of Probability and Its Applications*, **28** (1983), 278–302.

University of Tennessee
Knoxville, TN 37996-1300

Received September 5, 1996

Degree Theory on Wiener Space and an Application to a Class of SPDEs

A. Süleyman Üstünel and Moshe Zakai

Let T be a C^1 map from $\mathbb{R}^n$ to $\mathbb{R}^n$ and proper (i.e. the inverse image of each compact set is compact). The degree theorem for this map states that for any bounded real valued function $\varphi(x), x \in \mathbb{R}^n$ with compact support

$$\int_{\mathbb{R}^n} J(x)\varphi(Tx)dx = q \int_{\mathbb{R}^n} \varphi(x)dx \tag{1}$$

where q, the degree is an integer and does not depend on φ, $J(x)$ is the Jacobian of T : $\det\left(\frac{\partial T_i(x)}{dx_j}\right)_{n \times n}$. The degree q satisfies

$$q = \sum_{y \in T^{-1}\{x\}} \text{sign } \det\left(\frac{\partial T_i(y)}{\partial y_j}\right)_{n \times n} \tag{2}$$

for almost all $x \in \mathbb{R}^n$. The notion of degree was extended in several directions and in particular, applied to establishing the existence of a solution x to equations of the type $f(x) = x$. In 1934 the notion of degree was extended by Leray and Schauder to a class of transformations on Banach space and applied to the proof of existence of solutions to certain partial differential equations [De], [FG].

The possibility of extending the theory of degree to Wiener space was first pointed out by Eels and Elworthy. In 1986 E. Getzler [Ge] introduced the notion of degree for the shift transformations $Tw = w + u(w)$, where u is a Cameron-Martin space-valued random variable. The results were improved in [Ku] and [ÜZ]. The results derived in these papers were extensions of (1) and (2) under strong integrability assumptions.

In the present work we show the following result on the abstract Wiener space (W, H, μ):

Theorem A *Let $u(w)$ be an H-valued random variable, $Tw = w + u(w)$. Assume that*

(1) *for a.a.w, $h \mapsto u(w + h)$ is a compact map on H.*

(2) *for almost all w, for any $h_0 \in H$*

$$\sup\left\{|h| \ : \ h_0 = h + u(w + h)\right\} < \infty$$

(this condition is satisfied if, e.g., $\varliminf |h_n + u(w + h_n)|_H \longrightarrow \infty$ whenever $|h_n| \to \infty$).

Let D_n denote an increasing sequence of bounded subsets of H, $D_n \nearrow H$ as $n \to \infty$. Then, as n goes to $+\infty$

$$\deg\left(T(w + h) - w,\, D_n,\, h_0\right) \longrightarrow q$$

almost surely, where q is a non random constant. If $q \neq 0$, then $\{\nu : T(\nu) = w\}$ is a.s. non empty and the equation for $\nu : \nu + u(\nu) = w$ has a measurable solution $\nu(w)$.
If (2) is replaced by (2)':

(2)' *for a.a. w, $|u(w + h)|_H = o\left(|h|_H\right)$ as $|h|_H \to \infty$, then (2) is also satisfied and $q = 1$.*

Combining Theorem A with a Sard inequality and previously known results yields:

Theorem B *Assume that $u : W \to H$ is $H - C$-compact and $H - C^1_{loc}$ (i. e., there exists a positive random variable ρ such that $\mu\{\rho(w) > 0\} = \mu(Q) = 1$ and $h \mapsto u(w + h)$ is C^1 on the open set $\{h \in H : |h|_H < \rho(w)\}$) with $Q + H \subset Q$ (cf. [ÜZ]). Suppose also that $\sup(|h|_H : h + u(w+h) = h_0) < \infty$ almost surely for any given $h_0 \in H$ and suppose also that the degree of T corresponding to u is nonzero, then μ is absolutely continuous with respect to $T^*\mu$. Denote by S the measurable right inverse of T whose existence was proved in the previous theorem. Then μ is absolutely continuous with respect to $T^*\mu$. Letting L be the corresponding Radon-Nikodym derivative, we have the following Girsanov-type identity:*

$$E[f \circ T\, L \circ T] = E[f]$$

for any $f \in C_b^+(W)$. Moreover the following hold:

i)

$$\mu(A) \leq \int \mathbf{1}_A \circ T |\Lambda_u| d\mu,$$

in particular $\mu \ll T^*\mu|_M$ (hence $\mu \approx T^*\mu|_M$) with

$$\frac{du}{dT^*\mu|_M}(w) = \left[\sum_{y \in T^{-1}\{w\} \cap Q} \frac{1}{|\Lambda_u(y)|}\right]^{-1},$$

where

$$\Lambda_u = det_2(I_H + \nabla u) \exp -\delta u - \frac{1}{2}|u|_H^2$$

and $det_2(I_H + \nabla u)$ denotes the modified Carleman-Fredholm determinant.

ii)

$$u(S^{-1}(A)) \leq \int 1_A |\Lambda_A| \Lambda_u d\mu,$$

for an $A \in \mathcal{B}(W)$. In particular $S^*\mu \ll \mu|_M$ and we have

$$\frac{dS^*\mu}{d\mu|_M}(w) = |\Lambda_u(w)| 1_{S(W)}(w),$$

for almost all $w \in W$.

As an application to a class of SPDEs, consider first the following corollary to Theorem A:

Corollary A *Let K be a linear Hilbert-Schmidt operator on H and $G(h)$ a continuous function from H to itself such that $|G(h)|_H = o(|h|_H)$ as $|h|_H \to \infty$. Then the equation for y:*

$$y + KG(y) = \tilde{K}w \tag{3}$$

where $\tilde{K}$ is the natural extension of K to the abstract Wiener space, possesses a measurable H-valued solution.

Proof: Consider the shift:

$$Tw = w + G(\tilde{K}w). \tag{4}$$

Since K is compact, $G(\tilde{K}w)$ satisfies the requirements on $u(w)$ in Theorem A. Therefore the equation $T\nu = w$ has a measurable solution, say $\nu = T^{-1}w$. Replacing w with $T^{-1}w$ in (4) and operating with $\tilde{K}$ yields

$$\begin{aligned}\tilde{K}w &= \tilde{K}\left(T^{-1}w + G(\tilde{K}T^{-1}w)\right) \\ &= \tilde{K}T^{-1}w + KG(\tilde{K}T^{-1}w)\end{aligned}$$

comparing with (3) yields that $y = \tilde{K}T^{-1}w$ solves (3).

Let $\mathcal{D}$ be a bounded domain in $\mathbb{R}^d$, $\dot{w}$ a white noise on $\mathbb{R}^d$ and H_o will denote the Hilbert space of real valued functions on $L^2(\mathcal{D})$. Let $g(x,r)$, $x \in \mathcal{D}$, be real valued and such that for any $f \in H_o$, $g\left(x, f(x)\right)$, $x \in R^d$ is a continuous and bounded transformation from H_o to itself satisfying the assumption on G imposed in Corollary A with H replaced by H_o. We want to consider the stochastic partial differential equation

$$- \Delta \xi(x) + g\Big(x, \, \xi(x)\Big) = \dot{w}, \ x \in \mathcal{D} \tag{5}$$

where Δ is the Laplace operator on $\mathcal{D}$ with the boundary condition

$$\xi\Big|_{\partial \mathcal{D}} = 0 \, . \tag{6}$$

We restrict d to be 1,2 or 3 since, as shown by Buckdahn and Pardoux [BP], $K = (-\Delta)^{-1}$ subject to the boundary condition (6) is a strictly positive Hilbert-Schmidt kernel and this is needed later. Equations (5), (6) can, therefore, be written as

$$\xi(x) + K g\Big(x, \, \xi(x)\Big) = K \dot{w} \, . \tag{7}$$

Let R_x^d denote $\{y \, : \, y_i \leq x_i, \, \forall i \leq d\}$. Define the Cameron-Martin space H induced by H_o as follows:

$$\eta(x) = \int_{R_x^d \cap \mathcal{D}} \xi(y) \, dy \, , \qquad (\eta_1, \, \eta_2)_H = (\xi_1, \, \xi_2)_{H_o} \, .$$

Set

$$\Big(G_1(\eta)\Big)(x) = \int_{R_x^d \cap \mathcal{D}} g(y, \, \xi(y)) \, dy \, ,$$

and

$$(K_1 \eta)(x) = \int_{R_x^d \cap \mathcal{D}} \int_{\mathcal{D}} K(y_1, \, y_2) \, \xi(y_2) \, dy_2 dy_1 \, .$$

Then equation (7) is equivalent to:

$$\eta + K_1 G_1(\eta) = \tilde{K}_1 w \tag{8}$$

and the existence of a measurable solution to (8) follows from Corollary A. ∎

The reader is referred to [ÜZ] for details and further results.

References

[BP] R. Buckdahn and E. Pardoux, *Monotonicity Methods for White Noise Driven Quasi-Linear SPDEs*. Progress in Probability, M. Pinsky (Editor), **Vol. 2**, Birkhäuser, 1993.

[De] K. Deimling, *Nonlinear Functional Analysis* , Springer-Verlag, 1985.

[FG] I. Fonseca and W. Gangbo, *Degree Theory in Analysis and Applications*. Oxford Lecture Series in Mathematics and Its Applications 2. Clarendon Press, Oxford, 1955.

[Ge] E. Getzler, Degree Theory for Wiener Maps, *J. Funct. Anal.*, **Vol. 68**, (1996) 388–403.

[Ku] S. Kusuoka, Some Remarks on Getzler's Degree Theorem. In: *Proc. of 5th Japan USSR Symp.* Lect. Notes Math, **Vol. 1299**, 239–249), Berlin Heidelberg New York: Springer 1988.

[ÜZ] A. S. Üstünel and M. Zakai, Applications of the Degree Theorem to Absolute Continuity on Wiener Space, *Prob. Theory Rel. Fields*, **Vol. 95** (1993), 509–520.

[ÜZ] A. S. Üstünel and M. Zakai, Transformation of the Wiener Measure under Non-Invertible Shifts, *Prob. Theory Rel. Fields* **Vol. 99** (1994), 485–500.

[ÜZ] A. S. Üstünel and M. Zakai, Degree Theory of Wiener Space. To appear.

A.S. Üstünel
ENST, Dépt. Réseaux
46, rue Barrault
75013 Paris, France
ustunel@res.enst.fr

Moshe Zakai
Department of Electrical Engineering
Technion—Israel Institute of Technology
32000 Haifa, Israel
zakai@ee.technion.ac.il

Received September 29, 1996

On the Interacting
Measure-Valued Branching Processes

Xuelei Zhao[1]

1. Introduction

Let $(E, \mathcal{E})$ be a Polish space. Denote by $M(E)$ the family of finite measures on E, and equip $M(E)$ with the usual weak convergence topology. Let us first briefly introduce the approximating processes $(\mu_t)_{t>0}$, called interacting branching diffusion processes. For each time t, μ_t is a random measure which models branching and diffusing particles in the following way: $\mu_t = \sum_{i \in I_t} \delta_{x_t^i}$ where I_t is the set of indexes of particles alive at time t, δ_x is the Kronecker delta function at point x, and x_t^i are the locations of particles indexed i at time t. Their dynamics are the following: μ_0 is a finite measure on E, describing the initial configuration of the system. Each particle moves following a homogeneous conservative Feller process with generator $(A, \mathcal{D}(A))$, and after a certain lifetime, it vanishes at the location x with the death rate $\lambda(x, \mu_t)$ and is replaced by a random number of children. The reproduction law depends on the state of the system by μ_t and the location x. This kind of interaction is well known in infinite particle systems. Under suitable hypothesis this interacting diffusion process μ_t approximates to the measure-valued branching diffusion processes $(X_t, P^\mu)_{\mu \in M(E)}$ which uniquely satisfies the following martingale problem: $\forall F \in bC^2(R)$, $\forall f \in \mathcal{D}(A)$, $F_f(\mu) = F(\langle \mu, f \rangle)$,

$$
F_f(X_t) - F_f(X_0) - \int_0^t [\langle X_s, (A + b(\cdot, X_s))f \rangle F'(\langle X_s, f \rangle)
$$
$$
+ \langle X_s, \frac{c(\cdot, X_s)}{2} f^2 \rangle F''(\langle X_s, f \rangle)] ds \qquad (1.1)
$$

is a P^μ-local martingale. Where $c \geq 0$ and $b : M(E) \times E \longrightarrow R$ are measurable functions. When $b(x, \mu)$ and $c(x, \mu)$ are independent of μ (the global system), the corresponding processes are the classical DW-superprocesses,

[1]Research supported in part by China Postdoctoral Science Foundation and the National Natural Science Foundation of China.

which have been extensively studied in recent decades. However, in the population genetic model, the interaction is somehow essential according to Darwin's population evolution theory. Therefore, the more interesting case is that b may depend on the global system. From this viewpoint we call b the *interacting intensity*, and further assume $\sup_{\mu \in M(E),\ x \in E} b(\cdot, \mu) < \infty$ and c, a non-negative and bounded function, is referred to as *branching rate*. The details are found in Méléard and Roelly (1992), Dawson (1993), and Perkins (1992). We will directly work with this kind of processes.

Assumption 1.1 *Let ξ be the Feller process associated with $(A, \mathcal{D}(A))$ in E. Denote by P_t its semigroup. There is a σ-finite reference measure $\mathcal{L}$ on E such that ξ has transition density $p_t(x, y)$ with respect to $\mathcal{L}$. Assume that*

1. *For any $f \in bC(E)$, $\lim_{t \to 0+} P_t f(x) = f(x)$, $x \in E$.*

2. *There exists $0 < \beta < 1$ such that $\forall T > 0$, $\sup_{0 \le t \le T,\ x,y \in E} p_t(x, y) t^{\beta} < +\infty$.*

3. *The transition density $p_t(x, y)$ is A-symmetric, i.e. $A p_t(\cdot, y)|_x = A p_t(x, \cdot)|_y$.*

4. *The Comparison Lemma for A^n holds for some continuous function ρ in $(0, \infty) \times E$ and for any $n \ge 1$, i.e., if u, v satisfy*

$$
\begin{aligned}
\dot{u}(t; x_1, \ldots, x_n) &= Au(t; x_1, \ldots, x_n) + \rho(t; x_1, \ldots, x_n), \\
\dot{v}(t; x_1, \ldots, x_n) &\le Av(t; x_1, \ldots, x_n) + \rho(t; x_1, \ldots, x_n), \\
&\qquad (x_1, \ldots, x_n) \in E^n
\end{aligned}
$$

   and $u(0, x) \ge v(0, x)$, then $u(t, x) \ge v(t, x)$ for all $(t; x_1, \ldots, x_n) \in (0, +\infty) \times E^n$.

5. *For some $\delta, \gamma > 0$ and let $p_r(x, y) = 0$ for $r < 0$ by convention,*

$$
\int_0^t \int_E [p_{t-s}(z, x) - p_{r-s}(z, y)]^2 \mathcal{L}(dz) ds
$$
$$
\le C(K)(|t - r|^{\delta} + |y - x|^{\gamma}), \quad 0 < r, t \le K,\ x, y \in E. \tag{1.2}
$$

Clearly the assumption makes sense for Brownian motion, or more generally symmetric diffusions in R^d with Lebesgue measure as the reference measure (cf. Konno and Shiga (1988), Lee and Ni (1992)). In this paper, under the above assumption, we shall mainly concern ourselves with the moments and the absolute continuity of measure-valued branching processes with interactions relative to reference measure $\mathcal{L}$. We shall also present a generalized result of Etheridge and March (1991).

2. Moments

Moments of a stochastic process are prerequisite in studying the properties of such a process. The moments of a superprocess have been easily formulated by using analytic knowledge since it has so-called log-Laplacian property (see Dawson (1993)). However, generally speaking, the log-Laplacian property does not hold for interacting measure-valued branching processes. In this section we shall discuss the moments of X_t under Assumption 1.1 (1–4). We shall find out that the Comparison Lemma and the symmetry of the operator A have played a fundamental role, but these are not necessarily required for classical superprocesses. Define the nth-moment measure $M_n(t, \mu; dx_1 \cdots dx_n)$ on E^n by

$$\int_{E^n} f_1(x_1) \cdots f_n(x_n) M_n(t, \mu, dx_1 \cdots dx_n)$$
$$= P^\mu \langle X_t, f_1 \rangle \cdots \langle X_t, f_n \rangle, \quad f_i \in pB(E), \quad i = 1, \cdots, n$$

Lemma 2.1 *The nth-moment measure $M_n(t, \mu, dx_1 \cdots dx_n)$, $n = 1, 2, \cdots$, are well defined.*

Proof. It suffices to prove $P^\mu \langle X_t, 1 \rangle^n < \infty$. We first note that there exists a sequence of stopping time T_k, $k \geq 1$, $T_k \to \infty$ as $k \to \infty$, and for any fixed k,

$$P^\mu \langle X_{t \wedge T_k}, 1 \rangle^n = \langle \mu, 1 \rangle^n + P^\mu \int_0^{t \wedge T_k} [n \langle X_s, b(X_s) \rangle \langle X_s, 1 \rangle^{n-1}$$
$$+ n(n-1) \langle X_s, \frac{1}{2} c(X_s) \rangle \langle X_s, 1 \rangle^{n-2}] ds \qquad (2.1)$$

Because b and c have a common bound, say $C > 0$,

$$P^\mu \langle X_{t \wedge T_k}, 1 \rangle^n \leq \langle \mu, 1 \rangle^n + \int_0^t C[n P^\mu \langle X_{s \wedge T_k}, 1 \rangle^n + n(n-1) P^\mu \langle X_{s \wedge T_k}, 1 \rangle^{n-1}] ds.$$
$$(2.2)$$

From this and Gronwall's inequality we have that $P^\mu \langle X_{t \wedge T_k}, 1 \rangle \leq e^{Ct} \langle \mu, 1 \rangle$, and thereby let $k \to \infty$,

$$P^\mu \langle X_t, 1 \rangle \leq e^{Ct} \langle \mu, 1 \rangle, \qquad (2.3)$$

And this easily yields the desired result by the inductive argument. ∎

Lemma 2.2 *For any fixed $n \in Z$, $M_n(t, \mu, dx_1 \cdots dx_n)$ is absolutely continuous with respect to the n-multiple product of $\mathcal{L}$. That is, there exists a* *measurable function* $m(t, x_1, x_2, \cdots, x_n)$ *such that*

$$M_n(t, \mu, dx_1 \cdots dx_n) = m(t, \mu; x_1, x_2, \cdots, x_n) \mathcal{L}(dx_1) \mathcal{L}(dx_2) \cdots \mathcal{L}(dx_n), \; t > 0.$$

Proof. For any $h > 0$, we have from the martingale property that,

$$\int_{E^n} p_h(x_1, y_1) \cdots p_h(x_n, y_n) M_n(t, \mu, dx_1 \cdots dx_n)$$

$$= \langle \mu, p_h(\cdot, y_1) \rangle \cdots \langle \mu, p_h(\cdot, y_n) \rangle$$

$$+ P^\mu \int_0^t \sum_{i=1}^n [\langle X_s, (\triangle + b(\cdot, X_s)) p_h(\cdot, y_i) \rangle \prod_{j \neq i}^n \langle X_s, p_h(\cdot, y_j) \rangle$$

$$+ \sum_{j \neq i}^n \langle X_s, c(\cdot, X_s) p_h(\cdot, y_i) p_h(\cdot, y_j) \rangle \prod_{k \neq i, j}^n \langle X_s, p_h(\cdot, y_k) \rangle] ds \quad (2.4)$$

Set $C_1 := \sup_{\mu, x} b(\mu, x); \; C_2 := \sup_{\mu, x} c(\mu, x)/2$, and

$$g_h(t, \mu; y_1, \cdots, y_n) := \int_{E^n} p_h(x_1, y_1) \cdots p_h(x_n, y_n) M_n(t, \mu, dx_1 \cdots dx_n).$$

Notice that Assumption 1.1(3), we have that,

$$\frac{d}{dt} g_h(t, \mu; y_1, \cdots, y_n) \leq (\sum_{i=1}^n (A_i + C_1)) g_h(t, \mu; y_1, \cdots, y_n) \quad (2.5)$$

$$+ C_2 \int_{E^{n-1}} \sum_{i=1, \, j \neq i}^n p_h(x_j, y_i) p_h(x_j, y_j) \prod_{k \neq i, j}^n p_h(x_k, y_k) M_{n-1}(t, \prod_{j \neq i} dx_j)$$

with the initial value

$$g(0, \mu; y_1, \cdots, y_n) = \langle \mu, p_h(\cdot, y_1) \rangle \cdots \langle \mu, p_h(\cdot, y_n) \rangle$$

where A_i stands for the Laplace operator carrying out $g_h(t, \mu; y_1, \cdots, y_n)$ in the ith components. From Lemma 2.1 and Assumption 1.1(4) we have that,

$$g_h(t, \mu; y_1, \cdots, y_n) \leq e^{nC_1 t} \langle \mu, p_{t+h}(\cdot, y_1) \rangle \cdots \langle \mu, p_{t+h}(\cdot, y_n) \rangle$$

$$+ C_2 \int_0^t e^{nC_1(t-s)} \int_{E^{n-1}} \sum_{i=1, \, j \neq i}^n p_{t-s+h}(x_i, y_i) p_{t-s+h}(x_i, y_j)$$

$$\prod_{k \neq i, j}^n p_{t-s+h}(x_k, y_k) M_{n-1}(s, \prod_{j \neq i} dx_j) ds, \quad (2.6)$$

By Assumption 1.1 (1), we know that $1_B(x) = \lim_{h\to 0} \int_B p_h(x,y)\mathcal{L}(dy)$ if B is an open set, then for any open sets B_i, $i = 1, 2, \cdots, n$,

$$M_n(t, \mu; B_1, \cdots, B_n)$$

$$= \lim_h \int_{B_1 \times B_2 \times \cdots \times B_n} \prod_{i=1}^n p_h(x_i, y_i)\mathcal{L}(dy_i)M_n(t, \mu; dx_1 \cdots dx_n)$$

$$= \lim_{h\to 0} \int_{B_1 \times B_2 \times \cdots \times B_n} g_h(t, \mu; y_1, \cdots, y_n)\mathcal{L}(dy_1) \cdots \mathcal{L}(dy_n) \qquad (2.7)$$

In particular, when $n = 1$ (2.6) and (2.7) clearly show that $M_1(t, \mu, dx)$ is absolutely continuous relative to $\mathcal{L}$ for $t > 0$, and thus from the inductive reasoning, the nth moment is also absolutely continuous relative to $\mathcal{L}^n$ for all $n \geq 1$, and $m(t, \mu, x_1, \ldots, x_n) = \lim_{h\to 0+} g_h(t, \mu; x_1, \ldots, x_n)$, $t > 0$. ∎

Notice that $\mathcal{D}(A)$ is dense in $B_+(E)$ in the sense of pointwise convergence. We can prove that.

Corollary 2.3 *If* $C_1 := \sup_{\mu,x} b(\mu, x)$, $C_2 := \sup_{\mu,x} c(\mu, x)/2$, *then for* $f_i \in B_+(E)$, $i = 1, 2, \cdots, n$ *and* $t \geq 0$,

$$P^\mu \prod_{i=1}^n \langle X_t, f_i \rangle \leq e^{C_1 nt} \prod_{i=1}^n \langle \mu, P_t f_i \rangle$$

$$+ C_2 \sum_{i=1}^n \sum_{j=1, j\neq i}^n \int_0^t e^{C_1 n(t-s)} \int_{E^{n-1}} P_{t-s}f_i(x_j)P_{t-s}f_j(x_j)$$

$$\prod_{k\neq i,j} P_{t-s}f_k(x_k)M_{n-1}(s, \mu; \prod_{i\neq j} dx_i)ds. \qquad (2.8)$$

Where P_t is the semigroup of ξ, i.e. $P_t f(x) = \int_E p_t(x,y)f(y)\mathcal{L}(dy)$.

3. The absolute continuity

Absolute continuity is always an interesting topic for random measures. As a typical problem for measure-valued processes, it is not surprising that many authors have worked on it (see Konno and Shiga (1988), Sugitani (1989), Zhao (1994) etc.) and have completely solved, (if we may say so) this kind of question for the classical DW-superprocesses. Méléard and Roelly (1992) conjectured that measure-valued branching Brownian motion with interaction has a similar result. Indeed, we have,

Theorem 3.1 (Zhao (1997), Li (1996)) *Under Assumption 1.1, X_t is almost surely absolutely continuous with respect to the reference measure $\mathcal{L}$ for any $t > 0$. The Radon-Nykodym derivative $X(t, x)$ satisfies a partial stochastic differential equation:*

$$\frac{\partial}{\partial t} X(t, x) = \sqrt{\tilde{c}(X(t, x)) X(t, x)} \dot{W}_t(x) + AX(t, x) + \tilde{b}(X(t, x)) X(t, x),$$
$$(3.1)$$

where $\dot{W}(t, x)$ is a "time-space white noise" defined on an extension of the original probability space, and $\tilde{b}(X(t, x)) := b(x, X_t)$, $\tilde{c}(X(t, x)) := c(x, X_t)$. More precisely, the equation (3.1) can be interpreted as follows,

$$\int_E X(t, x) f(x) \mathcal{L}(dx) - \int_R f(x) X_0(dx)$$
$$= \int_0^t \int_E \sqrt{\tilde{c}(X(s, x)) X(s, x)} f(x) \dot{W}_s(x) \mathcal{L}(dx) ds$$
$$+ \int_0^t \int_E X(t, x) [Af(x) + \tilde{b}(X(s, x)) f(x)] \mathcal{L}(dx) ds \quad (3.2)$$

for all $t \geq 0$ and $f \in bC(E)$ almost surely.

The outline of proof. There is a continuous orthogonal martingale measure $\{M_t(A), A \in \mathcal{B}(E), t \geq 0\}$ with covariance measure $c(X_s, x) X_s(dx) ds$, such that for $f \in C^2(E)$,

$$X_t(f) = X_r(P_{t-s} f) + \int_r^t \int_E P_{t-s} f(x) M(ds, dx)$$
$$+ \int_r^t \int_E b(X_s, x) P_{t-s} f(x) X_s(dx) ds. \quad (3.3)$$

Let $Z_t^r(x) = \int_r^t \int_E P_{t-s}(z, x) M(ds, dz)$, and $Y_t^r(x) = \int_r^t \int_R b(z, X_s) p_{t-s}(z, x) X_s(dz) ds$, $t > 0$, $x \in E$. From estimates of moments in Section 2, we can prove in the same manner as in Konno and Shiga (1988) that $X_t(dx)$ is absolute continuity and by (3.3) its density $X(t, x)$ satisfies

$$X(t, x) = \int_E p_t(z, x) X_r(dz) + Z_t^r(x) + Y_t^r(x), \quad x \in E, \quad (3.4)$$

And then we can define a "time-space white noise" $\dot{W}_t(x)$ such that for any $f \in C_+^2$, $\int_0^t f(x) M(ds, dx) = \int_0^t \int_E \sqrt{\tilde{c}(X(s, x)) X(s, x)} f(x) \dot{W}_s(x) \mathcal{L}(dx) ds$. Notice that

$$\int_0^K \sup_{x, y \in E} p_s(x, y) ds < \infty, \quad K > 0, \quad (3.5)$$

and from (3.5) we can prove from the well known Kolmogorov's Lemma that both $Z_t^r(x)$ and $Y_t^r(x)$ are continuous in $(t, x) \in (0, \infty) \times E$ for any fixed $r \geq 0$, and equation (3.1) holds by a standard argument (see Walsh (1986), Konno and Shiga (1988), Li (1996)). ∎

4. A generalized result of Etheridge and March

It is well known that the (classical) DW-superprocess is closely associated with the (classical) FV-superprocess. This connection is first observed by Konno and Shiga (1988), rigorously proved by Etheridge and March (1991), and generalized by Perkins (1992). We here shall consider the same question for the measure-valued branching processes defined by (1.1) in which $c \equiv 1$ and b, a bounded continuous function, only depends on the global mass, i.e. $b(\cdot, X_s) = b(X_s(1))$. In term of the particle system, this means that interaction among particles depends upon only global mass in the system but geographical locations.

Theorem 4.1 (Zhao and Yang (1997)) *Let $P^{\mu,T,\epsilon}(\cdot) = P^{\mu}(\cdot \,|\, |X_t(E) - 1| < \epsilon, \ 0 \leq t \leq T)$. For any fixed $T > 0$, If $\epsilon_n \to 0_+$, $T_n \uparrow T < \infty$ and μ_n weakly converges to a probability measure μ, then the conditioned probability law $P^{\mu_n, T_n, \epsilon_n}$ weakly converges to the probability law Q^{μ} of FV-superprocess. That is, Q^{μ} is the unique solution for the following martingale problem:*

(1) $\quad Q^{\mu}(X_0 = \mu) = 1$

(2) $\quad \forall f \in \mathcal{D}(A), \ M_t(f) = \langle X_t, f \rangle - \langle X_0, f \rangle - \int_0^t \langle X_s, Af \rangle ds$

is a square integrable Q^{μ}–martingale with variance process $\frac{1}{2} \int_0^t [X_s(f^2) - (X_s(f))^2] ds$.

Skeleton of proof. This theorem can be proved by replacing the auxiliary function $u_\epsilon(t, x)$ in Etheridge and March (1991) by that define by

$$\begin{cases} \frac{\partial}{\partial t} u_\epsilon = \frac{1}{2} x \frac{\partial^2}{\partial x^2} u_\epsilon + b(x) x \frac{\partial}{\partial x} u_\epsilon(x, t), & |x - 1| < \epsilon, t > 0; \\ u_\epsilon(x, 0) = 1, & |x - 1| < \epsilon \\ u_\epsilon(1 \pm \epsilon, t) = 0, & t > 0. \end{cases} \tag{4.1}$$

From [BH], solution $u(t, x)$ satisfies: (1) $u(t, x)$ is continuous in $(0, \infty) \times \{x : |x - 1| < \epsilon\}$ and $0 \leq u(x, t) \leq 1$. (2) The limit $\lim_{t \to \infty} u(x, t)$ converges to 0 uniformly in $[1 - \epsilon, 1 + \epsilon]$. We can then prove our result in a similar manner to that in Etheridge and March (1991) despite some small obstacles that demand sweeping out. For the sake of brevity we omit the details.   ∎

References

[BH]   H. Begher, and G. C. Hsiao, *Nonlinear Boundary Value roblem for a Class of Elliptic Systems*, Lecture Notes of Mathematics, Springer-Verlag, Berlin, 1980.

[D]   D. Dawson, *Measure-Valued Markov Processes*, Lecture Notes of Mathematics, **1541**, Springer-Verlag, Berlin, 1993.

[EM]   A. Etheridge and P. March, A note on superprocesses, *Probab. Th. Rel. Fields*, **89** (1991), 141–147.

[KS]   N. Konno and T. Shiga, Stochastic partial differential equations for some measure-valued diffusions, *Probab. Th. Rel. Fields*, **79** (1988), 201–225.

[LN]   T. Y. Lee and W. M. Ni, Global existence , large time behavior and life span of solutions of a semilinear parabolic Cauchy problem, *Trans. Amer. Math. Soc.*, **333** (1992), 365–378.

[L]   Z. H. Li, On the absolute continuity of branching Brownian motion with mean field interaction, preprint, 1996.

[MR]   M. Méléard and S. Roelly, Interacting branching measure processes, in: *Stochastic Partial Differential Equations and Applications* (ed. G. Da Prato and L. Tubaro), **PRNM 268**, Longman Scientific and Technical, Harlow, 1992.

[P1]   E. Perkins, Measure-valued branching diffusions with spatial interactions, *Probab. Th. Rel. Fields*, **94** (1992), 189–245.

[P2]   E. Perkins, *Conditional Dawson-Watanabe processes and Fleming-Viot processes*, *Seminar on Stochastic Processes*, Birkhäuser, Boston, 1993.

[S]   S. Sugitani, Some properties for the measure-valued Branching diffusion processes, *J. Math. Soc. Japan*, **41**.3 (1989), 437–461.

[W] B. J. Walsh, *An Introduction to Stochastic Partial Differential Equations*, Lecture Notes of Mathematics, **1180**, Springer–Verlag, New York, 1986, 266–348.

[Z1] X. L. Zhao, Some absolute continuity of superdiffusions and superstable processes, *Stoch. Proc. Appl*, **50** (1994), 21–36.

[Z2] X. L. Zhao, The Absolute Continuity for Interacting Measure-Valued Branching Brownian Motion, *Chin. Ann. Math.*, **18B**.1 (1997), 47–54.

[ZY] X. L. Zhao and M. Yang, A Limit Theorem for Interacting Measure-Valued Branching Processes, *Acta Mathematica Scientia*, **17B**.1 (1997).

Institute of Mathematics
Shantou University
Shantou, 515063
China

Received September 19, 1996

Progress in Systems and Control Theory

Series Editor

Christopher I. Byrnes
Department of Systems Science and Mathematics
Washington University
Campus P.O. 1040
One Brookings Drive
St. Louis, MO 63130-4899

Progress in Systems and Control Theory is designed for the publication of workshops and conference proceedings, sponsored by various research centers in all areas of systems and control theory, and lecture notes arising from ongoing research in theory and applications control.

We encourage preparation of manuscripts in such forms as LATEX or AMS TEX for delivery in camera-ready copy which leads to rapid publication, or in electronic form for interfacing with laser printers.

Proposals should be sent directly to the editor or to: Birkhäuser Boston, 675 Massachusetts Avenue, Cambridge, MA 02139, U.S.A.

PSCT8   Analysis of Controlled Dynamical Systems: Proceedings of a Conference
in Lyon, France, July 1990
B. Bonnard, B. Bride, J. P. Gauthier and I. Kupka

PSCT9   Nonlinear Synthesis: Proceedings of a IIASA Workshop held in Sopron,
Hungary, June 1989
Christopher I. Byrnes and Alexander Kurzhansky

PSCT10   Modeling, Estimation and Control of Systems with Uncertainty: Proceedings
of a Conference held in Sopron, Hungary, September 1990
Giovanni B. Di Masi, Andrea Gombani and Alexander B. Kurzhansky

PSCT 11 Computation and Control II: Proceedings of the Second Conference at
Bozeman, Montana, August 1990
Kenneth Bowers and John Lund

PSCT 12 Systems, Models, and Feedback: Theory and Applications. Proceedings
of a U.S.-Italy Workshop in honor of Professor Antonio Ruberti, held in
Capri, Italy, 15-17, June 1992
A. Isidori and T.J. Tarn

PSCT 13 Discrete Event Systems: Modelling and Control. Proceedings of a Joint
Workshop held in Prague, Poland, August 1992
S. Balemi, P. Kozak, and R. Smedinga

PSCT 14 Essays on Control: Perspectives in the Theory
and its Applications
H. L. Trentelman and J. C. Willems

PSCT 15 Computation and Control III: Proceedings of the Third
Bozeman Conference, Bozeman, Montana, August 1992
K. L. Bowers and J. Lund

PSCT 16 Set-Valued Analysis and Differential Inclusions,
A Collection of Papers from a Workshop held in
Pamporovo, Bulgaria, September 1990
A. B. Kurzhanski and Vladimir M. Veliov

PSCT 17 Advances in Nonlinear Dynamics and Control:
A Report from Russia
A. B. Kurzhanski

PSCT 18 Modeling Techniques for Uncertain Systems,
Proceedings of a Conference held in Sopron,
Hungary, July 1992
A. B. Kurzhanski and Vladimir M. Veliov